G. Muselin

4 II 1985

DEVELOPMENT AND EVOLUTION

THE SIXTH SYMPOSIUM OF
THE BRITISH SOCIETY
FOR DEVELOPMENTAL BIOLOGY

Development and evolution

EDITED BY

B. C. GOODWIN

Reader in Developmental Biology, University of Sussex

N. HOLDER

Lecturer in Anatomy, King's College, London

C. C. WYLIE

Reader in Anatomy and Embryology, St George's Hospital Medical School, London

#6424

CAMBRIDGE UNIVERSITY PRESS

CAMBRIDGE

LONDON   NEW YORK   NEW ROCHELLE

MELBOURNE   SYDNEY

Published by the Press Syndicate of the University of Cambridge
The Pitt Building, Trumpington Street, Cambridge CB2 1RP
32 East 57th Street, New York, NY 10022, USA
296 Beaconsfield Parade, Middle Park, Melbourne 3206, Australia

First published 1983

Printed in Great Britain at the Pitman Press, Bath

Library of Congress catalogue card number: 82–14728

British Library Cataloguing in Publication Data
Development and evolution.
1. Developmental biology—Congresses
2. Biological patterns—Congresses
I. Goodwin, B. C.   II. Holder, N.
III. Wylie, C. C.   IV. British Society for
Developmental Biology
574.3   QH491
ISBN 0 521 24949 X

Contents

Introduction

The BSDB symposium on Development and Evolution was held at the University of Sussex from 31 March to 2 April 1982, the centenary year of Darwin's death. Some of the participants came to bury Darwin and others came to praise him, but such has been his influence on biological thought that all the dialogue was conducted in relation to Darwin's evolutionary vision and his conception of biology as a historical science.

All interesting scientific theories face problems or difficulties which can be resolved in one of two ways: either explanations are found within the existing theory, perhaps with some shift of emphasis but no basic change of principle so that the problem is assimilated into the accepted conceptual structure; or the problem may turn out to be an unassimilable challenge to orthodoxy, requiring revision or transformation of the theory before it can be satisfactorily explained. Current biological theory, having successfully consolidated an impressive array of observations within the framework of Darwin's evolutionary vision, may or may not be in the early stages of such a transformation. Certainly there have been rumblings in recent years. Some of these rumblings have undoubtedly arisen from differences of language and approach which are inevitable in a subject with such a diversity of disciplines as biology, making dialogue difficult. All the more reason, then, for symposia which bring together people from somewhat different traditions of biological research to attempt to resolve the superficial differences and to uncover any deeper divergences. The time seems now to be ripe for a concerted attempt to examine relationships between the research traditions of systematics, evolution theory, genetics, and developmental biology; and this volume sets out to do just this.

If major change is in the offing, then one of the chief conceptual difficulties facing neo-Darwinian orthodoxy is the problem of biological form. There are two main aspects of this problem, which is itself a part of the greater problem of biological organisation. One of these is concerned with taxonomy, the systematic relationships between organisms with different morphologies; while the other relates to development, the way in which organisms of specific form are generated. The former problem has been well publicised in the work of those, such as the (neo)-cladists, who believe in the existence of a natural taxonomy of organisms as opposed to a Darwinian genealogy: i.e. they believe that biological systematics is based upon a real and discoverable hierarchy of logically organised structures (forms) which must be known before we can decide if it is fully explicable as a historical sequence of organisms ordered by relationships of descent dependent upon contingencies. This view is represented in this volume by Patterson's wide-ranging and comprehensive paper. All other contributions address, from one perspective or another, the rather less articulated but equally important dialogue between those who see in the regularities of organismic form and transformation evidence of developmental constraints or generative rules which place significant limits on evolutionary change, and those who see such constraints as minor, essentially reducible to evolutionary contingencies. The most extreme form of the developmental constraint principle, linking it quite closely to Patterson's form of cladism, takes the view that a theory of organisation (of which morphogenesis and systematics are aspects) is the primary biological problem, whose solution is required before it is possible properly to understand what the evolutionary record is telling us. This view is presented in the paper by Goodwin & Trainor. The other extreme, in which evolutionary forces leaving their trace in 'genetic programmes' are seen as the major source of information in developing organisms, and intrinsic organisational constraints on morphogenesis are regarded as minor, is presented in Wolpert's paper. Maynard Smith allows for significant limitations on genetic variations resulting from properties of the developmental process such as hierarchical ordering of morphogenetic events, but sees them all arising by historical contingency rather than from developmental necessity. The implication of this is that there cannot be a theory of the organism and its transformations as a significant component of an evolutionary

theory, and that biology is ineluctably a historical science. This is
the conventional view, and may well turn out to be correct; but it is
a hypothesis which is put under explicit and implicit scrutiny by a
number of contributors to this volume.

The editors felt that, despite these two clear expositions emphasising the dominant developmental role of the genome and the
implicit acceptance of this view in several other papers, an explicit
statement of the geneticist's approach to the analysis of development was missing from the conference presentations. This is
therefore provided by Whittle's paper, which was a spontaneous
response to this deficiency.

The other papers in this volume constitute a very important
contribution to a rich and lively domain of enquiry. Although the
major emphasis is on the study of generative rules or developmental
constraints in insects and vertebrates, reflecting what are still the
chief preoccupations of developmental biologists, studies of other
taxa which have made major contributions to an understanding of
the relationship between ontogeny and phylogeny are represented
by papers of the highest quality, such as those of Frankel on the
Protozoa and Walbot on plants, with emphasis on maize. If there
are morphogenetic universals which constrain development, then
their discovery and description depend upon wide-ranging comparative studies of the type so ably represented in these papers.

The concept of homology, which has had a turbulent history, is
tackled in a number of important papers, especially those of
Hinchliffe & Griffiths, Nieuwkoop & Sutasurya, and Maden *et al.*
Whether this relation will remain the major conceptual tool in
comparative morphogenesis and systematics, or sink into the
historical abyss, remains to be seen. Multiple but related meanings
continue to make its use somewhat problematic, and we can see in
the highly interesting papers of García-Bellido and French rather
different uses of the concept in relation to 'ancestral patterns' and
segmental (serial) homology in insects. However, what is emerging
from these studies is a firm empirical basis for comparative
morphogenesis, the contemporary generative (i.e. developmental)
response to the problems left unsolved by comparative morphology. At a more conceptual level, Kauffman addresses the problem
of the developmental principles which define the intrinsic logic of
morphogenesis, giving us a clarification and development of
relevant ideas.

The final collection of papers deal with principles of tissue interactions involved in the development of vertebrates. These range from an analysis of epithelial–mesenchymal interactions, particularly with regard to differentiation of neural crest derivatives by Hall, to specific controls of spatial patterning in vertebrate limbs by Maden *et al.* and Holder. The vertebrate limb is the focus of five of the papers presented in this volume. This structure provides an ideal focal point for analysing the relationship between development and evolution because of our extensive knowledge of tissue level interactions controlling development of the limb and the readily available and richly diverse collection of limb anatomies in the various tetrapod taxonomic groups.

March 1982   B. C. Goodwin
                                                 N. Holder
                                                 C. C. Wylie

How does phylogeny differ from ontogeny?

COLIN PATTERSON

Department of Palaeontology, British Museum (Natural History),
London SW7, UK

HAECKEL
VON BAER
ONTOGENY

Introduction

The concepts evolution and development have always been very closely related; indeed, the two were synonymous until the 1840s (Bowler, 1975). The change in meaning of 'evolution', from development of the embryo to transmutation of species, was initiated by diffusion of von Baer's (1828) ideas of ontogeny – 'a process leading to the production of heterogeneity or complexity of structure' (Bowler, 1975, p. 100) – and the importance of von Baer's contributions, particularly his four laws, is a theme throughout this paper.

Just as Darwin had first promoted natural selection by analogy with artificial selection, so Haeckel (1866), who coined the word 'phylogeny', first promoted it by analogy with ontogeny. Haeckel was drawing on an older tradition, the parallel between ontogeny and the succession of fossils first proposed by Agassiz (1844, 1845; also Russell, 1916; Gould, 1977; Ospovat, 1976, 1981). Haeckel's biogenetic law, that ontogeny recapitulates phylogeny, was also a re-statement of a much older tradition, the ancient idea of recapitulation or progressive ascent of a ladder-like, uniserial chain of being. Von Baer's 1828 book was, in large part, an attack on that older doctrine of recapitulation (Ospovat, 1976; Gould, 1977).

But Haeckel clearly differentiated his evolutionary version of recapitulation, the biogenetic law, from the older, non-evolutionary version by pointing out that the ontogeny of the individual corresponds to, or recapitulates, only a part of phylogeny, 'a simple *unbranching* or graduated chain of forms; and so it is with that portion of phylogeny which comprises the palaeontological history of development of the *direct ancestors only* of an individual

organism. But *the whole of phylogeny* – which meets us in the
natural system . . . forms a *branching* or tree-shaped developmental
series, a veritable pedigree' (Haeckel, 1876, vol. 1, p. 314, italics
original). With 'the natural system' Haeckel introduces the third
term of what Agassiz (1859) called the three-fold parallelism,
between ontogeny, palaeontology, and comparative anatomy as
summarised in the natural hierarchy of systematics. According to
Haeckel, the aim of phylogenetic research was to reconstruct the
tree-shaped pedigree of the natural system, and tracks through that
were sketched, more or less accurately or fully, by individual
ontogenies, whereas fossils provided 'the infallible and indisputable
records which fix the correct history of organisms upon an irrefrag-
able foundation' (Haeckel, 1876, vol. 1, p. 55).

The metaphor of the branching tree, which Haeckel used so
graphically (see Figs. 1, 6, 7), was, as Ospovat (1981) shows,
becoming widespread among non-evolutionary biologists in the
1840s and 1850s. Ospovat also demonstrates that the impulse
behind adoption of the idea of branching was von Baer's laws of
ontogeny, particularly his first law: 'the more general characters of
a large group of animals appear earlier in their embryos than the
more special characters' (T. H. Huxley's (1853) translation). Von
Baer's laws are presented in the terms of systematics, groups and
their characters, illustrated by a hierarchical diagram (Fig. 2). And
it was chiefly among systematists that the branching-tree model was
discussed in the 1840s (Ospovat, 1976, 1981, p. 124 ff.).

Haeckel's biogenetic law fell into disfavour and disuse, and
Gould (1977) has chronicled its decline. Gould also points out
(*ibid.* pp. 2, 185) the 'monumental confusion that persists to this
day' between von Baer's and Haeckel's laws. Today, trees are
usually presented in Haeckel's sense, as pedigrees or images of
phylogeny, rather than in the sense of von Baer and pre-Darwinian
systematists, as images of hierarchical organisation, though cladist
systematists have learnt to distinguish phylogenetic trees from
cladograms (Nelson & Platnick, 1981). My aim here is to use trees
and similar images to explore the interrelationship between
ontogeny, systematics and reconstruction of phylogeny, particularly with regard to von Baer's contributions on ontogeny.

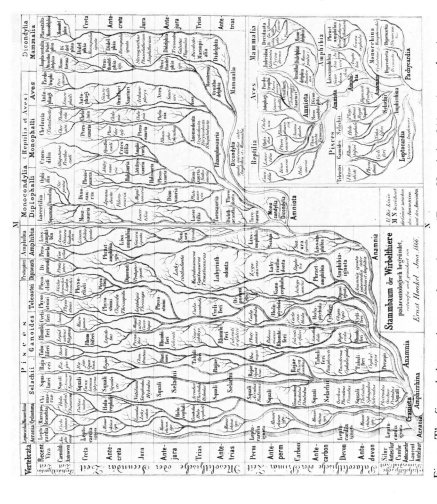

Fig. 1. The first phylogenetic tree of vertebrates. (From Haeckel, 1866, vol. 2, pl. 7.)

Scheme of the Progress of Development.

Lowest Grade of Development. — The animal rudiment is either:

- a germ-granule (itself germ),
- or an ovum with a germ. In this arises:
  - ? Radiate development .... ? *Animals of the peripheral type.*
  - Spiral development .... *Animals of the massive type.*
  - Symmetrical development...... *Animals of the elongated type.*
  - Doubly symmetrical development... *Vertebrata.* They have a chorda dorsalis, dorsal plates, visceral plates, nerve tubes, gill-clefts, and acquire...
    - *Gills*......
      - No true lungs formed.
        - The skeleton does not ossify .... *Cartilaginous fishes.*
        - The skeleton ossifies .... *Osseous fishes.*
      - Lungs formed... *Amphibia.* The gills
        - persist .... *Sirenidæ.*
        - do not persist
          - remain external .... *Urodela.*
          - become enclosed .... *Anura.*
    - A much-developed allantois.
      - No umbilical cord.
        - No wings nor air-sacs .... *Reptilia.*
        - Wings and air-sacs .... *Aves.*
      - An umbilical cord, *Mammalia,*
        - which falls off early,
          - without union with the parent ?......... *Monotremata.*
          - after a short union with the parent.... *Marsupialia.*
        - which persists longer. The yelk-sac ....
          - grows for a long time The allantois grows ......
            - very little .... *Rodentia.*
            - moderately .... *Insectivora.*
            - much .... *Carnivora.*
          - grows little; The allantois grows ......
            - little .... *Quadrumana.*
            - Umbilical cord very long. *Man.*
            - very long. Placenta
              - in scattered masses. *Ruminantia.*
              - evenly distributed. *Pachydermata.* *Cetacea.*

Highest Grade of Development.

Fig. 2. 'Scheme of the progress of development' in vertebrates. (From von Baer, 1828, facing p. 225, in Huxley's 1853 translation, p. 215.)

How shall we depict phylogeny?
(Gould, 1977, p. 212)

Gould's answer to this is: by a series of comparable or standardised stages of development, conventionally a sequence of adults. I want to look at his question from a different angle, and to consider depictions of phylogeny which employ conventions rather than organisms. There are three principal models in use, which I will call trees (e.g. Fig. 1), maps (e.g. Fig. 3) and spindle diagrams (e.g. Fig. 5). The last two examples, and other pre-Darwinian examples of all three models reproduced by Nelson & Platnick (1981), show that these models have no *necessary* phylogenetic implication.

I will concentrate first on Milne-Edwards's map (Fig. 3) and Agassiz's spindle diagram (Fig. 5), both published in 1844. With his map, Milne-Edwards was arguing, like von Baer, for the importance of embryology in systematics. He used the tree metaphor to explain his views, describing ontogenies as 'a multitude of . . . series which appear to branch from one another at different heights, or rather which are joined in a bundle at their base, and break up into secondary, tertiary and quaternary bundles, as in ascending towards the end of embryonic life they deviate from one another and take on distinct characters' (Milne-Edwards, 1844, p. 72, my translation). Milne-Edwards compared (*ibid.* p. 82) his diagram (Fig. 3) to the map of 'the islands of a vast archipelago' within which 'one could distinguish secondary archipelagos'. His map is readily translated into a hierarchical classification, resembling von Baer's diagram (Fig. 2), or into a branching diagram (Fig. 4*a*; cf. Nelson & Platnick, 1981, Fig. 2.59) which is a cladogram rather than a tree, since it carried no implication of descent for its author. Comparing the cladograms derived from Milne-Edwards's and von Baer's diagrams (Fig. 4), the two differ in that von Baer's unites birds and reptiles, whereas Milne-Edwards places those two groups in a trichotomy with mammals. However, the character von Baer used to unite birds and reptiles was 'no umbilical cord' (Fig. 2), that is, no placenta. Aristotle called such negative characters 'privative', and argued that they could not logically characterise groups, since they cannot be further subdivided: absence is merely absence (Nelson & Platnick, 1981, p. 71). In von Baer's scheme (Fig. 2), several other 'groups' are characterised in this way, by

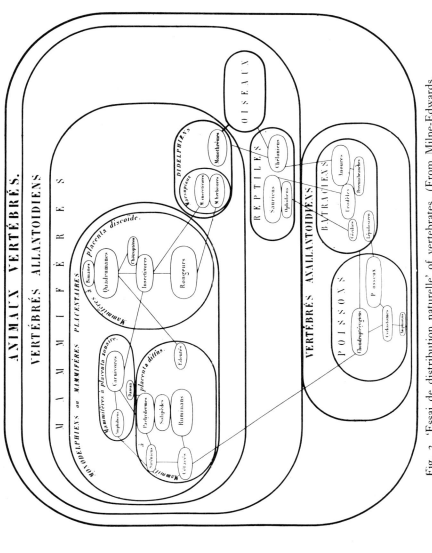

Fig. 3. 'Essai de distribution naturelle' of vertebrates. (From Milne-Edwards, 1844, facing p. 98.) The relative positions of the groups and the lines joining them symbolise 'the various degrees of affinity' between them.

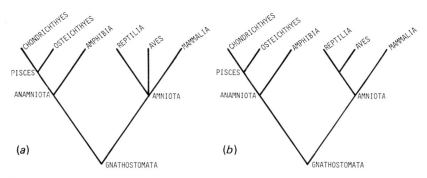

Fig. 4. Cladograms of the main groups of gnathostome vertebrates derived from
(*a*) Milne-Edwards's map (Fig. 3) and (*b*) von Baer's scheme (Fig. 2).

absence: fishes (no lungs), perennibranchiate amphibians (no
transformation of gills), and reptiles (no wings or air-sacs). Milne-
Edwards's group Anallantoïdiens (Anamniota in Fig. 4*a*) has the
same disadvantage, since its characters are 'ni allantoïde ni amnios'
(*idem* 1844, p. 89). Use of such groups in a phyletic context is
discussed on p. 18, below.

Agassiz's spindle diagram (Fig. 5) is called a 'généalogie'.
Whereas Milne-Edwards used his map to illustrate the importance
of von Baer's laws for systematics, Agassiz presented his diagram in
an essay on classification of fishes which began with a comparison
between systematists and mapmakers (Agassiz, 1844, vol. 1,
p. 165). Agassiz goes on to discuss the correlation he observed
between the sequence of features in ontogeny and the fossil record.
This parallel between palaeontology and embryology was generally
seen as an indication that von Baer's laws of divergence and
individualisation applied to both (Ospovat, 1976; 1981, p. 136).
Discussing ways of depicting the relations between species, Agassiz
(1844, vol. 1, p. 169) rejects linear series, and recommends
diagrams in which clear-cut divisions are each centred about 'les
types mieux connus', around which can be grouped other 'types'
which may, in turn, serve as centres for groups of subsidiary rank.
Palaeontological information may be added to these zoological maps
('cartes') to produce genealogical trees ('arbres'), showing the time
of appearance of each group, and its importance in each geological
period. Thus Agassiz's diagram can be seen as a version of
Milne-Edwards's map (création actuelle' in Fig. 5), with the time
dimension added as a summary of the fossil record. Unlike

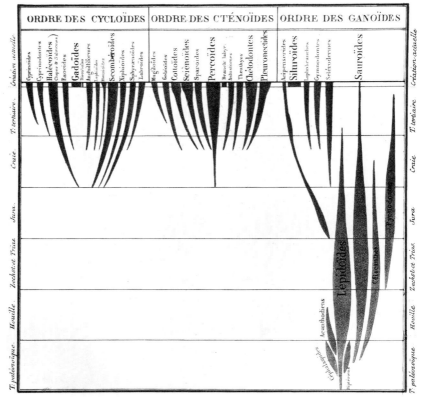

Fig. 5. Agassiz's 'généalogie de la classe des poissons' (1844, vol. 1, facing p. 170), modified by omitting the chondrichthyans.

Milne-Edwards's diagram, Agassiz's cannot be readily converted into a tree, except perhaps by continuing the converging lines downwards until they meet. Agassiz avoided this, for the implication of common descent would be inevitable, and Agassiz expressly denied it. Diagrams like Agassiz's are frequently modified in this way today (examples in Romer, 1966, and many other modern works).

Common descent was added to the fossil record in Haeckel's (1866, 1868) 'palaeontologisch begründet' trees (Figs. 1, 6). In these early trees the trunk and main branches are thick (Fig. 7), as in Milne-Edwards's simile of the bundle ('faisceau') at the base, and are labelled with names or numbers using five different con-

9

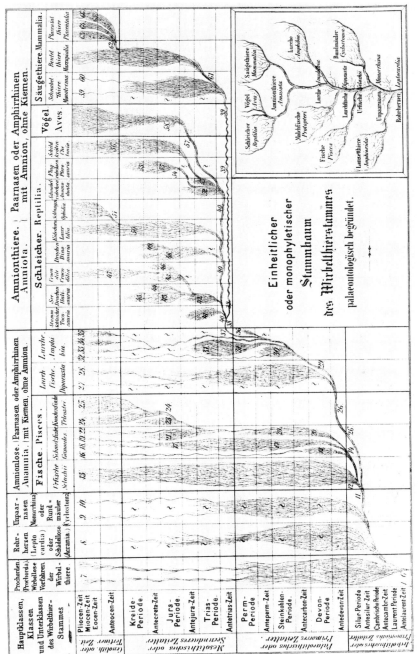

Fig. 6. Haeckel's monophyletic tree of vertebrates (1868, pl. 6).

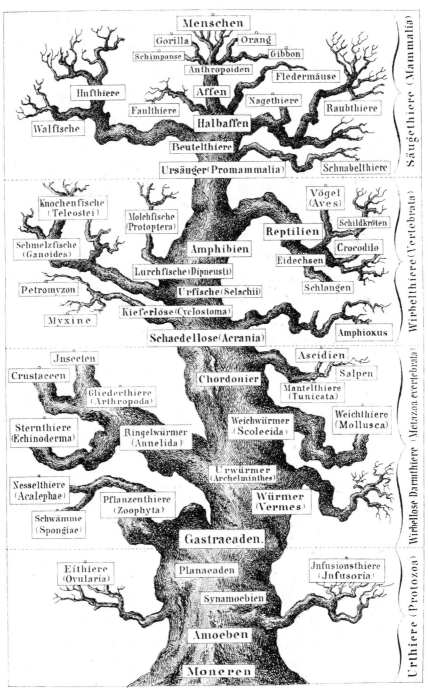

Fig. 7. Haeckel's phylogenetic tree of man (1874, pl. 12).

ventions of ancestry. The first is names of higher taxa whose rank decreases as one ascends the tree (e.g. Craniota, Amphirrhina (= Gnathostomata), Amniota in Fig. 1): the convention here is that a group gives rise to its subgroups. Secondly, there are groups which give rise to taxa of equal or higher rank (e.g. Acrania to Craniota, Anamnia to Amniota in Fig. 1, both of equal rank; Amphibia to Amniota in Figs. 1, 7, of higher rank). Thirdly, there are unknown or hypothetical members of taxa of the first two types (e.g. *Dipnoi ignoti, Amphibia ignota, Anamnia ignota* in Fig. 1; in Fig. 6, nos. 11, 26, 29, 37 are 'unbekannte Zwischenformen' or 'unbekannte Uebergangsformen', no. 38 is 'Protamnien – gemeinsame Stammform' and no. 39 is 'Stammsäuger (Prom- ammalia)'). These unknown or hypothetical forms are really equiva- lent to the groups to which they belong, since 'hypothetical forms are themselves generalised classification groups' (Russell, 1916, p. 294). The fourth type includes extinct groups of rank lower than those to which they are ancestral (e.g. no. 12 in Fig. 6 is the genus *Onchus*, Silurian and Devonian finspines ancestral to cartilaginous fishes; no. 14 in Fig. 6 is the genus *Pteraspis*, ancestral to bony fishes; no. 30 in Fig. 6 is the order Ganocephala, including the Carboniferous amphibians *Archegosaurus* and *Dendrerpeton*, ancestral to amphibians and amniotes). And the fifth type is groups inferred only from early ontogeny (e.g. Moneren, Amoeben, Synamoebien, Planaeaden, Gastraeaden in Fig. 7). As Gould (1977, p. 170) points out, these 'groups' relied on no other evidence than early ontogeny, Monera being anucleate primordia, Amoeben the nucleate zygote, Synamoebien the morula, Planaeaden the planula and Gastraeaden the gastrula. Like unknown and hypothe- tical forms (type 3), these ancestral embryonic forms are equivalent to the first two types, ancestral groups whose rank is higher than or equal to that of their descendants, since we could replace the ancestral embryos by the names of extant taxa without violating current concepts of phylogeny: Moneren = kingdom Monera (prokaryotes), Amoeben = kingdom Protista, Synamoeben + Planaeaden = Metazoa, and Gastraeaden = Diploblastica (Hansen, 1977; cf. Tribe *et al.*, 1981).

The five different conventions that Haeckel used to denote ancestry do not exhaust the possibilities, for there is a sixth, ancestral species. There are no ancestral species in Haeckel's trees, yet this type of ancestry is the only one compatible with evolutionary theory, since species and groups of species are descended from species, not groups (Wiley, 1979).

These six forms of ancestry may be categorised in different ways, but the distinction between phylogeny and ontogeny is demonstrated most clearly by splitting them into those which are consistent with von Baer's law of decreasing generality, or with a map like Fig. 3, or with a cladogram (Fig. 4), and those which are inconsistent with von Baer's law and with a cladogram, and can only be accommodated in a phylogenetic tree of historical descent. The first set, which may be called von Baerian conventions of ancestry, includes groups whose rank decreases as one ascends the tree and unknown or hypothetical members of such groups. The second set, which can be placed only in trees, may be called Haeckelian conventions of ancestry. They are extant or extinct taxa whose rank is equal to or lower than that of their descendants and unknown or hypothetical members of such groups. Named ancestral species are components of trees, but they may also occur in cladograms, where an ancestral species will appear in a trichotomy or polychotomy as equal in rank to its descendant groups or species.

Von Baerian and Haeckelian ancestral groups differ in one significant way. Examples of von Baerian ancestry are the taxa named at the nodes in the cladograms in Fig. 4, and taxa like Craniota, Amphirrhina (= Gnathostomata), Amniota and Mammalia in Fig. 1. The convention is that a group gives rise to its subgroups, and the pattern of character distribution is for the ancestral group to have certain characters, and the descendant groups to have certain additional characters. Examples of Haeckelian ancestry are Acrania, Anamnia and Amphibia in Fig. 1, various extinct groups in Fig. 6, and Moneren (= prokaryotes) and Amoeben (= protists) in Fig. 7. The convention here is that a group gives rise to another of equal or higher rank, and the pattern of character distribution is that the ancestral group is always characterised only by lack of the features which distinguish the descendant group, i.e. by absent or privative characters. In terms of the hierarchical structure of the tree, rather than character distribution, von Baerian ancestry is consistent with the hierarchical structure, which is one of decreasing generality, and Haeckelian ancestry is inconsistent with that structure.

Embryos and ancestors

With the theory of evolution established, and the groundwork of phylogeny reconstruction laid by Haeckel's biogenetic law, the work of building genealogies forged ahead in the late nineteenth century, especially amongst embryologists, and then gradually fell into disrepute. By 1893, an editor could write 'three weeks at the seaside with borax-carmine and a rocker microtome, no longer result in an epoch-making paper containing a hypothetical ancestor, and a brand new pedigree' (Anon., 1893, p. 82). Gould (1977, p. 184) points out that one neglected factor in the decline of the biogenetic law was von Baer's demolition of recapitulation. In the nineteenth century, the only von Baerian criticism of Haeckelian recapitulation was Hurst's (1893*a–d*): 'The two views – von Baer's and the Recapitulation Theory – are irreconcilable. Von Baer's is applicable only to certain animals, but its applicability to these disproves 'recapitulation' in these' (Hurst, 1893*b*, p. 365). Significantly, the only replies to Hurst came from palaeontologists, Bather (1893*a*, 1893*b*) and Buckman (1893). Bather (1893*a*, p. 275) first demanded that 'ontogeny recapitulates phylogeny' be disproved by contrary examples 'from the actual history of extinct beings', that is, by 'direct comparison of the ontogeny with the phylogeny in a large number of cases', as Hurst put it (1893*b*, p. 365). Hurst went on 'that has hitherto proved impossible', though he did not say why. But what Bather demanded was phylogenies established without the help of the biogenetic law, and the only alternative method of establishing phylogenies was stratigraphic succession of fossils (e.g. Huxley, 1870; cf. Patterson, 1981). Indeed, Bather referred Hurst to 'the brilliant researches of Hyatt and Buckman on [fossil] Cephalopoda . . . which . . . have been inspired by the Recapitulation Theory' (1893*a*, p. 280). Bather went on to accuse Hurst of the 'main fallacy' of substituting 'contemporary relations for ancestors', pointing out that 'there is all the difference in the world between filial and fraternal relationship'.

This exchange illuminates the trend of phylogenetic research in the late nineteenth century, and Russell's superb survey (1916) takes the story 20 years on: 'There was a close spiritual affinity between the speculative evolutionists and the transcendentalists' (*ibid.* p. 303); 'Phylogenetic method differed in no way from transcendental – except perhaps that it had learned from von Baer

and from Darwin to give more weight to embryology' (*ibid.* p. 304).
Phylogenetic research fell out of favour because, under the influence of recapitulation theory, it led to the proposal of innumerable
irreconcilable and uncriticisable speculations. Russell adds (*ibid.* p.
357) that 'It was a curious, but very typical characteristic of
evolutionary morphology that its devotees paid very little attention
to the positive evidence accumulated by the palaeontologists, but
shut themselves up in their tower of ivory and went on with their
work of constructing ideal genealogies.' Summarising the work of
post-Darwinian palaeontologists, Russell notes first that intermediates between major groups were notably lacking (cf. Olson, 1981),
but that close study of a few limited but complete series had
confirmed the law of recapitulation, and had established 'the actual
"phyletic series"' (Russell, 1916, p. 360), which, when generalised,
showed 'an immense number of phyletic lines which evolve
parallel to one another' (*ibid.* p. 361). And so speculative genealogies, built on embryos, fell into disrepute and were replaced by
concrete phylogenies based on fossils – Haeckel's 'infallible and
indisputable records which fix the correct history of organisms'
(Haeckel, 1876, vol. 1, p. 55).

The new view of phylogeny reconstruction that Russell was
summarising – fossils as the arbiters, parallelism and polyphyly as
the rule – was prevalent until the 1960s or later (e.g. Grassé, 1977).
However, some biologists eventually came to regard fossils and
their stratigraphic succession as no more foolproof than the
biogenetic law (Hennig, 1969; Schaeffer *et al.*, 1972; reviews of
particular groups in Patterson, 1977, 1981; Rosen *et al.*, 1981).
The problem is that epitomised in the 1893 exchange between
Hurst, who asserted that the biogenetic law is false, and Bather,
who demanded that Hurst provide actual phylogenies that disprove
it. How were these phylogenies to be made out? Stratigraphic
succession is the only source that has ever been offered for actual
ancestor–descendant series. The method has recently been made
more explicit as stratophenetics (Bretsky, 1979; Gingerich, 1979).
But referring to the general argument from stratigraphy, Nelson &
Platnick (1981, p. 333) write 'This . . . is simply fallacious;
stratigraphic sequence alone cannot indicate that two fossils belong
to the same lineage . . . . Fossils must be ordered on the basis of
systematic hypotheses, and since these hypotheses may always be
incorrect, fossils so ordered cannot be said to show the truth . . . of

evolution.' Indeed, the possibility that several of the classic 'phyletic series' of fossils may be incorrect has come up in the recent debate over gradualism (Gould & Eldredge, 1977; Stanley, 1979). Riedl (1979, p. 256) criticises the view of fossils as arbiters from a different angle: 'construction of phylogenies without fossil evidence . . . has been condemned as methodological sin . . .. However . . . the condemnation turns the matter upside down. The sole authority is in fact morphological theory.' Where Riedl proposes 'morphological theory' as the authority, Nelson & Platnick (above) offer 'systematic hypotheses'. But the two prescriptions are really no different. For Riedl, 'the so-called homology theorem . . . is the essential part of the principles of morphology' (1979, p. 32), whereas for Nelson & Platnick 'systematics in general consists of the search for defining characters of groups' (1981, p. 304). The alternative arbiters, homology and characters of groups, coincide if 'homology is the relation characterizing natural groups' (Patterson, 1982, p. 65), an opinion also held by Eldredge & Cracraft (1980, p. 36) and Nelson & Platnick (1981, p. 158). Riedl's book, and those of Eldredge & Cracraft and Nelson & Platnick, are signs of a recent resurgence of interest in hierarchical order in nature, and in methods of discerning and representing that order (also Gould, 1980; Lauder, 1981, 1982). Aspects of this are discussed in the next three sections.

Hierarchy versus continuum

The change in systematic or comparative biology which accompanied general acceptance of evolution can be summed up as a shift from an attempt to discover and represent natural order, to a world view in which life was known to be a historical continuum on which biologists must impose order (Griffiths, 1974, p. 87). One illustration of this change is the contrast between Milne-Edwards's (1844) and Agassiz's (1844) use of the map simile (p. 5 above), and Haeckel's post-Darwinian use of the tree (Figs. 1, 6, 7). Milne-Edwards and Agassiz viewed nature as imperfectly known territory which biologists were exploring and charting. But for Haeckel and his successors the *form* of the territory is known – it is a tree – and the biologist's task is to discover the unknown creatures (e.g. '*Amphibia ignota*'; 'unbekannte Uebergangsformen') which will complete the continuum.

✳ Not classes!

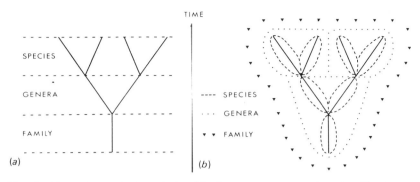

Fig. 8. Two views of the relationship between a tree and a classification, or between time and relationship. (After Simpson, 1961, Fig. 19.)

For pre-Darwinian biologists, species were generally thought to be real – discrete and immutable. For Darwin, they ceased to be real and became arbitrary segments of a continuum (Eldredge, 1979; Cracraft, 1979, 1981). The change is epitomised in an indirect exchange between Agassiz and Darwin: 'If species do not exist at all, as the supporters of the transmutation theory maintain, how can they vary?' (Agassiz, 1860, p. 142) – 'How absurd that logical quibble . . . . As if anyone doubted their temporary existence' (Darwin, 1887, vol. 2, p. 333).

Given the continuous tree composed of evanescent species, it was inevitable that the aims of systematics should change. Discovery of the natural hierarchy was replaced by the imposition of order, more or less arbitrarily, on a continuum. This change is exemplified in Fig. 8, which contrasts a pre-Darwinian and a neo-Darwinian view of the relation between time and hierarchy. For von Baer, the ontogeny of vertebrates is exactly congruent with the natural hierarchy of those animals: the characters of Vertebrata appear before those of vertebrate classes, and so on through the hierarchy. Thus ontogeny is hierarchical, just as is the natural system. Agassiz and Richard Owen extended these ideas into the fossil record: the oldest fossil vertebrates were just vertebrates, and fishes did not assume their characters until tetrapods appeared, nor did reptiles become reptiles until birds and mammals appeared (Agassiz, 1859, p. 166). Owen drew similar conclusions from fossil mammals (Ospovat, 1976). So in Fig. 8a, the time axis may be read as ontogenetic time (von Baer) or geological time (Agassiz, Owen, extending von Baer's laws into the fossil record). The neo-Darwinian view (Fig. 8*b*) is entirely different. Because the tree is a continuum, composed of species throughout, the concept of relationship becomes problematical, and some arbitrariness is necessary to form higher groups (Simpson, 1961; Mayr, 1969, p. 98). 'The only problem was where to apply the ax. The same tree could be chopped up in many different ways, and it often was' (Nelson & Platnick, 1981, p. 126). So it became orthodox to view classification as an art reflecting personal taste or discrimination rather than the hierarchical structure of the tree itself. And since tastes vary, there are many possible classifications, no one necessarily better than another.

This thesis, species and higher categories arbitrarily imposed on a continuum, has far-reaching consequences in biology, resulting in an approach named (by its critics) 'transformational' (Eldredge, 1979), characterised by concentration on process, by smooth extrapolation from population genetics to the origin of higher taxa (Gould, 1980), and so by reductionism which sees evolution and phylogeny as nothing more than changes in gene frequency. The alternative approach, christened 'taxic' by Eldredge (1979), is marked by an interest in pattern before process, or hierarchy rather than continuum, by anti-reductionism which denies that higher hierarchical levels have properties reducible to or predictable from the lowest level, by an interest in testable historical hypotheses rather than those consistent with any observation (Lauder, 1981, 1982), and by the belief that taxa are real and may be discovered (Ball, 1976, p. 426). If taxa are real, it follows that there is only one real or natural hierarchy, one correct or true classification. In systematics, these ideas originated with a new definition of relationship, which led to a new conception of types of group.

Ancestry and types of group

Haeckel used five different conventions to denote ancestry in his trees, and there is a sixth possibility, ancestral species. I suggested above that these could be divided into two: first, those which invoke only the decreasing generality of von Baer's law (cf. Fig. 8*a*); and secondly, those which concern actual historical descent (cf. Fig. 8*b*). As phylogeny, documented by fossils, and the historical viewpoint came to dominate higher classification (Simp-

son, 1945), and since ancestral species remained elusive, groups of
the second type occupied the trunks of trees and played a major part
in phylogenies. These ancestral groups, or Haeckelian conventions
of ancestry, may be extinct, containing fossils only, or extant. For
example, Romer's (1966, Fig. 316) spindle diagram of placental
mammals shows an extant order, Insectivora, as ancestral to all
other orders, and an extinct order, Condylarthra, as ancestral to all
the ungulate orders.

    With this view of phylogeny, groups giving rise to groups, the
concepts of monophyly and polyphyly, which Haeckel had introduced (1866), were much discussed, and 'minimal monophyly' was
proposed (Simpson, 1945, 1961): a group may be considered
monophyletic if it is descended from a group of equal or (prefer-
ably) lower rank. This criterion allowed descendant groups to be
polyphyletic, in the sense that they were derived by more than one
lineage from the ancestral group, and during the 1950s and 60s
much paper was used in discussions of the mode of origin of various
groups. These discussions ceased when the consequences of Hennig's (1966, p. 74) definition of relationship became widely dif-
fused. Hennig defined relationship by recency of common ancestry: given three species or groups, those two are more closely
related which share an ancestral species not shared by the third.
One consequence of this definition is that groups of related species
exactly match the hierarchical structure of the tree. In other words,
Hennig used evolutionary theory to rediscover a congruence between groups and the hierarchy like that used by pre-Darwinians
such as von Baer, Milne-Edwards and Agassiz (Figs. 2–5).

    A second consequence of Hennig's definition of relationship is
that there are not two types of group, monophyletic and polyphyletic, as evolutionists from Haeckel to Simpson had supposed, but
three. Hennig named the third type paraphyletic: a monophyletic
group includes an ancestor and all its descendants, a paraphyletic
group includes an ancestor and some but not all of its descendants,
and a polyphyletic group does not include a common ancestor
(those statements paraphrase later discussions of Hennig's original
definitions – Holmes, 1980). Figure 8*b* illustrates the three concepts as applied to a theory of relationships: the whole family would
be monophyletic, since it includes the common ancestral species
and all its descendants; the extinct genus would be paraphyletic,
since it includes the common ancestral species and some but not all

of its descendants; and the two extant genera would be polyphyletic, since neither includes the species ancestral to its members. The three concepts may also be defined in practical rather than theoretical terms: monophyletic groups are characterised by shared derived characters (Hennig's synapomorphies), paraphyletic groups by shared primitive characters (Hennig's symplesiomorphies), and polyphyletic groups by convergent characters. Again, these definitions are illustrated by Fig. 8*b*. For example, the two extant genera can only be characterised by independently evolved (convergent or non-homologous) characters, since whatever they uniquely share was absent in their common ancestor.

The concepts of monophyly, paraphyly and polyphyly have since been much discussed (see Holmes, 1980; Wiley, 1981), but there is general agreement that Hennig's discovery of paraphyly was new and important. This is evident from Fig. 8*b*, taken from Simpson (1961), who regarded all the groupings as valid monophyletic taxa. Hennig recommended (1966, p. 146) that paraphyletic groups should be avoided, since they have no independent history, reality, individuality, or exclusive common ancestry and point of origin; 'The concept of "extinction" has thus a different meaning for them than it does for monophyletic groups.' For a paraphyletic group can become extinct only by 'transformation' – this is true of all ancestral groups. Hennig's references to reality and individuality reflect his belief that higher taxa are real, and 'exist whether or not there are any systematists around to perceive or name them' (Wiley, 1981, p. 72). This idea is diametrically opposed to the prevailing neo-Darwinian view that taxa are artifacts, invented or imposed on nature according to the skill and taste of systematists. The theme of the reality of monophyletic taxa, in Hennig's sense, has come to be one of the central tenets of cladistics, the school of systematics that he founded. And the equation of monophyletic groups with natural groups (e.g. Farris, 1980; Nelson & Platnick, 1981; Wiley, 1981) is another sign of convergence between modern systematics and the pre-Darwinian systematists' attempt to map or mirror a discoverable natural hierarchy.

Hennig's objections to paraphyletic (= ancestral) groups were theoretical. An additional theoretical objection to such groups is that they are inconsistent with evolutionary theory: species, not groups of species, are the ancestral units in evolution (Wiley, 1979). Practical objections are that paraphyletic groups can only be

defined arbitrarily or by authority, since an arbitrary line must be
imposed to separate the ancestral group from the group or groups
descended or detached from it, and this residual cluster has no
characters of its own (Patterson, 1982) – it is an 'almost meaning-
less set' (Ball, 1976, p. 426). To be consistent with evolutionary
theory, evolutionary trees must consist only of species. Those
species may be hypothetical or named, but in neither case is it
legitimate to combine them in ancestral groups.

Once paraphyletic groups are recognised as systematists' inven-
tions, the trunks of phylogenetic trees melt away into hypothesis.
Nevertheless, Hennig's concepts of monophyly and paraphyly were
firmly tied to phylogeny, both in theory, by common ancestry, and,
in practice, by derived (= advanced) versus primitive (= ancestral)
characters, which necessitate determination of character phylogenies, morphoclines, or 'transformation series', as Hennig called
them.

Character phylogeny

Methods of distinguishing primitive and derived characters, or
determining the polarity of morphoclines, have recently been
reviewed from several angles. De Jong (1980) distinguished no less
than 21 different arguments, and recommended seven, particularly
outgroup analysis. Stevens (1980) discussed seven groups of
criteria (ten criteria in all), and also recommended outgroup
analysis. Wiley (1981) reviewed six methods and recommended
two, outgroup comparison and ontogenetic sequence, which he
regarded as 'complementary to . . . out-group comparison'. Rieppel
(1979) discussed, and rejected, the ontogenetic criterion. Watrous
& Wheeler (1981) analysed outgroup comparison, and commended
it above all other criteria; and Bishop (1982) reviewed five criteria
and rejected all save functional morphology. Thus most authors
choose outgroup comparison as the decisive criterion, and reject
ontogenetic sequence. The reasons given for rejecting ontogeny are
varied. De Jong (1980, p. 18) writes 'the ontogenetic argument is of
little use for the distinction between apomorphous and plesiomorphous character states' because 'parallels between ontogeny and
phylogeny . . . can only be observed if the phylogeny is known
. . .. Observation of gill arches in adult fishes and embryonic
mammals does not *per se* tell us that the mammals evolved from a

fish-like ancestor. Without the help of fossils and other evidence (i.e. without a presupposed phylogeny) the fishes could have been regarded as derived from mammals.' In short, the argument from ontogeny is 'based on tendencies observed in established phylogenies' (*ibid.* p. 21). Like Wiley (1981), Stevens (1980, p. 341) argued that the ontogenetic criterion must be used in a comparative context, 'i.e. in some form of out-group analysis', and that it depends on 'the implicit assumption that evolution leads to an increase of complexity.' Both de Jong and Stevens comment that the ontogenetic argument is invalid whenever neoteny or paedomorphosis occurs. Rieppel (1979) criticised the ontogenetic criterion by giving examples from reptiles (vertebrae, palatobasal articulation, trabeculae) to show that ontogenetic criteria fail to match phylogenies established by outgroup comparison. Rieppel accounts for that failure by heterochrony, deletion, deviation or paedomorphosis.

All those criticisms seem to me to miss the point. Hennig (1966, p. 88) introduced the terms 'apomorphous' and 'plesiomorphous' with a discussion of evolution as transformation, with the equation of 'derived' (as in 'shared derived characters') and 'transformed', and with the expression 'transformation series' to describe the difference between symplesiomorphy and synapomorphy. The notion of character phylogeny is therefore one of transformation, or of direction of transformation.

Phylogeny is generalised transformation, but we have no empirical experience of phylogeny; the only transformations of which we have empirical evidence are those of ontogeny. When de Jong suggests that, on the evidence of gill arches in the embryo, we might as well derive fishes from mammals as the reverse, ontogenetic transformation shows him to be wrong. For we know organisms (anurans, urodeles) in which we observe transformation of serially similar, gill-bearing gill arches, separating open clefts (the fish condition), into serially differentiated, cleftless, gill-less arches (the mammal condition), and we know no examples of the reverse transformation. This does not show that mammals are descended from fishes, for the same observations were available to von Baer and Agassiz. Rather it shows, as it showed von Baer, that there is a general condition, seen in embryos of fishes, amphibians and mammals, and a special condition in adults of each of these types; and that the transformations in amphibians are more extensive than

Genealogies of domesticated animals with known pedigrees.

in fishes, and those in mammals are more extensive than in amphibians. Empirical transformation is, I believe, what Nelson (1973, 1978) meant when he differentiated the ontogenetic criterion of polarity from all others as a direct argument 'not involving other species or groups' (1973, p. 87), or not requiring 'a prior assumption of relationships (or a higher level phylogeny)' (1978, p. 325). All other criteria, including outgroup comparison, are indirect, and require prior assumptions.

The mistake made by those who criticise or neglect the ontogenetic criterion is, in my view, that their training forces them to see it only in a phylogenetic, or evolutionary, or Haeckelian context. For example, Wiley (1981, p. 154) writes, 'The ontogenetic criterion assumes that ontogenetic transformation toward a particular character reflects the phylogenetic development of that ontogeny.' But no such assumption is necessary. All one need assume is that, since ontogenetic transformation is consistently observed to be in one direction, and never in the reverse, we have direct evidence of transformation, and may rate the untransformed state as more general ('primitive') than the transformed state. Hence, Nelson's (1978, p. 327) re-statement of Haeckel's biogenetic law in von Baerian terms: 'given an ontogenetic character transformation, from a character observed to be more general to a character observed to be less general, the more general character is primitive and the less general advanced.' Just as the transformation is empirical, so is the generality, in terms of numbers of species. Wiley (1981, p. 156) rephrased the law, replacing 'observed to be more general' by 'found in the outgroup', and replacing 'observed to be less general' by 'found only in the group considered', and so argued that the ontogenetic criterion is complementary to outgroup comparison. But Wiley's modification brings in the problem with outgroup comparison, and all other indirect methods; what criterion is to be used to establish the ingroup and the outgroup? (Patterson, 1982, p. 52). In other words, without some prior knowledge of character phylogeny, how can we investigate phylogeny? More to the point is that Wiley's reformulation obscures the fact that the general condition is not 'found in the outgroup', but found both in the outgroup *and* the ingroup (in early ontogeny), so that no prior establishment of ingroup and outgroup is necessary. Without the transformations of ontogeny, which by themselves define nested sets (outgroup + ingroup → ingroup), systematics would be impossible, as Cartmill (1981), who neglects this criterion, suggests it is.

Von Baerian and Haeckelian recapitulation

Løvtrup (1978) distinguished these two types of recapitulation. He epitomised von Baer's laws as 'Ontogeny recapitulates the taxonomic course and the mechanism of phylogeny' (*ibid.* 351), and contrasted that with Haeckel's biogenetic law. Løvtrup differentiated the two theories by the Haeckelian demand that all phylogenetic novelties be terminal, so that terminal features of ancestors are shunted back in descendants to become non-terminal, and the von Baerian tolerance of non-terminal novelties. Løvtrup concluded (*ibid.* p. 352) 'All instances of ontogeny are accounted for by von Baerian recapitulation', whereas Haeckelian recapitulation is excluded during morphogenesis (the creation of form, as opposed to growth, the modification of form), and can concern only terminal or subterminal features.

Leaving aside the propriety of citing von Baer as an advocate of recapitulation (Gould, 1979), von Baerian and Haeckelian recapitulation can also be sharply differentiated in phylogenetic rather than ontogenetic terms: von Baerian recapitulation is a guide to *common* ancestry, and Haeckelian recapitulation is a guide to *direct* ancestry. Since direct ancestry is a special case of common ancestry (Nelson, 1974), this distinction agrees with Løvtrup's (1978) proposition that Haeckelian recapitulation is a special case of von Baerian recapitulation. The distinction between the two types of recapitulation in terms of ancestry recalls the differentiation (p. 12) between von Baerian conventions of ancestry (a group gives rise to its subgroups) and Haeckelian conventions – ancestral groups of rank equal to or lower than that of their descendants.

There is no space here to summarise Riedl's (1979) analysis of terminal and non-terminal features (which he calls metaphenes and interphenes). Riedl does not distinguish between Haeckelian and von Baerian recapitulation, regarding the biogenetic law as true and all recapitulation as Haeckelian, but his discussion of recapitulatory and caenogenetic features in terms of his concept of burden ('the responsibility carried by a feature or decision', quantified in numbers of dependent homologues) is relevant and illuminating.

Does phylogeny differ from ontogeny?

In seeking answers to this question, we must bear in mind Haeckel's distinction between the two aspects of phylogeny, the chain of direct ancestors which he saw as recapitulated in the individual's ontogeny, and the branching phylogenetic tree, which he saw in the natural hierarchy of systematics, and which is expressed in ontogenetic terms in von Baer's laws. The difference between the two aspects is shown in the contrast between Haeckel's biogenetic law of serial transformation and von Baer's laws of hierarchical individualisation. Further, in order to specify where or how phylogeny differs from ontogeny, we need some knowledge of phylogeny. Regarding the first aspect, the chain of ancestors in any particular lineage, we are hardly better informed than in Haeckel's time, for no general method of identifying ancestors has yet been found. Instead the chain of ancestors has been approximated by what I have called Haeckelian conventions of ancestry, chopping the systematic hierarchy horizontally into extinct ancestral groups, or phenetic envelopes surrounding certain known fossils. Such groups are paraphyletic, or unreal, or invented, and tell us nothing more about phylogeny than that monophyletic groups possessing certain characters have evolved from ancestors which lacked those characters. Since this is no more than a rephrasing of von Baer's laws, it can hardly be inconsistent with ontogeny, in which adults possessing certain characters evolve from embryos which lack those characters.

With the branching-tree aspect we seem to be on firmer ground, for the systematic hierarchy has been refined since Haeckel's time. Figure 9 compares the cladogram of gnathostomes derived from von Baer's 1828 scheme (Fig. 4b) with one modern version. The significance of the differences between the two is simple. Just as Milne-Edwards's diagram (Fig. 4a) differed from von Baer's in lacking one 'group' (reptiles + birds) whose only character in von Baer's scheme was a lack (of the placenta), so the modern clado- gram of gnathostomes differs from von Baer's in having fragmented other 'groups' which were uncharacterised. Von Baer's Pisces, characterised by lack of lungs (Fig. 2) is broken up into four groups, each the sister-group of everything to its right: chondrich- thyans, actinopterygians, and two groups undiscovered in von Baer's time, coelacanths and lungfishes. Von Baer's Anamniota,

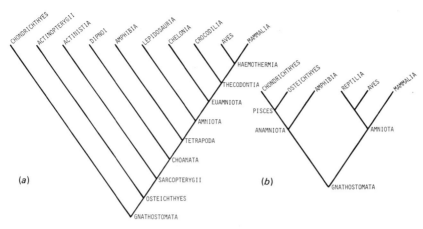

Fig. 9. Cladograms of gnathostome vertebrates from (*a*) Gardiner (1982, Fig. 2 – tetrapods) and Rosen *et al.* (1981, Fig. 62 – non-tetrapods), and (*b*) von Baer (1828: cf. Fig. 4*b*).

uniting fishes and amphibians and characterised by gills, has also disappeared; it comprises the four successive groups of fishes and the amphibians. And von Baer's Reptilia, characterised by lack of wings or air-sacs, is broken up into three groups, lepidosaurs, chelonians and crocodiles, each the sister of everything to its right. The change comprises the elimination by subdivision of all the uncharacterised, or paraphyletic, or potentially ancestral groups.

Von Baer's laws can be tested against the cladogram in Fig. 9*a*. His first law — more general characters appear earlier in the embryo than more special – predicts that the characters of the groups named up the axis of the cladogram should develop in sequence. They do not: the amnion and allantois, characterising Amniota, a subgroup of Tetrapoda, appear before any feature of the latter group, and even before features of Gnathostomata such as jaws and limb girdles. Von Baer's first law therefore fails as a sequential guide to hierarchical ranking of features (cf. Müller, 1864, on Crustacea). Von Baer's second law – less general characters are developed from the most general, and so on – is not presented as a guide to sequence in the whole organism, but to sequence in structural relations. The embryonic membranes of amniotes, as developments (outgrowths) or more widely distributed structures, are consistent with this law, as is, so far as I know, every other

observation in vertebrate morphology. Von Baer's third and fourth
laws are denials of recapitulation of adult stages. He summed up his
work – 'the most general result' – as 'The history of the development
of the individual is the history of its increasing individuality in all
respects' (Huxley's (1853) translation, p. 237). That statement
seems to be applicable to Fig. 9a, and to any more detailed analysis
of vertebrate relationships.

Von Baer's conclusions seem to be generally true for the systematic hierarchy. But are they true of phylogeny? First, is Fig. 9a a
phylogeny? By any definition, phylogeny concerns history, and the
diagram may easily be converted from an atemporal cladogram into
a historical tree by mentally adding an implication of common
ancestry to the bifurcations. This makes what has been called an
'X-tree', a tree containing no ancestors (each ancestor is an
unknown 'X') but carrying the implication of a time-scale and of
descent and speciation. Von Baer's second law and his general
conclusion must apply to this tree just as they applied to the
cladogram, for the mental operation of converting it from tree to
cladogram cannot have modified its content. How then does
phylogeny differ from ontogeny?

There is one class of phylogenies, or X-trees, that does differ
from ontogeny: molecular phylogenies, derived from nucleotide or
protein sequence data. The transformations postulated in molecular phylogenies are gene duplications and changes in nucleotides
that are nowhere matched in ontogeny. Yet, paradoxically, this
difference brings out one of the similarities between phylogeny and
ontogeny. Molecular sequences are linear, or one-dimensional, and
do not themselves display hierarchic organisation. In building
phylogenies from homologous sequences, methods such as parsimony (minimum evolution), compatibility or likelihood (Felsenstein, 1979) must be used to link the sequences by hypothetical
ancestral sequences. Different methods (assumptions) may give
different phylogenies from the same data (e.g. Romero-Herrera *et
al.*, 1978); that is, the tree topology and the ancestral sequences
hypothesised are assumption dependent. This means that hierarchical structure is not inherent in comparative sequence data
(except with very limited samples), but is imposed by the investiga-
tor. Ontogeny also imposes hierarchical organisation on linear
sequence: the genome is one-dimensional, but ontogeny is hierarchical. So in comparative morphological data, hierarchical orga-

nisation is inherent, for it is imposed by the transformations of ontogeny (von Baer's laws). Where ontogeny is well known, as it is generally in vertebrates, an almost infinite number of alternative X-trees (all of which are equally plausible given sequence data alone) is reduced to manageable proportions, or constrained, by groupings dictated by the known, unreversed (and therefore inviolate) transformations of ontogeny. As Jarvik wrote (1980, p. 230), 'it is to be required that phylogeny be in accord with ontogeny'. That this is so is indicated by the fact that most evolutionary 'laws' or 'rules' seem to be descriptions of ontogeny: this is true, for example, of Dollo's law of irreversibility, of Cope's rule of increasing size, of Cope's 'law of the unspecialised', that new types of construction start from small, unspecialised forms, of Williston's rule of increasing individuality of initially similar parts, and so on. This congruence recalls Nelson's (1978, p. 336) conclusions 'that systematics and comparative anatomy (applied to fossils, too) are possible only to the extent that ontogeny is orderly' and 'that the concept of evolution is an extrapolation, or interpretation, of the orderliness of ontogeny'. If so, phylogeny can hardly differ from ontogeny.

Of course, phylogeny differs from ontogeny in that it is a sequence of ontogenies, or life cycles, but those life cycles are lost to us, and remain conjectural. Deductions about them come from diagrams like Fig. 9a, or from more detailed X-trees including those fossil forms which are sufficiently well known to be placed unambiguously. Fig. 9a demands, for reasons of parsimony, that birds and mammals are descended from ancestors in whose ontogenies median fins with horny and bony finrays were developed; several generations of teeth developed on the jaw bones and on all the dermal bones of the buccal cavity and pharynx; enamel and dentine developed on the entire body surface rather than on the teeth alone; dermal bones developed all over the body, and were more numerous in the skull; enclosed sensory canals with neuromasts developed on the head and trunk; and so on. Transformations from these conditions to those found in adult birds and mammals are not seen in ontogeny, and so phylogeny differs from ontogeny. Yet if the transformations required are generalised, they comprise only loss or suppression (e.g. of median fins and finrays, of teeth in birds), and restriction or individualisation (e.g. of dentine and enamel formation, of dentition, of dermal bones). And

of course, loss or suppression, restriction and individualisation are
well known in ontogeny.

Finally, phylogeny also differs from ontogeny in that it is
conceived as a random walk, whereas ontogeny is the epitome of
economy, or purposefulness. However, Riedl asserts (1979, p. 235)
'that evolution is excluded from accident to a much greater degree
than hitherto supposed and that this shielding is the necessary
result of a selection dictated . . . by the functional systemic
conditions in the organization of organisms themselves'. Which
again sounds like a description of ontogeny.

References

Agassiz, J. L. R. (1833–44). *Recherches sur les poissons fossiles.* 5 vols. Petit-
pierre: Neuchâtel. Dates of publication of parts are given by W. H. Brown in
Woodward, A. S. & Sherborn, C. D. (1890) *A catalogue of British fossil
Vertebrata*, pp. xxv–xxix, Dulau: London.

Agassiz, J. L. R. (1845). *Monographie des poissons fossiles du vieux grès rouge.*
Petitpierre: Neuchâtel.

Agassiz, J. L. R. (1859). *An essay on classification.* Longman, Brown, Green,
Longmans & Roberts, and Trübner & Co.: London.

Agassiz, J. L. R. (1860). Professor Agassiz on the origin of species. Individuality
and specific differences among Acalephs. *Am. J. Sc.* (2) **30**, 142–55.

Anon. (1893). Vitalism. *Nat. Sci. Lond.* **2**, 82–3.

Ball, I. R. (1976). Nature and formulation of biogeographical hypotheses. *Syst.
Zool.* **24**, 407–30.

Bather, F. A. (1893*a*). The recapitulation theory in palaeontology. *Nat. Sci. Lond.*
2, 275–81.

Bather, F. A. (1893*b*). Phylogeny *v.* ontogeny. *Nat. Sci. Lond.* **3**, 238–9.

Bishop, M. J. (1982). Criteria for the determination of character state changes.
Zool. J. Linn. Soc. **74**, 197–206.

Bowler, P. J. (1975). The changing meaning of 'Evolution'. *J. Hist. Ideas*, **36**,
95–114.

Bretsky, S. S. (1979). Recognition of ancestor–descendant relationships in
invertebrate paleontology. In *Phylogenetic analysis and paleontology*, eds. J.
Cracraft and N. Eldredge, pp. 113–63. Columbia University Press: New York.

Buckman, S. S. (1893). The recapitulation theory in biology. *Nat. Sci. Lond.* **3**,
138–9.

Cartmill, M. (1981). Hypothesis testing and phylogenetic reconstruction. *Z. zool.
Syst. Evolutionsforsch.* **19**, 73–96.

Cracraft, J. (1979). Phylogenetic analysis, evolutionary models, and paleontology.
In *Phylogenetic analysis and paleontology*, eds. J. Cracraft and N. Eldredge,
pp. 7–39. Columbia University Press: New York.

Cracraft, J. (1981). Pattern and process in paleobiology: the role of cladistic
analysis in systematic paleontology. *Paleobiology*, **7**, 456–68.

Darwin, F. (1887). *The life and letters of Charles Darwin.* 3 vols. John Murray: London.

Eldredge, N. (1979). Alternative approaches to evolutionary theory. *Bull. Carnegie Mus. Nat. Hist.* **13**, 7–19.

Eldredge, N. & Cracraft, J. (1980). *Phylogenetic patterns and the evolutionary process.* Columbia University Press: New York.

Farris, J. S. (1980). The information content of the phylogenetic system. *Syst. Zool.* **28**, 483–519.

Felsenstein, J. (1979). Alternative methods of phylogenetic inference and their interrelationship. *Syst. Zool.* **28**, 49–62.

Gardiner, B. G. (1982). Tetrapod classification. *Zool. J. Linn. Soc.* **74**, 207–32.

Gingerich, P. D. (1979). The stratophenetic approach to phylogeny reconstruction in vertebrate paleontology. In *Phylogenetic analysis and paleontology*, eds. J. Cracraft and N. Eldredge, pp. 41–77. Columbia University Press: New York.

Gould, S. J. (1977). *Ontogeny and phylogeny.* Belknap Press: Cambridge, Mass.

Gould, S. J. (1979). On the importance of heterochrony for evolutionary biology. *Syst. Zool.* **28**, 224–6.

Gould, S. J. (1980). Is a new and general theory of evolution emerging? *Paleobiology*, **6**, 119–30.

Gould, S. J. & Eldredge, N. (1977). Punctuated equilibria: the tempo and mode of evolution reconsidered. *Paleobiology*, **3**, 115–51.

Grassé, P. P. (1977). *Evolution of living organisms.* Academic Press: New York.

Griffiths, G. C. D. (1974). On the foundations of biological systematics. *Acta biotheor.* **23**, 85–131.

Haeckel, E. (1866). *Generelle Morphologie der Organismen.* 2 vols. G. Reimer: Berlin.

Haeckel, E. (1868). *Natürliche Schöpfungsgeschichte.* G. Reimer: Berlin.

Haeckel, E. (1874). *Anthropogenie oder Entwickelungsgeschichte des Menschen.* W. Engelmann: Leipzig.

Haeckel, E. (1876). *The history of creation.* 2 vols. H. S. King. London.

Hansen, E. D. (1977). *The origin and early evolution of animals.* Wesleyan University Press: Middletown, Conn.; and Pitman: London.

Hennig, W. (1966). *Phylogenetic systematics.* University of Illinois Press: Urbana, Ill.

Hennig, W. (1969). *Die Stammesgeschichte der Insekten.* W. Kramer: Frankfurt am Main.

Holmes, E. B. (1980). Reconsideration of some systematic concepts and terms. *Evolut. Theory*, **5**, 35–87.

Hurst, C. H. (1893*a*). The recapitulation theory. *Nat. Sci. Lond.* **2**, 195–200.

Hurst, C. H. (1893*b*). The recapitulation theory: a rejoinder. *Nat. Sci. Lond.* **2**, 364–9.

Hurst, C. H. (1893*c*). 'Recapitulation' and 'earlier inheritance'. *Nat. Sci. Lond.* **3**, 239.

Hurst, C. H. (1893*d*). Phylogeny and ontogeny. *Nat. Sci. Lond.* **3**, 316.

Huxley, T. H. (1870). The anniversary address of the President. *Q. J. geol. Soc. Lond.* **26**, xlii–lxiv.

Jarvik, E. (1980). *Basic structure and evolution of vertebrates*, vol. 2. Academic Press: London.

de Jong, R. (1980). Some tools for evolutionary and phylogenetic studies. *Z. zool. Syst. Evolutionsforsch.* **18**, 1–23.

Lauder, G. V. (1981). Form and function: structural analysis in evolutionary morphology. *Paleobiology*, **7**, 430–42.

Lauder, G. V. (1982). Introduction to *Form and function* by E. S. Russell, 2nd edn, pp. xi–xlv. University of Chicago Press: Chicago.

Løvtrup, S. (1978). On von Baerian and Haeckelian recapitulation. *Syst. Zool.* **27**, 348–52.

Mayr, E. (1969). *Principles of systematic zoology*. McGraw-Hill: New York.

Milne-Edwards, H. (1844). Considérations sur quelques principes relatifs à la classification naturelle des animaux. *Annls Sci. nat.* (3) **1**, 65–99.

Müller, F. (1864). *Für Darwin*. W. Engelmann: Leipzig.

Nelson, G. J. (1973). The higher-level phylogeny of vertebrates. *Syst. Zool.* **22**, 87–91.

Nelson, G. J. (1974). Classification as an expression of phylogenetic relationships. *Syst. Zool.* **22**, 344–59.

Nelson, G. J. (1978). Ontogeny, phylogeny, paleontology, and the biogenetic law. *Syst. Zool.* **27**, 324–345.

Nelson, G. J. & Platnick, N. I. (1981). *Systematics and biogeography: cladistics and vicariance*. Columbia University Press: New York.

Olson, E. C. (1981). The problem of missing links: today and yesterday. *Q. Rev. Biol.* **56**, 405–42.

Ospovat, D. (1976). The influence of Karl Ernst von Baer's embryology, 1828–1859. *J. Hist. Biol.* **9**, 1–28.

Ospovat, D. (1981). *The development of Darwin's theory*. Cambridge University Press: Cambridge.

Patterson, C. (1977). The contribution of paleontology to teleostean phylogeny. In *Major patterns in vertebrate evolution*, eds. M. K. Hecht, P. C. Goody and B. M. Hecht, pp. 579–643. Plenum Press: New York.

Patterson, C. (1981). Significance of fossils in determining evolutionary relationships. *A. Rev. Ecol. Syst.* **12**, 195–223.

Patterson, C. (1982). Morphological characters and homology. In *Problems of phylogenetic reconstruction*, eds. K. A. Joysey and A. E. Friday, pp. 21–74. Academic Press: London.

Riedl, R. (1979). *Order in living organisms: a systems analysis of evolution*. John Wiley: Chichester.

Rieppel, O. (1979). Ontogeny and the recognition of primitive character states. *Z. zool. Syst. Evolutionsforsch.* **17**, 57–61.

Romer, A. S. (1966). *Vertebrate paleontology*, 3rd edn. University of Chicago Press: Chicago.

Romero-Herrera, A. E., Lehmann, H., Joysey, K. A. & Friday, A. E. (1978). On the evolution of myoglobin. *Phil. Trans. R. Soc. Lond. B* **283**, 61–163.

Rosen, D. E., Forey, P. L., Gardiner, B. G. & Patterson, C. (1981). Lungfishes, tetrapods, paleontology, and plesiomorphy. *Bull. Am. Mus. nat. Hist.* **167**, 159–276.

Russell, E. S. (1916). *Form and function.* John Murray: London.
Schaeffer, B., Hecht, M. K. & Eldredge, N. (1972). Phylogeny and paleontology. *Evolut. Biol.* **6,** 31–46.
Simpson, G. G. (1945). The principles of classification and a classification of mammals. *Bull. Am. Mus. nat. Hist.* **85,** 1–350.
Simpson, G. G. (1961). *Principles of animal taxonomy.* Columbia University Press: New York.
Stanley, S. M. (1979). *Macroevolution.* W. H. Freeman: San Francisco.
Stevens, P. F. (1980). Evolutionary polarity of character states. *A. Rev. Ecol. Syst.* **11,** 333–58.
Tribe, M., Morgan, A. & Whittaker, P. (1981). *The evolution of eukaryotic cells,* Institute of Biology's Studies in Biology, No. 131. Edward Arnold: London.
Von Baer, K. E. (1828). *Ueber Entwickelungsgeschichte der Thiere. Beobachtung und Reflexion.* Bornträger: Königsberg.
Von Baer, K. E. (1853). Fragments relating to philosophical zoology. (Selected and translated by T. H. Huxley.) *Sci. Mem.* 1853, 176–238.
Watrous, L. E. & Wheeler, Q. D. (1981). The out-group comparison method of character analysis. *Syst. Zool.* **30,** 1–11.
Wiley, E. O. (1979). Ancestors, species, and cladograms – remarks on the symposium. In *Phylogenetic analysis and paleontology,* eds. J. Cracraft and N. Eldredge, pp. 211–25. Columbia University Press: New York.
Wiley, E. O. (1981). *Phylogenetics.* John Wiley: New York.

Better get this !

#6 42C0

Evolution and development

J. MAYNARD SMITH

School of Biological Sciences, University of Sussex, Brighton,
Sussex BN1 9QG, UK

EMBRYOLOGY - MECHANISMS
HOMOLOGY

It is difficult to discuss the relationship between evolution and development because, although we have a clear and highly articulated theory of evolution, we have no comparable theory of development. I therefore start by outlining the kind of theory of development which, however tentatively, most of us hold. I will then discuss what relevance, if any, such a theory has for evolution.

I must first explain what I shall mean by 'development'. Essentially, I mean the processes leading to an increase in geometrical complexity between egg and adult. Adult organisms may synthesize a greater variety of proteins than do eggs, but the difference is neither large nor, in principle, hard to understand. What is hard to understand, however, is how a spatially heterogeneous adult can develop from a relatively homogeneous egg.

A classification of developmental processes

Many different kinds of processes have been suggested which could lead to an increase in geometrical complexity. I start by offering a classification of such processes. It is not exhaustive, nor am I convinced that it is arranged in the most illuminating way. It is offered in the hope of provoking others into giving a more complete and rational classification.

We have a good idea of how cells can produce sets of specific proteins. Since Monod & Jacob (1961), we can imagine how different cytoplasmic constituents of cells can activate different sets of genes, thus amplifying initially small differences. Although such differentiation must be both more complex and more stable in eukaryotes than in prokaryotes, there are no particular conceptual difficulties in imagining how it might happen. The major problem lies in how different kinds of cells come to be arranged in space in particular ways. I suggest the following classification of the processes responsible for such spatial arrangement:

A. Molecular jigsaws.
B. Cellular mechanisms.
    1. Cell division and determination coincident in time.
    2. Cell sorting.
    3. Determination subsequent to cell division:
        (i) causes external to tissue (e.g. induction),
        (ii) causes internal to tissue:
            (*a*) positional information
            (*b*) prepattern and competence
            (*c*) cell shape generates tissue shape.

A. Molecular jigsaws

We have a rather good understanding of how the primary structure of proteins is determined, and some reason to think that the primary structure determines the three-dimensional shape of the whole molecule. It is then possible that larger-scale structures arise because proteins fit together, like pieces of a jigsaw. If so, the final structure would require a specification of which proteins were to be made (i.e. which genes were to be switched on), in approximately what proportions. It could also depend on the sequence in which proteins are made, and possibly on the pre-existence of some structure analogous to the seed of a crystal. It would require that physical conditions (e.g. temperature, pH) be kept within certain bounds.

It seems certain that such processes play a role in the formation of viral structures, and probably also of subcellular organelles. I doubt if they are important on a larger scale, except insofar as the existence of such structures as ribosomes is a necessary precondition for the development of larger-scale structures. However, in those cases in which there is a large amount of cytoplasmic localization in the egg before cleavage, we cannot rule out that some of the localization arises by a jigsaw-like process.

B. Cellular mechanisms

1. *Cell division and determination coincident in time*

The fate of two cells may become different at the time of cell division because of pre-existing cytoplasmic localizations. An alternative possibility is that the state of activation of genes becomes different at the time of chromosome replication. Cytoplasmic localization is known to be important in most eggs, and is probably imposed from outside by the female laying the egg. Determination occurring at the time of cell division is also involved in the differentiation of, for example, the haemoipoetic system.

If all division merely partitions, between cells, cytoplasmic localizations which already existed, the increase in geometric complexity generated would be small. Alternatively, we can suppose that new localizations are generated within daughter cells, which in turn specify new planes of division and new differentiations. If so, there could be a continued increase in complexity. The difficulty lies in imagining how successive differentiations could be coordinated on a global scale; consequently, such a process would be unlikely to be able to regulate.

2. *Cell sorting*

Suppose that cells of two or more kinds are produced, and subsequently move relative to one another. If the cells have differing degrees of affinity for one another, analogous to the interlocking of jigsaw pieces, then they will arrange themselves in regular patterns, regardless of their initial arrangement. There remains the question of how appropriate numbers of different kinds of cells arise in the first place; cell sorting will merely arrange them in space once they are formed.

3. *Determination subsequent to cell division*

I now consider cases in which a population of cells – e.g. a sheet of cells – are initially identical in their potentialities, and become different only later.

(i) *Causes external to the differentiating tissue*
In embryonic induction, a set of initially similar cells become different because some of them come into contact with another cell

layer. In effect, the folding of one cell layer can induce increased spatial complexity in another. This raises the possibility of serial induction: each group of cells, once differentiated, imposing differentiation on the next group.

(ii) *Causes internal to the differentiating tissue*
Since Turing (1952), it has become possible to imagine how an initially homogeneous region can become heterogeneous, without the need of a localized external influence. Turing imagined this happening through chemical reaction and diffusion, without flow, active transport, or electrical phenomena, and without need for cellularization. However, we can, in principle, imagine a variety of pattern-forming processes. In effect, dynamical processes in space rather readily give rise to wave-like inhomogeneities; if the wave-length is large relative to the size of the field, the resulting pattern will be gradient-like rather than wave-like.

I will not attempt to list the various physical and chemical processes which have been proposed. However, some general points are worth making. First, the pattern-generating process need not depend on cellularization. Secondly, the patterns which arise are likely to be simple – e.g. waves, gradients. Thirdly, there is no need to postulate any initial inhomogeneity, other than the presence of boundaries to the developing fields. But some models do assume the existence of some privileged region or boundary, which may, for example, act as a source or sink.

The interactions between the morphogenetic field and the cells it influences can take several forms (Fig. 1).

(*a*) *Positional information.* Wolpert (1969) imagines a process giving rise to a monotonic gradient (or to two orthogonal gradients). The value of some scalar quantity (or quantities) uniquely specifies each position in the field. Cells then respond to the local value of this scalar by differentiating appropriately. All the complexity of the final pattern, over and above that manifest in a monotonic gradient, is then generated by the capacity of cells to respond differently to different values of the positional scalar.

(*b*) *Prepattern and competence.* Stern (1954) and Maynard Smith & Sondhi (1961) were primarily concerned to account for patterns composed of one kind of element, or of a few kinds, arranged in

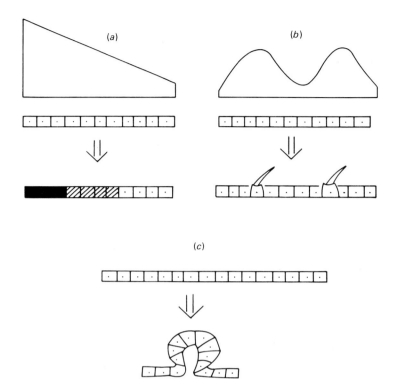

Fig. 1. Three kinds of field process: (*a*) positional information (Wolpert, 1969); (*b*) prepattern and competence (Stern, 1954); (*c*) elastic instability (Oster *et al.*, 1980). Shading indicates monotonic gradient.

repeated patterns. It was therefore natural to imagine that some field process gave rise to a standing wave of values of some scalar quantity, and that cells responded by giving rise to (e.g.) a bristle if the value rose above some threshold. Thus less of the final complexity was supposed to reside in the differential response of cells, and more in the field itself (prepattern). Logically, there is no reason why a row of bristles should not be generated by a monotonic gradient with cells responding by forming a bristle in response to (say) odd number values, and by not doing so in response to even values, but it seems implausible.

(*c*) *Cell shape generates tissue shape.* Both positional information and prepattern models postulate the existence of some invisible spatial inhomogeneity, which elicits differentiation in cells locally, presumably by activating different sets of genes. Oster *et al.*, 1980 has proposed a model of pattern formation in which all the cells composing a sheet have the same tendency to change shape. However, since they are stuck to one another and to a basement membrane, changes of shape in one cell cause changes in others to which it is attached. A homogeneous sheet of cells may be mechanically unstable, and may generate patterns such as placodes or invaginations, without further gene activation. In such a mechanism, there is no distinction between positional information and competence to respond.

The hierarchical nature of development

The processes listed above are not mutually exclusive. I see no reason to think that only one basic kind of developmental process occurs, any more than the occurrence of nerve conduction rules out hormonal mechanisms in physiology. It seems to be well established that the increase in geometrical complexity occurs in a stepwise fashion. By considering the accuracy with which fixed numbers of parts can be produced, I have argued (Maynard Smith, 1960) that one reason why this must be so is that there is a limit to the amount of additional complexity that can be generated in one step; the limit is set by the accuracy with which scalar quantities can be specified.

Our picture of development, then, is one of a succession of qualitatively different kinds of process, arranged hierarchically in time, with the products of one step forming the starting point for the next. For reasons of accuracy and repeatability, only a limited increase in spatial complexity is achieved in each step. The accuracy of the process as a whole depends on the accuracy of gene replication, and the accuracy with which specific sets of genes can successively be associated.

Evolution and development

Evolutionary biologists see evolution in a variety of ways. To a palaeontologist, evolution typically presents itself as a succession of adult forms; to a geneticist, evolution is a change in the frequencies

* Won't work — some stepwise some not.

of genes in populations, leading to the substitution of one allele by another, at a locus. These are complementary views of the same process. However, if we are interested in mechanisms rather than outcomes, I think the geneticist's view is more fundamental, for the reason stated by Weismann. If a gene in a fertilized egg is altered, that will lead to a change in the resulting adult, and to the transmission of a similarly changed gene to the next generation of eggs. In contrast, if the structure of an adult is changed, that will not lead to changes in the eggs produced which are of a kind to cause the development of a new generation of adults with the changed structure. The difference is crucial; I will discuss in a moment whether it is always true.

The issue is essentially about the nature of heredity. Heredity is a property of entities which multiply, and which vary. Thus, suppose there are entities of kinds A, B, C etc. If, when they multiply, with high probability A's give rise to A's, B's to B's, and so on, the entities possess heredity. Only such entities can evolve by natural selection. Hence, to an evolutionist, the interesting question about, for example, cytoplasmic localization is not whether such localization is essential for proper development, but whether, if the localizations are changed, the result will be an adult which produces eggs with similarly altered localizations.

In general, the answer to such questions is no. There are a few well-established exceptions, of which the phenomenon of 'cortical inheritance' in ciliates is perhaps the most important. Neo-Darwinists should not be allowed to forget these cases, because they constitute the only significant experimental threat to our views. However, the overwhelming majority of inherited differences between similar organisms are caused by differences between chromosomal genes. Further, most cases of 'cytoplasmic' inheritance turn out to depend on non-nuclear DNA or RNA, a fact which encourages a still more gene-centred and less organism-centred view of evolution.

Provisionally, then, I continue to hold a Weismannist and gene-centred view of evolution. It is worth asking *why* heredity should be Weismannist and not Lamarckian. The following answers are possible.

(i) Weismannism is true because the 'central dogma' of molecular biology is true; i.e. because information does not pass from proteins to nucleic acids. I think this is correct. However, it leaves open the question of why a hereditary mechanism with a one-way flow of information has been retained. After all, the hereditary mechanism itself is subject to evolutionary change, and I see no reason in principle why 'reverse translation' should be impossible; some information channels can work both ways.

(ii) Most 'acquired characters' are the results of injury, disease and old age. Therefore, a mechanism able to transmit such changes would lower the fitness of offspring. I think this is also correct, and may be the 'functional' explanation of why inheritance is rarely Lamarckian. If an organism had some way of telling which of its acquired characteristics were adaptive, a mechanism for transmitting them would be favoured. Of course, this is precisely what happens with learning and cultural inheritance, but such a mechanism could hardly be primitive.

(iii) Development is 'epigenetic' rather than 'preformationist', and this makes it hard to see how Lamarckian inheritance could work. I think this was the main reason why Weismann himself was a Weismannist, and the argument has recently been reformulated by Dawkins (1982). Thus, if development consisted merely of the growth in size of a preformed structure, gamete formation could be merely a reversal of this process; it is true that a gamete could hardly be formed merely by shrinking an adult, but some process resembling Darwin's 'pangenesis' might have the same effect. If so, when an adult structure was altered, the alteration might be transmitted. But development is not like that. It is much harder to imagine how, for example, the hypertrophied muscles of a black-smith could cause him to produce altered sperm which, although themselves not containing anything directly corresponding to muscle size, would cause his children to have larger muscles. As Weismann remarked, if one were to come across a case of the inheritance of an acquired character, it would be as if a man were to send a telegram to China, and it arrived translated into Chinese.

This argument is, of course, not decisive. We do not understand how genetic information is translated into adult structure, but are happy to assume that it is. We cannot, therefore, rule out the possibility of reverse translation merely because we cannot see how it could work. It is best to regard the Weismannist nature of inheritance as a contingent rather than as a necessary fact about nature.

[handwritten margin note:] Quite the opposite!

[handwritten note at bottom:] ✗ But Weisman considered himself a preformationist!

Genes, development, and the conservation of form

Comparative anatomists have taught us to regard similarities of form as a better guide to relationship than similarities in habits, and the fossil record demonstrates, at least for vertebrates, that the relationships being revealed are in fact phylogenetic relationships. It follows that anatomical patterns are more conservative in evolution than are ways of life: the pentadactyl limb is more conservative than the habit of running, climbing, flying or swimming. Why should this be so?

It is easy to see how changes in genes could cause changes in the structures produced by any of the processes listed above. For example, Stern's concepts of prepattern and competence were formulated with the specific aim of asking whether genes causing particular differences in *Drosophila* acted by changing the prepattern or the cellular response. He concluded that the differences were usually of competence; Sondhi and I agreed, but argued that some of the changes we observed were more plausibly interpreted as arising from changes in the prepattern. There is, in any case, no difficulty in principle in seeing how gene changes could cause prepattern changes. If, for example, one imagines the formation of chemical waves as proposed by Turing (1952), then any gene coding for a protein affecting either the reaction or diffusion rates could alter the prepattern. For the positional information model, in its extreme form, hereditary change can arise only by changing the genes which affect the responses of cells to unchanging positional information; it is for this reason that, as an evolutionist, I find the model less appealing, although I am ready to agree that prepatterns may often take the form of monotonic gradients.

If development is hierarchical, as most of us would think, then evolutionary changes could occur at any stage. However, in practice it is hard to see how viable alterations could occur except as terminal alterations or additions (if development passes through a free-living larval stage, neotenic changes are also a possibility). Consider the hierarchy in Fig. 2, in which each arrow represents a process and each letter a resulting structure. An alteration early in the hierarchy would be unlikely to be compatible with life. It is easier to imagine an alteration or addition at the end being compatible with survival. In the case of organisms with free-living

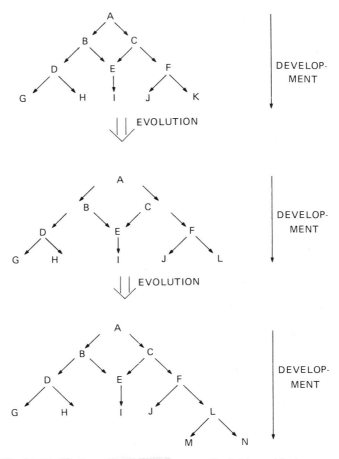

Fig. 2. The hierarchical nature of development. Each hierarchical tree represents the development of an individual, the three trees representing successive generations. Arrows within each hierarchy represent developmental processes, and letters represent resulting structures.

larval stages, the omission of later stages (i.e. neoteny) is also a possibility.

If so, the conservation of morphological patterns becomes comprehensible. Conservation requires the accuracy of DNA replication, but by itself that is not a sufficient explanation, because it cannot explain the conservation of pattern despite change of function.

If we say that adult structure A in species 1 and structure B in species 2 are 'homologous', we may mean any one of a number of different things. For example,

  (i) structures A and B have similar topological relationships with other structures,
  (ii) the modes of development of A and B are similar,
  (iii) structures A and B are connected by a series of small gradations to a structure C in an common ancestor of species 1 and 2.

It is rare for us to have direct evidence of C. More commonly, (iii) is a hypothesis, for which (i) and (ii) provide supporting evidence. It might seem that (i) and (ii) are circular. For example, a bone is called an articulator if it articulates with a quadrate, but to decide whether it does one must first know what is a quadrate. This objection does not seem to me particularly serious. In practice we recognize the isomorphism between two sets of objects, and regard as homologous objects occupying corresponding positions in the two sets. Similarly, we recognize the hierarchical patterns of development of two organisms as being isomorphic, and then treat as homologous structures derived from corresponding elements of the hierarchy (in Fig. 2, $K \equiv L \equiv M + N$).

On this view, anatomical conservatism is a consequence of the conservation of the hierarchical structure of development, which in turn is a consequence of the fact that alterations in the early stages of such a hierarchy are almost certain to be incompatible with life. It follows that there are, indeed, constraints on the kinds of structures which can evolve from any given ancestor. However, these constraints are historical and contingent in nature, arising from the existing pattern of development. The alternative to this view would be that there are absolute 'laws of form', not historically contingent, determining what kinds of organisms can and cannot exist. I can see little to recommend such a view.

Conclusions

It has never been part of Darwinism to claim that phenotypic variation is random or unconstrained. In any given taxon, some variations occur commonly and others rarely or not at all. Darwin himself noted this phenomenon, and coined for it the term 'analogous variation'. There are, however, two senses in which a

contemporary Darwinist would claim that variation is unconstrained. First, at the molecular level, we know of no constraints on
the kinds of changes in DNA sequence that can arise by mutation.
Secondly, the phenotypic changes resulting from a mutation do not
in general adapt the organism to withstand the agent which caused
the mutation. There is, therefore, nothing 'un-Darwinian' about
the claim that there are developmental constraints. Our difficulty is
that we can rarely predict the nature of those constraints from a
knowledge of development.

The question of sudden versus gradual changes in evolution is
more controversial. It is known that single mutational events can
and do give rise to large phenotypic changes, as a visit to any
genetics laboratory will establish. What is at issue is how often such
mutations are incorporated in evolution. The question is ultimately
an empirical one. The most fruitful approach is through a genetic
analysis of related species which differ morphologically. Such
analyses do little to suggest that mutations of large effect have been
important (Charlesworth *et al.*, 1982), but it would be wise to keep
an open mind. However, two assertions can be defended on
theoretical grounds. First, a mutation of large phenotypic effect is
unlikely to be incorporated in evolution unless it increases the
fitness of its carriers. Secondly, 'systemic mutation' as envisaged by
Goldschmidt (1940) – i.e. complex new adaptations arising without selection from a restructuring of the genome – can be ruled out
on probabilistic grounds.

I cannot see that developmental constraints can help to explain
'stasis' (Gould & Eldredge, 1977) – i.e. the morphological constancy of species over long geological periods. Thus it may be that
variation occurs in some characteristics more readily than others, or
that, in some cases, variation is discontinuous, permitting the
existence of only a discrete subset of the conceivable phenotypes.
However, it is manifestly *not* true that heritable variation is absent
altogether. Indeed, many traits which show stasis in the fossil
record vary in contemporary populations. One of the few general-
izations on which geneticists agree is that, if a sexual population is
exposed to selection for a phenotypically variable trait, then the
population will change (for a rare exception, see Maynard Smith &
Sondhi, 1960). Hence, if such traits remain constant for long
periods, it must either be because no selection operates on them, or
because selection is normalizing. Whatever one's view may be on

electrophoretic variability of proteins, no naturalist would accept that morphological variations are selectively neutral.

Thus we must distinguish between the idea that developmental constraints are responsible for the conservation of major morphological patterns, which is plausible, and that such constraints are responsible for the constancy in time of traits which vary in contemporary populations, which is not.

This brings me to my final point. If we are to understand evolution, we must remember that it is a process which occurs in populations, not in individuals. Individual animals may dig, swim, climb or gallop, and they also develop, but they do not evolve. To attempt an explanation of evolution in terms of the development of individuals is to commit precisely that error of misplaced reductionism of which geneticists are sometimes accused. *Substitute "selection."*

References

Charlesworth, B., Lande, R. & Slatkin, M. (1982). A Neo-Darwinian commentary on macroevolution. *Evolution* (in press).

Dawkins, R. (1982). *The extended phenotype*. W. H. Freeman: Oxford.

Goldschmidt, R. B. (1940). *The material basis of evolution*. Yale University Press: New Haven, Conn.

Gould, S. J. & Eldredge, N. (1977). Punctuated equilibria: the tempo and mode of evolution reconsidered. *Palaeobiology* **3**, 115–51.

Maynard Smith, J. (1960). Continuous, quantized and modal variation. *Proc. R. Soc. Lond. B* **152**, 397–409.

Maynard Smith, J. & Sondhi, K. C. (1960). The genetics of a pattern. *Genetics* **45**, 1039–50.

Maynard Smith, J. & Sondhi, K. C. (1961). The arrangement of bristles in *Drosophila*. *J. Embryol. exp. Morph.* **9**, 661–72.

Monod, J. & Jacob, F. (1961). Teleonomic mechanism in cellular metabolism, growth, and differentiation. *Cold Spr. Harb. Symp. quant. Biol.* **26**, 389–401.

Oster, G., Odell, G. & Alberch, P. (1980). Mechanics, morphogenesis and evolution. *Lect. Math. Life Sci.* **13**, 165–255.

Stern, C. (1954). Two or three bristles? *Am. Sci.* **42**, 213–47.

Turing, A. M. (1952). The chemical basis of morphogenesis. *Phil. Trans. R. Soc. Ser. B*, **237**, 37–72.

Wolpert, L. (1969). Positional information and the spatial pattern of cellular differentiation. *J. theor. Biol.* **25**, 1–47.

Constancy and change in the development and evolution of pattern

L. WOLPERT

Department of Anatomy and Biology as Applied to Medicine,
The Middlesex Hospital Medical School, London W1P 6DB, UK

#0427

Patterns of cellular differentiation and overall form change during evolution. What is it that changes? Are all the developmental processes involved equally likely to change, or are some processes more conserved? If so, why is this? Again, what restrictions are there on change? And how should we consider these changes in relation to the genome? Does the genome contain a programme for the development of pattern?

I wish to argue that in relation to pattern and form the major changes are not in the cellular activities themselves but in their spatial and temporal organization. In relation to changes in form, the cellular forces that bring these about, such as localized contractions or cell movements, as, for example, in sea-urchin morphogenesis (Gustafson & Wolpert, 1967) and neural tube formation (Karfunkel, 1974), have probably not changed significantly during evolution. However, the change in their spatio-temporal organization will produce quite different forms. Similarly with pattern formation. In, for example, the evolution of the vertebrate limb, the changes in cellular differentiation leading to cell types such as cartilage or muscle are probably of physiological rather than developmental significance. What has changed in evolution is the spatial organization. How then has this change occurred? We can examine this in terms of the concept of positional information.

Positional fields

In certain systems the development of the spatial organization of cells can be understood in terms of positional information. Cells are assigned positions within a coordinate system and interpret this

according to their genetic constitution and developmental history
(Wolpert, 1971, 1981). I suggest that, in general, the coordinate
system remains the same in evolution and that what changes is the
interpretation. Let us examine the coordinate system first.

There are two aspects to setting up a positional field. First, there
are the cell-to-cell interactions required for specifying position,
and, secondly, there is the mechanism whereby position is recorded
in the cell – the specification of positional value. (It is thus
important to distinguish between a positional signal and a positional
value.) There is quite good reason to believe that both are quite
well conserved during evolution. The justification for this will be
briefly summarized.

(i) The positional field in different insect imaginal discs is the
same. The evidence is that they have the same sets of positional
values and respond to similar positional signals. Thus, in genetic
mosaics of homoeotic mutants, cells behave according to their
position and genetic constitution (Postlethwait & Schneiderman,
1974). Again, position-appropriate intercalation occurs when fragments from different discs are combined (Bryant, 1979). (For
further discussion, see French, this volume.)

(ii) The positional signal from the polarizing region is the same
in forelimb and hindlimb and is common to amniotes (Fallon &
Crosby, 1977). Thus, mouse polarizing region grafted to the
anterior margin of chick limb bud, specifies the formation of chick
digits (Tickle et al., 1976).

(iii) The induction of epidermal structures seems to be the same
in urodeles and anurans. This can be interpreted as the transfer of
positional information from mesoderm to the overlying ectoderm
(Wolpert, 1981). Thus, anuran ectoderm grafted to the ventral
region of a urodele embryo, forms suckers in the appropriate
position (Holtfreter & Hamburger, 1955).

(iv) Sengel (1976) has shown that mesenchymal epidermal
interactions are similar in amniotes.

(v) In general, both the rules for pattern formation and the
phenomenology are strikingly similar in very different animals
(Huxley & de Beer, 1934; Wolpert, 1971). The best evidence for
this is the polar coordinate model for epimorphic regulation
(Bryant et al., 1981) which highlights the similarities between
insects and amphibians.

If the positional signal and positional values are conserved during evolution then it is the interpretation that must change. It is as if signals remain the same but the response alters. Having suggested that positional signals and positional values remain constant in evolution, we must nevertheless ask how similar these are in any one developmental system. Is the same set of signals and positional values used in, for example, the primary body axis of the chick and the limb buds? Separate mechanisms are probably required for specifying and recording positional value along each dimension. Thus separate signals and sets of positional values are needed for the antero-posterior and proximo-distal axes of the limb. It also seems likely that the positional values specified in the limb must be different from those in the main body axis in order to avoid ambiguity, but the signals could be the same.

Central to the above ideas is the concept of positional value and non-equivalence (Lewis & Wolpert, 1976). It is because cells have different positional values even if they belong to the same differentiation class, such as cartilage, that localized changes in cellular activities, such as growth, can occur in evolution. This can be illustrated by growth in the chick limb.

Growth of skeletal elements

The growth of skeletal elements once they have been laid down, is considerable. In the chick wing, the elements are not much more than 0.3 mm when laid down, but grow to 300 times that length. Almost all of that growth is within the growth plates at the epiphyses. As Kember (1978) has shown for rats, the key feature in determining the rate of growth of a growth plate is the size of the proliferative zone. It is this that underlies the difference in growth rates of different growth plates, since the proliferative rate of the cells seems to be the same in different plates. Thus the length of an element will be determined largely by the size of the proliferative zone in the growth plate and for how long growth continues, since different growth plates cease growing at different times. From this it can be seen that the size of the initial growth plate, rather than the size of the rudiment as a whole, is crucial in determining its final size. Thus, what may appear to be heterochrony could be the result of a change in spatial organization (Wolpert, 1982b).

Very interesting examples of differential growth, illustrating non-equivalence, can be seen in the chick wing and leg. The wrist elements of the wing are similar in size to the humerus and radius and ulna when they are first laid down, but they grow very much less (Summerbell, 1976). By 10 days of incubation they are only one-tenth of the length of the adjacent elements. This early lack of growth results partly from these elements not having anything corresponding to the diaphyseal region of the adjacent long bones. The cells do not undergo hypertrophy, which plays a major role, together with matrix secretion and cell division, in increasing the volume of the long bones. This dilation is directed along the long axis because of the restraint provided by the perichondrium in the diaphyseal region (Wolpert, 1982*a*).

A further example concerns the tibia and fibula. When these are initially laid down, they are not very different in either diameter or length, though the fibula is slightly narrower than the tibia (C. Archer, P. Rooney & L. Wolpert, unpublished). However, by 10 days of incubation the fibula is nearly half the length of the tibia. We have found that the reason for this is that the distal end of the fibula, the distal epiphysis, becomes detached from the fibula and joined to the distal end of the tibia to form part of the tibio-tarsus. Thus, virtually lacking a distal half, the fibula becomes much shorter than the tibia. The details are particularly interesting. There is a pronounced bulge from the distal end of the tibia towards the fibula, possibly reflecting a weak perichondrium in that region, and, quite early on, the region where separation will occur can be recognized in the fibula. Thus there are well-defined localized changes in form which must reflect localized changes in cellular properties. These have profound effects on the overall form of the limb.

A further implication of our findings on the fibula is in relation to Hampé's (1960) well-known experiments. Hampé found that operations on the limb bud, such as inserting a barrier between the early tibia and fibula, resulted in the fibula becoming much longer and resembling the fossil *Archaeopteryx*. This experiment has been interpreted by some (e.g. Gould, 1982) as involving heterochrony and induction. It is clear of course that it involves neither and that the operation merely prevents detachment of the distal epiphysis of the fibula. However, we now have some insight into the cellular changes that might have occurred during evolution.

As a final example, I will consider the particular case of the evolution of the one-toed horses from three-toed horses (Robb,

1936). The central digit is the longest of the three toes and the two lateral digits are slightly shorter. As the horse goes up onto its toes in evolution, the three digits can still touch the ground because the two lateral digits are slightly posterior. The growth of the digits is probably linear with time (because of the nature of the growth plate) and the central digit grows faster than the lateral ones. A plot of the length of the central digit against the lateral ones shows that it grows 1.4 times faster (Robb, 1937). As the size of the animal increases during evolution the side digits will no longer be able to contact the ground. This is probably what happened in the evolution of the three-toed horses and could be completely accounted for by a smaller proliferative zone in the growth plates of the lateral digits, and an increased period of growth. If the size of the proliferative zone of these digits is further reduced and the distal ones lost, one-toed horses will result, with the lateral digits reduced to splints. This must involve localized changes in the growth of the lateral digits.

This last example, together with those from the chick, emphasize the importance of localized changes in cellular properties affecting growth. That such changes are possible is the result of the different cartilaginous regions being non-equivalent. It is because they have different positional values that local cellular properties can change, by alterations in interpretation.

Interpretation

This analysis of pattern places the core of pattern formation and change on the process of interpretation. The vast majority of changes in pattern will reflect localized changes in differentiation or growth or some other cellular activity and the examples of limb development given above, such as the localized changes in tibia and fibula, are typical. This is not to say that patterns cannot be altered by, for example, truncation of a coordinate system, and this is just what occurs in limb evolution when elements are lost. This always involves the loss of distal structures (Lande, 1978) and it is as if the proximo-distal axis were truncated by removal of the apical ridge and thus deletion of the progress zone.

The central problem in pattern formation is to understand the nature of positional value and its interpretation. Unfortunately, we remain largely ignorant. Not only do we not know how positional value is encoded within the cell, but we do not know how this might be interpreted so as to give well-defined patterns. Given a set of positional values, how could this be used to specify an appropriate pattern of cellular differentiation? There are almost no models that even address this problem, though MacWilliams & Papageorgiou (1978) have shown how gradients can be used to generate ellipses of different forms, and Meinhardt (1978) has shown how a gradient could turn on specific genes at particular concentrations. However, there are no models which, in a general way, show how positional values could interact with the cell's genetic information to generate a pattern.

The system about which we have the most information and which might throw light on this process is the set of *bithorax* genes in early insect development. Here it appears that a gradient activates additional genes along the body axis so that each segment has an additional gene activated as we progress antero-posteriorly (Lawrence, 1981). This might be regarded as recording the positional value of each segment by gene activation, though the possibility that positional value is encoded by some other means, and that the *bithorax* genes represent the first phase of interpretation, cannot be excluded. If we accept that the positional values of each segment is represented by particular combinations of the *bithorax* genes, we still have no idea as to how this leads to different spatial patterns. For example, mutations in the homoeotic gene *aristapaedia* lead to antennal structures being replaced by leg ones. How are we to think about this? Does the *aristapaedia* gene activate a whole set of other genes, and does this lead to the positional values in the antennal structure being interpreted differently and forming antenna rather than leg? Or does the *aristapaedia* gene code for RNA molecules or proteins which modify the interpretation of leg so that it forms an antenna? We need to know what processes and molecules are held in common in interpreting leg and antennal structures. Are mutations that affect different structures in similar ways acting on the positional field? One such example may be *decapentaplegic* which seems to be involved in the elaboration of distal positional values in a number of different discs (Spencer, Hoffmann & Gelbart, 1982).

Similar problems are to be faced when we consider the wing and leg of the chick. Several experiments show that the positional field and positional signals in the wing and leg bud are the same.

Moreover, some mutations affect both in a similar manner. *Ameta-podia* (Cole, 1967) results in the loss of metapodeal regions in both wing and leg. Again, there are, in man, brachydactylous anomalies, such as A2 (Temtamy & McKusick, 1978), which result in the shortening of the middle phalanges of the index finger and second toe. How then do leg and wing differ? Do they have largely independent systems for interpretation or do they share a common one that is modified? Are mutations that affect them similarly those that affect just the positional values rather than the interpretation?

We may ask why it is, as it appears to be, that positional values and signals are conserved in evolution and it is the interpretation that changes. It might be said that it is because, Nature, having found a convenient way of making patterns, has stuck to it. But this is not an adequate explanation. The essential point being that it is much easier when making something to alter what is being made with the existing machinery than to bring in quite new machinery. It is not that the new machinery might not be better, but it requires, in effect, starting from the beginning again, and this would be particularly difficult in evolution. Another way of looking at the same problem is to see that positional value and cell differentiation and behaviour have a one-to-many relationship. That is, the positional value of a cell can affect many cellular processes, particularly the synthesis of many different proteins. To change the positional value would thus alter the synthesis of many different processes and would produce gross changes, usually deletions. It seems far better to alter the interpretative mechanism and thus affect only a selected number of proteins.

Constraints and the genetic programme

To what extent does the mechanism of pattern formation described here provide a developmental constraint? At present, the concepts of positional information do not provide an understanding of some aspects of constancy in evolution. It does not provide any obvious constraints on change (Wolpert, 1982*b*) and thus fails to account for the conservation of the basic skeletal pattern during limb evolution. It may well be that this basic skeletal pattern is laid down by some other mechanism along the lines suggested by Ede (1982), and it is on this that the system of positional values is superimposed.

We can nevertheless consider the problem of constraints in a more general way. Consider, for example, whether, with a limitless population of mice, and as much time as required, it is possible to design a selection programme so that feathered flying mice evolved (Wolpert & Stein, 1982). It is evident at once that selection, and thus evolution, can only operate on existing variability. Even if we knew the detailed cellular mechanisms underlying the development of feather and hair, it might be very difficult to design an appropriate selection programme, unless there was appropriate variation in the relevant cellular processes. The essential point is that we require variability in order to select, and, even more important, a means for selection. We cannot select for two humeri if we never find them nor any intermediates. However, because of the pleiotropic effect of many genes, it may be possible to select for some other character such as long noses, which could result in the development of an extra humerus. It is quite clear, for example, that a mirror image limb duplication could be obtained if an additional polarizing region developed at the anterior margin of the limb. But even if there were the required variation in the early distribution of polarizing region activity, how could we select for it? Thus it seems that selection is as important a constraint as developmental variability. If we had freedom to manipulate the DNA of an animal we would have enormously greater potential for evolution. What we lack is a selection procedure so that the appropriate changes in the DNA are made and selected.

This, then, raises the question to what extent the DNA can be regarded as providing the programme for development. We wish to view development in terms of a generative programme contained within the DNA of the fertilized egg. The information content within the DNA is what specifies the processes for making the organism. The three-dimensional structure of a protein (for example, an enzyme) is uniquely determined by its linear sequence of the amino acids, which, in turn, is uniquely specified by the linear sequence of nucleotides in the DNA. In this sense, the genetic information for the structure of the protein can be thought of as being programmed by the DNA. This view is contrary to that of Stent (see Bonner, 1982), who wishes to reserve the term programme strictly for situations where there is a clear isomorphism between the programme and the outcome of the programme. In this view, the DNA contains the programme for the amino acid sequence of the enzyme, but the folding up of the chain is not in the programme and is something else. I feel, however, that this is too restrictive a concept, and misses the most interesting part of the programme concept (Wolpert & Stein, 1982). It also could well be that there is such an isomorphism in development but that we do not yet understand the relationship between the programme and the process. We do not know the code, for example, for interpretation. That such an isomorphism might well exist comes from studies on sporulation in bacteria. Mandelstam (1975) suggests that the sequence of events in bacterial sporulation can in part be understood in terms of a sequence of genes being activated. The activity of one gene leading eventually to the activity of the next gene, which initiates the next step. It is clear here that the genes are providing a genetic programme for development during sporulation. There is good reason to believe that a similar situation exists in development. Again, Stein (1982) has suggested a mechanism for differentiation in which the DNA directly specifies the sequence of binary decisions involved. For pattern formation, the crucial question is how positional value and the genetic programme are linked.

I would like to thank Miss M. Maloney for helping with the preparation of this paper.

References

Bonner, J. T. (ed.) (1982). *Evolution and development*. Heidelberg: Springer-Verlag.

Bryant, P. J. (1979). Pattern specification in imaginal discs. In *Determinants of spatial organization*, eds. S. Subtelny and I. R. Konigsberg, pp. 295–316. Academic Press: New York.

Bryant, S. V., French, V. & Bryant, P. J. (1981). Distal regeneration and symmetry. *Science* **212**, 993–1002.

Cole, R. K. (1967). *Ametapodia*: a dominant mutation in the fowl. *J. Hered.* **58**, 141–6.

Ede, D. A. (1982). Levels of complexity in limb mesoderm culture systems. In *Differentiation* in vitro, eds. M. M. Yeoman and D. E. S. Truman, pp. 207–30. Cambridge University Press: Cambridge.

Fallon, J. F. & Crosby, G. M. (1977). Polarizing zone activity in limb buds of amniotes. In *Vertebrate limb and somite morphogenesis*, eds. D. A. Ede, J. R. Hinchliffe & M. Balls, pp. 55–71. Cambridge University Press: Cambridge.

Gould, S. J. (1982). Change in developmental timing as a mechanism of macroevolution. In *Evolution and development*, ed. J. T. Bonner, pp. 333–46. Springer-Verlag: Heidelberg.

Gustafson, T. & Wolpert, L. (1967). Cellular movement and contact in sea urchin morphogenesis. *Biol. Rev.* **42**, 442–98.

Hampé, A. (1960). Le competition entre les elements osseux de zeugopode de Poulet. *J. Embryol. exp. Morph.* **8**, 241–5.

Holtfreter, J. & Hamburger, V. (1955). Amphibians. In *Analysis of development*, eds. B. N. Willier, P. A. Weiss and V. Hamburger, pp. 230–96. Saunders: Philadelphia.

Huxley, J. S. & de Beer, C. R. (1934). *The elements of experimental embryology*. Cambridge University Press: Cambridge.

Karfunkel, P. (1974). The mechanisms of neural tube formation. *Int. Rev. Cytol.* **38**, 245–71.

Kember, N. F. (1978). Cell kinetics and the control of growth of long bones. *Cell Tiss. Kinet.* **11**, 477–85.

Lande, R. (1978). Evolutionary mechanisms of limb loss in tetrapods. *Evolution* **32**, 79–92.

Lawrence, P. A. (1981). The cellular basis of segmentation in insects. *Cell* **26**, 3–10.

Lewis, J. H. & Wolpert, L. (1976). The principle of non-equivalence in development. *J. theor. Biol.* **62**, 479–90.

MacWilliams, H. K. & Papageorgiou, S. (1978). A model of gradient interpretation based on morphogen binding. *J. theor. Biol.* **72**, 385–411.

Mandelstam, J. (1975). Bacterial sporulation: a problem in the biochemistry and genetics of a primitive developmental system. *Proc. R. Soc. Lond. Ser. B* **193**, 89–106.

Meinhardt, H. (1978). Space-dependent cell determination under the control of a morphogen gradient. *J. theor. Biol.* **74**, 307–21.

Postlethwait, J. H. & Schneiderman, H. A. (1974). Developmental genetics of *Drosophila* imaginal discs. *A. Rev. Genet.* **7**, 381–433.

Robb, R. C. (1936). A study of mutations in evolution. III. The evolution of the equine foot. *J. Genet.* **33**, 267–73.

Robb, R. C. (1937). A study of mutation in evolution. IV. Ontogeny of the equine foot. *J. Genet.* **34**, 477–86.

Sengel, P. (1976). *Morphogenesis of skin*. Cambridge University Press: Cambridge.

Stein, W. D. (1982). Replicating RNA duplexes and the epigenetic address. *J. theoret. Biol.* **94**, 869–89.

Spencer, F. A., Hoffmann, F. M. & Gelbart, W. M. (1982). Decapentaplegic: a gene complex affecting morphogenesis in *Drosophila melanogaster Cell* **28**, 451–61.

Summerbell, D. (1976). A descriptive study of the rate of elongation and differentiation of the skeleton of the developing chick wing. *J. Embryol. exp. Morph.* **35**, 241–60.

Temtamy, S. A. & McKusick, V. A. (1978). The genetics of hand malformation. In *Birth defects: original article series*, XIV, No. 3. Alan R. Liss: New York.

Tickle, C., Shellswell, G., Crawley, A. & Wolpert, L. (1976). Positional signalling by mouse limb polarizing region in chick wing bud. *Nature* **259**, 396–7.

Wolpert, L. (1971). Positional information and pattern formation. *Curr. Top. devl. Biol.* **6**, 183–224.

Wolpert, L. (1981). Positional information and pattern formation. *Phil. Trans. R. Soc. Ser. B* **295**, 441–50.

Wolpert, L. (1982a). Cartilage morphogenesis in the limb. In *Cell behaviour*, eds. R. Bellairs, A. Curtis and G. Dunn, pp. 359–72. Cambridge University Press: Cambridge.

Wolpert, L. (1982b). Pattern formation and change. In *Evolution and development*, ed. J. T. Bonner, pp. 169–88. Springer-Verlag: Heidelberg.

Wolpert, L. & Stein, W. D. (1982). Evolution and development. In *Learning, development and culture*, ed. H. C. Plotkin, pp. 331–42. John Wiley: Chichester.

Litany and creed in the genetic analysis of development

School of Biological Sciences, University of Sussex,
Brighton, Sussex BN1 9QG, UK

#0428

Developmental puzzles and disasters attend mutations in the genome of an organism. These heritable changes offer an opportunity to developmental biologists to pursue local perturbation analysis of any developmental process so as to identify the 'rules' or 'logic' of ontogeny. The conventional wisdom includes the assertion that the common denominator between development and evolution is the genetic system. By itself, genetic methodology applied to development cannot provide the whole story, yet I wish to re-state and review the conceptual gains that emerge from this methodology, and to show that it is a very sharp tool with which to dissect development.

Our starting material is heritable phenotypic variation, and, unless we wish to be very parochial, an operational definition of heritable variation will include all manner of diverse phenomena. It will include corticotypes in ciliates (Aufderheide *et al.* 1980) mating-types in yeasts (Hicks *et al.*, 1979), and a plethora of processes in which cell phenotypes are perpetuated stably through cell proliferation in Metazoa. Cell determination in different imaginal disc types in *Drosophila* (Gehring, 1980), somatic variants in maize (Peterson, 1977) and X-chromosome inactivation in mammals (Lyon, 1974) all belong in this category. Conventionally, all these examples have been classified as non-genic or epigenetic, but we have to be careful not to dismiss them too prematurely from the province of genetics. The basic operation of a geneticist is to challenge one variant with another by bringing them together in the same cell or individual, recording what is seen, then tearing them apart to look again at the phenotypes of the products. This describes in other words sexual fusion, complementation and segregation analysis. I will concentrate upon the heritable differences the manipulation of which is achieved by sexual reproduction and through the production of genetic mosaics, although 'genetic' handling of the other variants is, in principle, possible, and I will return to this topic below (p. 70).

Let us be clear about the technical basis of the method. Mutagenic procedures of various kinds will generate heritable changes at high frequency, although with the surprising impression that induced mutations are frequently not like spontaneous mutations. Genetic variants are compared one with another for functional identity by complementation tests. The same procedure can be used systematically to uncover other lesions in the same functional unit or in an area of the genome defined only by structural criteria, for example by the limits of a genetic duplication or deletion. Recombination mapping will generate a summary map which is usually a close approximation to the actual physical disposition of the sites on the chromosome. Conventional cytogenetic rearrangements of gene order or cloning vehicles can be used to alter the number of copies of particular areas of the genome (Capdevila & García-Bellido, 1981). The genome of a cell can be changed in recognisable and controllable ways during phases of cell proliferation, thereby extending the analysis of genetic segregation to within-organism dimensions, so as to define the timing and site of expression of genetic units, as well as to reveal the pattern of cell lineage in general. Genetic chimaeras, the juxtaposition of genetically different cells from more than one embryo, define the autonomy of cellular functions as well as cell lineage (Mintz, 1981). By changing the conditions from permissive to non-permissive and vice versa, genotypes with conditional expression can be made to reveal the 'execution points' of particular elements (Hirsh & Vanderslice, 1976). Genomic cloning libraries are available within which, given a DNA transformation system, the DNA sequence complementary to any given genome-based variant can be retrieved. These clones can be used to 'probe' for the presence of transcripts from particular sequences. But what then is the litany associated with this methodology?

All genetic perturbations provide information on systematic properties

The most cautious view of the relevance of genetic methodology in the analysis of development is the empirical one which says (*a*) that, irrespective of the nature of the disturbance, the final phenotype and the route taken to that end-point, together demonstrate a property of the system and (*b*) that the phenotype is the response of the *system* to the insult (Kacser & Burns, 1981). For instance, *Drosophila* females homozygous for a recessive mutation *short egg* (Wieschaus *et al.*, 1981) form oocytes only 60 % of the normal length, but embryogenesis and the remainder of development are normal. What 'patterning' the embryo experiences, expressed by the embryonic fate map or manifest in the process of segmentation, is not sensitive to changes of this magnitude in the long axis of the embryo. Therefore we conclude that the patterning is unlikely to hinge on an exact template of fixed molecular dimensions. The *lin-5* mutation in the nematode *Caenorhabditis elegans* interferes with post-embryonic cellular and nuclear divisions (Albertson *et al.* 1978), but several other developmental events, normally concomitant with these divisions, nevertheless proceed in this mutant in the absence of cell or nuclear division and thus cannot be dependent upon the completion of those cell cycles. The essential point is that valuable conclusions can be drawn from the spectrum of the possible phenotypes that can be produced, merely by noting the existence of these phenotypes. Adherents of this empirical position would strongly resist the temptation to make statements about the nature of the mutable units in the genome.

Geneticists can draw inferences about the 'role' of the genetic elements being manipulated

This inference is made in the following way. By employing difference analysis, which is the comparison of mutant and normal phenotypes, we can decide what has changed as a result of the mutational event. We then infer that the genetic element prior to mutation was an essential component in the process showing disruption. Without further explanation, this logic would be naïve reductionism; we would discover 'wing genes' from observing strains of wingless flies or 'tail genes' by looking at tailless mice.

Three quite distinct and important issues arise from this carica-
ture of the way 'roles' are assigned to genetic elements. The first is
that, in common with other developmental biologists, geneticists
concede that development is likely to be organised hierarchically
(see p. 65), and that there are interactions between components and
between events. In a hypothetical system where phenotypic
descriptions could only be made at the level of the gene product and
between which products there were no interactions, the 'role' of
each 'gene' could be specified exactly, rather as if the variants were
an array of identical jigsaw puzzles each with a different missing
piece (J. Cohen, personal communication). The one gene–one
character fallacy faded out as an early misunderstanding of genetic
analysis, and, in the real world, regulative properties of the system
and hierarchical structure conspire to force us to a more circum-
spect description of the 'role' of a gene.

Secondly, we should not be misled by the nomenclature used as a
convenience by geneticists – what I have previously called the
'naming of parts' (Whittle, 1976). The heritable units or 'genes' are
initially defined by difference analysis between mutant and normal
individuals. Names are usually chosen prior to analysis and describe the appearance of the mutant (*eyegone, kinked, decapentaplegic, uncoordinated, wee, dwarf*). Names are preferred to numbers and the more bizarre the name the easier it is to remember in
discussion. One should not think that geneticists are trapped into
describing genes for abnormalities – the names are only a working
shorthand.

The third caveat about the assigning of 'roles' to genes requires
explanation of a point of 'internal logic' in genetic analysis. If the
mutation has altered rather than abolished a catalytic function of
that gene product, then our conclusion about the relationship
between phenotypic change and gene function might be unwit-
tingly very different from the true situation. This would also be the
case if the mutation turned out to be a genetic transposition
resulting in a bona fide gene product being formed in response to a
totally irrelevant or foreign induction signal (Errede *et al.* 1980).
What is altogether more powerful is that having located the
mutation on the chromosome, deletions of that area are used,
because the most unequivocal failure of a gene must surely be its
absence from the genome (García-Bellido, 1981). I conclude that
there exists both an experimental procedure and rules of descrip-

tion for considering that a given 'gene' (see p. 67) can be an essential component in a particular process or at least 'allows' it to occur (Holder, this volume). It is never quite clear whether such inferences will be illuminating or banal; lesions in the *rudimentary* gene in *Drosophila* starve the embryo of pyrimidines and so arrest it, with hindsight very understandable, but the comparable maternal effect mutation *dicephalic* (Lohs-Schardin, 1982) would seem to offer us more information on polarity in the oocyte.

Genetic lesions permit 'controlled' large scale repetition of an identical operation or subtle stepwise variations in perturbation

The introduction of a dominant mutation from the male parent into an embryo of a diploid species or the generation of a genotype homozygous for a recessive mutation by two heterozygous parents is a very subtle alteration in the starting conditions of that embryo (García-Bellido, 1980). The consequences can be examined in many individuals, and perhaps even at many alternative levels of description, by repeating the mating. Whether the mutation is typical for a failure of that genetic unit can be investigated by collecting a series of non-complementing mutations which, depending upon the criteria for recovery, can be arranged to include very minor changes as well as gross genetic errors in that unit (Shearn, 1980).

The combination of two independently arising complementary mutations in the same genome is another new experiment. Geneticists are therefore able to make not just single but pairwise and more complex combinations of alterations in starting conditions. Putting pairs of mutations together certainly has curiosity value for the empiricists but the greater value comes when the two separate 'genes' have been assigned functions. We can then validate their postulated functions by predicting the phenotype of the combination between the two mutations. Alternatively, the relationship between the processes affected, in respect of their hierarchy, their sequence in time or in the cellular focus (Hotta & Benzer, 1972), can be deduced by looking at the phenotype of the double mutant.

Suppose we have several distinct phenotypes, each associated with a particular mutation in 'genes' A, B and C, and we examine the phenotype of the double homozygote for two recessive mutations in genes A and B. If it resembles mutation A alone then it is reasonable to argue that gene A 'operates' prior to gene B, or alternatively that gene B is dependent upon normal function in gene A. We would expect all such pairwise tests to be internally consistent; that is, if the result above were represented $A > B$, then it would not be internally consistent to find $C > A$, and $B > C$ in these other two double mutant tests. These orderings of gene function can also be compared with any descriptive sequence of phenotypic changes seen during the process as it occurs normally. The individual phenotypic abnormalities associated with failure of each gene may then be classifiable as 'early' or as 'late' functions (Wood, 1980). Illustrations of this routine are legion, the paradigm being the dissection of metabolic pathways using mutations (Wagner & Mitchell, 1964). Gene functions in the elaboration of external sex-dimorphic characters in *Drosophila* have been ordered by their epistatic interactions (Baker & Ridge, 1980). Dissection of events in the cell cycle in yeasts have also been possible (see Hartwell *et al.*, 1974; Nurse, 1981) and Nüsslein-Volhard & Wieschaus (1980) have shown that the process of segmentation in the *Drosophila* embryo can be uncoupled from the establishment of the precise segment 'quality' in each individual segment.

Genetic evidence supports the concept that development is organised hierarchically

'Completed' organisms are structured and function hierarchically; there are functional 'domains' in the nervous systems of animals, and the organisation of cells into tissues which in turn are integrated into organ systems shows hierarchy. Now, is the process itself that generates this functional system arranged hierarchically? A descriptive account of animal development, as if it were taken from time-lapse microcinematography, reinforces this view. Furthermore, the 'information' or specification needed for a non-hierarchical system of development in a typical eukaryote would be immense (Abbott, 1977). Does genetic analysis provide a perspective on this idea?

The discovery of mutants with the same phenotype in different genes (many genes mapping to one phenotype) can be explained by hierarchical organisation. We could imagine the phenotype to reflect a biosynthetic activity in a particular tissue. Changes in mitotic rate, in gene copy-number, in messenger RNA stability, or cell size in this tissue would then be functionally equivalent (Abbott, 1977). The converse to the above is also true; a mapping relationship of one gene to many phenotypes (what geneticists call pleiotropy) may be easily explained by function required at a low level in a hierarchy (for example the *Mottled-brindled* mutation in mice (Hunt, 1974)). The compartment hypothesis in insects (Crick & Lawrence, 1975) is predicated upon cell lineage analysis, but receives strong reinforcement from the analysis of mutations which reveal genes affecting early branch points in 'decision trees'.

Mutational dissection has many times demonstrated that two previously tightly coupled events or processes can be uncoupled. Independence of the registration of correct segment-type for its position in the embryo from establishment of correct total segment number has already been mentioned. The welter of sex-change mutations reported in vertebrates, *Drosophila*, and nematodes testify to a hierarchy of events that are normally but not invariably coupled to various chromosomal triggers.

The obvious conclusion which emerges from an inspection of mutants, for example mutations affecting the central nervous system of the nematode *Caenorhabditis elegans*, is that not all mutations affect all events, so that interactions between genes or gene products must be limited to fewer than all the possible ones (Brenner, 1973).

Genetic evidence suggests that the cell is an important level in the hierarchy

There are very reasonable grounds on which we would expect this to be true. First, many molecular species never move between cells and, amongst these molecules, protein species reflect a very close mapping relationship with DNA sequence. Thus cell properties which are dependent upon such proteins indicate that the genes in question *must* be expressed *within* each of these cells. Secondly, all cells, including the product of sexual fusion, the zygote cell, derive from existing cells and not *de novo* by assembly from their components. Included within this continuity of cells is continuity by descent of the genomes of cells related by mitosis.

When cell autonomy of gene expression is tested by examining a genetic mosaic or chimaeric system, a considerable proportion of

genes are found to behave cell-autonomously; in mixtures, the
phenotypes of each cell genotype are unaffected by the presence of
other cells with differing genotype. Cellular determination affords
the most obvious example. In many species the result of removing
cells to 'foreign' sites during development indicates that cell
'potency' is restricted and that this cell state is refractory to
alteration despite severe traumas. Cell lineage analysis strongly
reinforces this conclusion, particularly in *Drosophila*. The implication is that there must be cell-autonomous perpetuation of determined states through cell proliferation (Gehring, 1980), and in this
species evidence is accumulating for the existence of gene functions
necessary for the maintenance of determination. These genes have
been termed selector genes by García-Bellido (1977). Struhl (1981)
has identified a gene whose product is necessary for 'registering' or
initiating any one of a number of alternative fates amongst cells in
the early stages of embryogenesis. There are also a number of
genetic elements required repeatedly or continuously through the
cell proliferation phase to maintain any one of a number of
alternative cell fates (Capdevila & García-Bellido, 1978; Ingham &
Whittle, 1980; Ingham, 1981; Duncan & Lewis, 1982). Finally
there are identifiable elements only required for the perpetuation
through cell proliferation of *certain* cell fates (the bithorax complex: Lewis, 1982) and not required at all for others.

The flurry of reports of cell-autonomous gene function implicated in determination events must not be mistaken for a claim that
development is not interactive at the cellular level or that field
properties cannot be countenanced as part of the process. However,
one of the preconditions for inductive phenomena and cell-group
phenomena must be that the cells involved in the interaction must
display a stable 'bias' of some sort, a commitment or competence
property. The bias may be dependent upon several gene functions
but it has to be a cell-limited property. Interactions between two
cell populations having different biases may generate further
differences in response, which may then be 'remembered' via the
induction of further cell-autonomous functions. Whilst elements of
the bithorax complex genetic element cell-autonomously constrain
cell fate, one of the consequent characteristics of this system is that
groups of cells of the same state show regulation in their total cell
population, which must imply cell–cell interaction. Cell-limited
gene functions, i.e. those not capable of being supplied by neigh-

bouring cells, are here conditioning those cells to be responsive to other members of the same group but not to other cells.

Genetic methodology provides underpinnings for the concept of binary combinatorial coding

The essence of this idea is that the many alternative cell states patently generated during development are specified by the states of a small number of two-state switch elements. On the one hand Kauffman particularised his model of switch networks (1971) to *Drosophila* (Kauffman *et al.*, 1978), supporting it with evidence from the 'permitted' transformations of cell state, in other words from the observed transdetermination sequences observed in *Drosophila* imaginal disc culture (Hadorn, 1980). García-Bellido (1977) in contrast bases his model on the genetic analysis of homoeotic mutants affecting thoracic and abdominal segment quality in *Drosophila*.

This parsimonious mechanism for specifying alternative determination states in groups of cells has strong formal links with the binary on–off genetic circuitry defined in prokaryotes. That geneticists, continually exercising the algebra of genetic segregation, should come up with dichotomous models should not be thought surprising, but there is considerable experimental evidence in support of the picture from genetic analyses. Evidence pointing to binary decisions in cell lineages is also accumulating from non-genetic techniques of analysis in mice (Snow, 1981) and in nematodes (Kimble, 1981).

Alterations in the genome can be resolved into functional units showing their own network properties

Up to this point I have equivocated regarding the use of the word gene, having usually written it as 'gene'. The formal definition of a gene as a mutable, mappable entity survives, particularly in the study of complex functional and morphological characteristics in higher eukaryotes. Genetic units defined functionally have considerable value in these *systemic* assays of complementation. However, there are other operational definitions of a gene.

Present investigations of the structure and function of particular genes defined by their specific gene products (for example, the *ovalbumin* gene in the chick, Heilig *et al.*, 1980) which use the increasing arsenal of DNA technology, have shown that the central dogma has many clauses in it. I will not presume to review this burgeoning area of molecular genetics but only to point out that different molecular techniques identify 'domains' in the genome that are capable of resolution into a network. Exact DNA sequences can be determined and related to mRNAs, revealing the existence of transcribed and non-transcribed regions of DNA, interspersed with sequences of codons. To understand transcription we must not ignore regions between and around transcribed sequences, nor overlook the role of DNA conformation and the proteins associated with this property. To understand the generation of an mRNA from a transcript requires identification of the coding sequence in the transcript and the interaction of the latter with another set of proteins, including those with catalytic functions in splicing and translocation. Phenotypic change could be generated by heritable alteration in DNA through the methylation of individual bases, or lack of integrity of 'TAATAA' sequences, changes in splicing recognition sequences or in regions recognised by other proteins with affinities for particular DNA sequences, or by changes in the codon sequences.

With the formalism of the *Cro* system in lambda phage (Ptashne *et al.*, 1980) or that of bacteriophage T4 assembly (Wood, 1980) as an inspiration, hypotheses of genome circuitry can be investigated once a gene can be 'assayed' at a level of organisation closer to its immediate gene product. Circuitry is suspected in many systems under genetic analysis (ecdysterone-mediated transcription in insect metamorphosis (Ashburner *et al.* 1973), and in spore and stalk formation in *Dictyostelium discoideum* (Loomis, 1975)), in addition to the processes quoted as examples in earlier sections.

Genetic variation facilitates both ascending and descending analysis of hierarchies in developmental processes

Both 'bottom up', or ascending, and 'top down', or descending, analyses of a developmental process (Brenner, 1975) are predicated upon the discovery of 'classical' heritable variation, which will have indicated the worth of pursuing these routes. The traditional route has been the descending one (Hadorn, 1960), the pursuit of difference analysis from a global or whole-organism phenotype to the physiological level and thence to a cellular description. Ascending analysis has been given an enormous impetus by the technology developed to analyse DNA, which *inter alia* can provide the DNA sequence within which the mutation lies. DNA transformation using genomic clones can effect rescue of genetic function in mutant individuals (D. Beach & P. Nurse, personal communication). This is extremely valuable in situations in which it is not clear whether there is any transcript involved at all, or in situations where the nature of the gene product is unknown. Cloned gene sequences can be used to 'probe' for the appearance of their messengers. In their turn, identified mRNAs may be translated to provide material with which to generate monoclonal antibodies recognising that gene product. Ascending and descending analyses are most productive when used together, in part because advances on each front can be tested with the complementary method to see whether they are coherent. One illustration will suffice. Cell lineage analysis in *Drosophila* has brought to light the phenomenon of compartmentation (García-Bellido *et al.*, 1973, Crick & Lawrence, 1975). Genetic analysis of the mutation *engrailed* (Lawrence & Morata, 1976; Kornberg, 1981) suggests that this gene affects one of the compartmentation steps and indicates the cell population in which, and the period during which, the gene acts. This makes strong predictions about the results to be expected from probes for transcription from the *engrailed* gene sequence in particular compartments.

If algorithms can be used to describe developmental processes (Kauffman, this volume), then analysis that proceeds stepwise up or down the hierarchy is valuable. The precise level at which an acceptable account is made, perhaps in terms of properties of groups of proteins or of groups of cells, will depend upon the phenotype under scrutiny.

Genetic methodology can be successfully extended to heritable phenomena which do not show regular segregation at sexual reproduction

In ciliates, corticotypes are alternative arrangements of the complex surface structure, which can be perpetuated stably in clones having identical nuclear genotypes. Microsurgery or traumas delivered at conjugation can generate new corticotypes, their segregation patterns can be followed within vegetative cell clones and their transition one into another can be watched (Nanney, 1977). We can record their 'replication' at binary fission, their stability through differentiated states like the cyst (Grimes, 1973), and these operations together have provided a substantial picture of the mechanisms at work (Aufderheide *et al.*, 1980; Frankel, this volume). Again the point is made that the heritable unit (this time having its manifestation in microtubular protein assemblies and not DNA) does figure in the final account.

The stability of alternative cell 'states' in cells having the same nuclear genome in imaginal disc dauer-culture has been used to define the properties of these states, suggesting that their properties are a function of positional relationships with neighbouring cells (Strub, 1979), and showing that changes, called transdetermination, are cell-group and not individual cell events (Hadorn, 1980). The conditions for change of determination in these cells are subtle, however, because the mixing together of cells from two different discs certainly does not automatically lead to transdetermination. No one has yet achieved the fusion of cells from two distinct discs, a situation which would permit questions to be posed about the 'dominance' hierarchy of the various cell states. In *Nicotiana*, F. Meins (personal communication) has illustrated how the basis of the tumorous state in the crown gall tumour can be analysed by examining the stability of its heritability in somatic tissue culture under a number of carefully defined environmental conditions.

These examples illustrate that the same formal logic and procedure that is employed with DNA-based heritable systems can yield information on the mechanism of perpetuation, the stability and transformation between alternative phenotypes and on the relationship of these variants to the global or gestalt systemic phenotype.

Conclusions

Virtues of the 'genetic method' have been characterised in this paper. Proponents of this position do not deny the validity of field concepts and indeed the latter are essential unless one subscribes to the view that the developmental process is no more than a complex 'crystallisation' event. Even so many geneticists will maintain that the very elements they use to perturb the system will have a position in the final picture.

I am grateful to colleagues and students who have helped to show me what I practise. This contribution was written after the conference as my response to it and has the advantages of hindsight.

References

Abbott, L. A. B. (1977). A biological and mathematical analysis of wing morphogenesis in *Drosophila*. D. Phil. Thesis, University of Colorado.

Albertson, D. G., Sulston, J. E. & White, J. G. (1978). Cell cycling and DNA replication in a mutant blocked in cell division in the nematode *Caenorhabditis elegans*. *Devl. Biol.* **63**, 165–78.

Ashburner, M., Chihara, C., Meltzer, P. & Richards, G. (1973). Temporal control of puffing activity in polytene chromosomes. *Cold Spr. Harb. Symp. quant. Biol.* **38**, 655–62.

Aufderheide, K. J., Frankel, J & Williams, N. E. (1980). Formation and positioning of surface-related structures in protozoa. *Microb. Rev.* **44**, 253–95.

Baker, B. S. & Ridge, K. A. (1980). Sex and the single cell. I. On the action of major loci affecting sex determination in *Drosophila melanogaster*. *Genetics* **94**, 383–423.

Brenner, S. (1973). The genetics of behaviour. *Br. Med. Bull.* **29**, 269–71.

Brenner, S. (1975). Closing remarks: the genetic outlook. In *Cell patterning*, pp. 343–5. Ciba Foundation Symposium No. 29. Elsevier: Amsterdam.

Capdevila, M. & García-Bellido, A. (1978). Phenocopies of *bithorax* mutants. *Wilhelm Roux Arch. Entw Mech. Org.* **185**, 105–126.

Capdevila, M. & García-Bellido, A. (1981). Genes involved in the activation of the *bithorax* complex of *Drosophila*. *Wilhelm Roux Arch. EntwMech. Org.* **190**, 339–50.

Crick, F. H. & Lawrence, P. A. (1975). Compartments and polyclones in insect development. *Science* **189**, 340–7.

Duncan, I. & Lewis, E. B. (1982). Genetic control of body segment differentiation in *Drosophila*. In Developmental Order: Its Origin and Regulation, 40th Symp. of Soc. for Devel. Biol. (ed. S. Subtelny). Alan Liss, N.Y.

Errede, B., Cardillo, T. S., Sherman, F., Dubois, E., Deschamps, J. & Wiame, J-M. (1980). Mating signals control expression of mutations resulting from insertion of a transposable element adjacent to diverse yeast genes. *Nature* **286**, 353–6.

García-Bellido, A. (1977). Homoeotic and atavic mutations in insects. *Am. Zool.* **17**, 613–29.

García-Bellido, A. (1980). Concluding remarks. In *Development and neurobiology in Drosophila*, eds. O. Siddiqi, P. Babu, L. M. Hall and J. C. Hall, pp. 1–2, Plenum: New York.

García-Bellido, A. (1981). From the gene to the pattern: chaeta differentiation. In *Cellular controls in differentiation*, eds. C. W. Lloyd, and D. A. Rees, pp. 281–304. Academic Press: New York.

García-Bellido, A., Ripoll, P. & Morata, G. (1973). Developmental compartmentalisation of the wing disk of *Drosophila. Nature New Biol.* **245**, 251–3.

Gehring, W. (1980). Imaginal disc determination. In *The genetics and biology of Drosophila,* vol. 2d, eds. M. Ashburner and T. R. F. Wright, pp. 511–54. Academic Press: New York.

Grimes, G. (1973). Morphological discontinuity of kinetosomes during the life cycle of *Oxytricha fallax. J. Cell Biol.* **57**, 229–32.

Hadorn, E. (1960). *Developmental genetics and lethal factors.* Methuen: London.

Hadorn, E. (1980). Transdetermination. In *The genetics and biology of* Drosophila Vol. 2c, eds. M. Ashburner and T. R. F. Wright, pp. 555–617. Academic Press: New York.

Hartwell, L. H., Cullotti, J., Pringle, J. R. & Reid, B. J. (1974). Genetic control of the cell division cycle in yeast. *Science* **183**, 46–51.

Heilig, R., Perrin, F., Gaunon, F., Mandle, J-L. & Chambon, P. (1980). The ovalbumin gene family: structure of the X gene and evolution of duplicated split genes. *Cell* **20**, 625–37.

Hicks, J., Strathern, J. N. & Klar, A. J. S. (1979). Transposable mating type genes in *Saccharomyces cerevisiae. Nature* **282**, 478–83.

Hirsh, D. & Vanderslice, R. (1976). Temperature-sensitive developmental mutants of *Caenorhabditis elegans. Devl Biol.* **49**, 220–35.

Hotta, Y. & Benzer, S. (1972). Mapping of behaviour in *Drosophila. Nature* **240**, 527–35.

Hunt, D. M. (1974). Primary defect in copper transport underlies mottled mutants in the mouse. *Nature* **249**, 852–4.

Ingham, P. W. (1981). *Trithorax*: a new homoeotic mutation of *Drosophila melanogaster.* II. The role of *trx*+ after embryogenesis. *Wilhelm Roux Arch. Entw Mech. Org.* **190**, 365–9.

Ingham, P. W. & Whittle, J. R. S. (1980). *Trithorax*: a new homoeotic mutation of *Drosophila melanogaster* causing transformations of abdominal and thoracic imaginal segments. *Mol. gen. Genet.* **179**, 607–14.

Kacser, H. & Burns, J. A. (1981). The molecular basis of dominance. *Genetics* **97**, 639–66.

Kauffman, S. A. (1971). Gene regulation networks: a theory for their global structure and behaviours. *Curr. Top. devl Biol.* **6**, 145–82.

Kauffman, S. A., Shymko, R. & Trabert, K. (1978). Control of sequential compartment formation in *Drosophila. Science* **199**, 259–270.

Kimble, J. E. (1981). Strategies for control of pattern formation in *Caenorhabditis elegans. Phil. Trans. R. Soc. Lond.* B **295**, 539–51.

Kornberg, T. (1981). Compartments in the abdomen of *Drosophila* and the role of the *engrailed* locus. *Devl Biol.* **86**, 363–72.

Lawrence, P. A. & Morata, G. (1976). Compartments in the wing of *Drosophila*: a study of the *engrailed* gene. *Devl Biol.* **50**, 321–37.

Lewis, E. B. (1982). Control of body segment differentiation in *Drosophila* by the bithorax gene complex. In *Embryonic development: genes and cells*, ed. M. Burger, Alan Liss: New York, (in press).

Lohs-Schardin, M. (1982). *Dicephalic* – a *Drosophila* mutant affecting polarity in follicle organisation and embryonic patterning. *Wilhelm Roux Arch. Entw Mech. Org.* **191**, 28–36.

Loomis, W. F. (1975). Dictyostelium discoideum, *a developmental system.* Academic Press: New York.

Lyon, M. F. (1974). Mechanisms and evolutionary origins of variable X-chromosome activity in mammals. *Proc. R. Soc. B* **187**, 243–68.

Mintz, B. (1981). Tracking genes in developing mice. *Science*, **215**, 44–7.

Nanney, D. (1977). Molecules and morphologies: the perpetuation of pattern in the ciliated protozoa. *J. Protozool.* **24**, 27–35.

Nüsslein-Volhard, J. & Wieschaus, E. (1980). Mutations affecting segment number and polarity in *Drosophila*. *Nature* **287**, 795–801.

Nurse, P. (1981). Genetic analysis of the cell cycle. In *Genetics as a tool in microbiology*, eds. S. W. Glover and D. A. Hopwood, Symposium of the Society for General Microbiology No. 31, pp. 291–315. Cambridge University Press: Cambridge.

Peterson, P. A. (1977). The position hypothesis for controlling elements in maize. In DNA: *insertions, elements, plasmids and episomes*, eds. A. I. Bukhari, J. A. Shapiro, and S. L. Adhya, pp. 429–60. Cold Spring Harbor Laboratory: New York.

Ptashne, M., Jeffrey, A., Johnson, A. D., Maurer, R., Meyer, B. J., Pabo, C. O., Roberts, T. M. & Sauer, R. T. (1980). How the lambda repressor and *Cro* work. *Cell* **19**, 1–11.

Shearn, A. (1980). Mutational dissection of imaginal disc development. In *The genetics and biology of* Drosophila, vol. 2c, eds. M. Ashburner and T. R. F. Wright, pp. 443–510. Academic Press: New York.

Snow, M. L. H. (1981). Autonomous development of parts isolated from primitive-streak-stage mouse embryos. Is development clonal? *J. Embryol. exp. Morph.* **65**, 269–87.

Strub, S. (1979). Heteromorphic regeneration in the developing imaginal primordia of *Drosophila*. In *Cell lineage and stem cell determination,* ed. N. le Douarin, pp. 311–24. North-Holland: Amsterdam.

Struhl, G. (1981). A gene product required for correct initiation of segmental determination in *Drosophila*. *Nature* **292**, 36–41.

Wagner, R. P. & Mitchell, H. K. (1964). *Genetics and metabolism*, 2nd edn. Wiley: New York.

Whittle, J. R. S. (1976). Mutations affecting the development of the wing. In *Insect development*, ed. P. A. Lawrence, pp. 118–31. Blackwell: Oxford.

Wieschaus, E., Audit, C. & Masson, M. (1981). A clonal analysis of the roles of somatic cells and germ line during oogenesis in *Drosophila*. *Devl Biol.* **88,** 92–103.

Wood, W. B. (1980). Bacteriophage T4 morphogenesis as a model for assembly of subcellular structure. *Q. Rev. Biol.* **55,** 353–67.

The ontogeny and phylogeny of the pentadactyl limb   #6429

B.C. GOODWIN* AND L.E.H. TRAINOR†

*Developmental Biology Group, School of Biological Sciences, University of Sussex, Brighton, Sussex BN1 9QG, UK and †Department of Physics and Medicine, University of Toronto, Toronto, Canada

This guy did a paper in Ho + Saunders.

Introduction

Ever since Richard Owen's demonstration that the limbs of tetrapod vertebrates can be understood as transformations of one another, the whole set of variations defining the typical form known as the pentadactyl limb, the phylogeny and the ontogeny of this basic vertebrate structure have been matters of considerable interest and dispute. Darwinians, reifying the abstract concept of a typical form, sought a common ancestral tetrapod limb structure in fossil amphibians such as *Eryops*, and antecedents in crossopterygian relatives such as *Sauripterus*, the limbs of later vertebrates being regarded as adaptive modifications of the ancestral form. Embryologists have looked for evidence of a basic pattern of cartilage condensation in the limb buds of developing amphibians, birds, and mammals prior to secondary modifications such as loss or fusion of carpal and tarsal elements. They were following the expectation, derived from the theory of recapitulation, that the ancestral form may appear transiently in the early ontogenesis of the limb. However, no consensus on an archetypal limb form has emerged.

This should not surprise us: the quest was initiated by an error of conceptualisation, the generalised form of which unfortunately runs through most of contemporary biology. This was Darwin's suggestion that the concept of the pentadactyl limb should not be treated as an abstract generalised form, as it was defined by Owen, but as an actual, materialised form which had (and perhaps still has) a historical existence in an ancestor common to the line of terrestrial vertebrate descent. This is explanation of form by historical contingency, not by morphogenetic law, the latter being the goal of the rational morphologists, such as Geoffroy St Hilaire, Cuvier, and Owen. The contemporary descendant of explanation by historical continuity and contigency is to be found in the concept that embryogenesis is a result of the operation of a genetic programme which is a historical record of adaptive success in a particular phylogenetic sequence. In this paper, we will not pursue this line of reasoning, but return instead to the concept that organismic morphology, and in particular vertebrate limbs, are to be understood in terms of general principles of spatial organisation, together with particular constraints (including the effects of specific gene products) which we develop in terms of the classical concept of the morphogenetic field.

Our objective is to demonstrate the simple proposition that tetrapod limb morphogenesis may be understood in terms of some basic generative principles capable of producing a great variety of limb forms which are all transformable one into the other under modifications of the limb-generating process. None of these forms is special in any way, none intrinsically more 'primitive' than any other, so that none can be selected as an ancestral form. Indeed, one of our purposes is to show that the study of principles or laws of biological form must be clearly separated from historical considerations, since the search for the invariants which define homology (equivalence of structure under a transformation which leaves certain spatial relations unchanged or invariant) depends solely on form and not on ancestry. Historical conservation of forms in particular lineages may also occur, but this observation does not help one to understand problems of biological organisation, such as the nature of the generative principles underlying the transformational equivalence of tetrapod limbs to which we now turn.

The generalised tetrapod limb form

It is convenient, with a generalised tetrapod limb form, to start by simply defining the essentials of its structure. For this we use a pattern similar to those found in texts on comparative anatomy such as Smith's *Evolution of chordate structure* (1960), though we emphasise again that, whereas he interprets this as a primitive ancestral limb form, the so-called 'cheiropterygium', we use the form shown in Fig. 1 to illustrate the general structures to be considered and the relationships of their components to the dif-

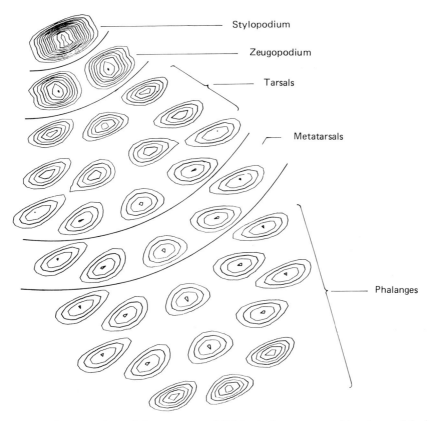

Stylopodium

Zeugopodium

Tarsals

Metatarsals

Phalanges

Fig. 1. Contour plots of the sequence of field solutions generated by the model of pattern formation for a hind limb with five digits. Each closed curve represents a particular level on a hill which describes a part of the field solution at some proximo-distal position. Thus the five metatarsal elements correspond to a single field solution with five peaks of the function which is selected by a particular 'energy' level at this position in the proximo-distal axis.

ferent parts of the tetrapod limb. The Figure is from a contour plot produced by a computer programme based upon a model of limb morphogenesis which is discussed briefly below and in detail in the Appendix (pp. 94–8). Each of the enclosed elements is a contour of a hill with a central maximum of a field function $u(x)$, as shown in a three-dimensional perspective in Fig. 2, and is taken to represent a domain where cartilage condensation will subsequently occur in a developing limb. Figure 1 thus describes an early stage in the formation of a 'pentadactyl' limb with (i) the usual number of

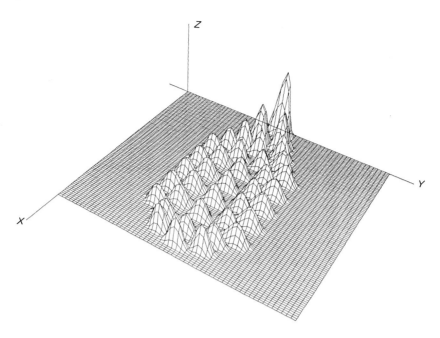

Fig. 2. A three-dimensional view of the pattern shown in Fig. 1 (but lacking the three terminal phalanges). Each hill represents a presumptive centre of cartilage formation in the developing limb bud. For variables, see Appendix p. 97.

elements in the stylopodium and the zeugopodium, where there is the greatest uniformity of structure in tetrapods, and (ii) an autopodium of 12 carpals or tarsals, 5 metacarpals or metatarsals, and a total of 13 phalanges arranged in 5 digits whose phalange formula is 23332. Since our objective is not to model the detailed development of any particular tetrapod limb but to explore the generative principles common to this type of limb and thus to gain some insight into the concept of homology from a developmental perspective, we have concentrated on the proximo-distal sequence of limb elements and ignored the secondary modifications which occur in most limbs with respect to the antero-posterior axis. Thus all the limb patterns shown have antero-posterior symmetry. Models incorporating these spatial details can quite readily be developed, but we have deliberately avoided these asymmetries in the present analysis in order to focus on the notion of transformational equivalence on which homology depends.

The principles used to model the morphogenetic field for limb formation in this paper correspond closely to those used in a previous publication dealing with a field description of the embryonic cleavage process (Goodwin & Trainor, 1980). We assume the existence of an order parameter, $u(x)$, which defines the state of the limb field at point x in terms of some relevant biological variable which describes that aspect of the limb-forming process which is the immediate causal antecedent of the emergent order resulting in a specific pattern of chondrogenic sites. This order parameter is assumed to obey a 'smoothness' principle of the general type used in describing the spatial variation of fields which is familiar to biologists from such examples as diffusion-like processes (e.g. Lawrence's (1970) 'sand model') or the intercalation process used in the clock model of limb regeneration (French *et al.*, 1976). This together with an assumption that the field values are everywhere bounded (limited in size) leads to a field equation for $u(x)$ whose solutions depend upon the geometry of the field and the values of the function on the boundaries of this domain, as well as on the specific value of a quantity generally analogous to an energy within the domain. In our model, a semicircular 'domain of competence' is assumed to propagate proximo-distally, starting at a particular stage in limb bud development, and to lay down a sequence of field solutions as shown in Fig. 1, each antero-posterior array of elements being generated by a particular field solution within the domain. This assumption is similar to that used by Ede (1976) in his model of limb morphogenesis. Thus, the presumptive centre for the humerus or femur is laid down first by a solution that produces a single maximum of $u(x)$ within the first hemispherical domain of competence. Then a solution with two maxima is established in a more distal domain generated by the advancing wave of competence; then a solution with three maxima, then four, then a series of four solutions with five maxima, and finally a solution with three maxima are laid down in proximo-distal order. The particular sequence thus obtained is, in this model, dependent upon a proximo-distal gradient in the limb-bud which may be compared to 'energy' levels stabilising or selecting particular solutions of the field equation. Thus to get the solutions generated in Fig. 1, the gradient is assumed to decrease progressively from a value which selects the solution with one period, giving the stylopodium, through values selecting solutions of periods 2, 3, 4, and 5, and finally rising to a value selecting the solution with period

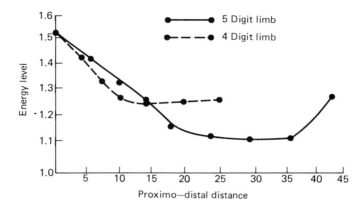

Fig. 3. The proximo-distal 'energy' profile which selects or stabilises a particular series of solutions of the field equation in successively more distal domains of competence in the limb bud. The solid curve gives rise to the sequence shown in Fig. 1, each solid circle on the curve defining an 'energy' level which corresponds to a particular eigensolution of the field equation, with a specific number of peaks across the antero-posterior axis. The dashed curve generates the pattern shown in Fig. 4, the curve flattening out at an 'energy' level which gives eigensolutions of period 4. The pattern generated is not dependent on limb bud size.

3 again. The gradient is assumed to be continuous, since solutions are stable only at particular 'energy' levels in this type of model; i.e. we are dealing with an eigenvalue problem. Repetition of the same solution, such as 5, occurs if the energy level is relatively constant so that solutions are set up as soon as a large enough domain is established by the advancing wave. To get the absence of a solution such as 4 between the distal 5 and 3 solutions, we assume a rather rapid change in the gradient so that a solution of period 4 has insufficient space or time to get established as the wave of competence propagates distally.

The sequence in Fig. 1 corresponds to the 'energy' profile shown in Fig. 3, which falls through a series of values generating the increasing series of elements, then levels off to give a sequence of four 5-unit solutions, and finally rises to give a terminal 3-unit solution. For comparison, a profile of a four-digit limb is included, which will be considered later.

Evidently it is the shape and the height of the proximo-distal 'energy' gradient which determine the sequence of elements which is laid down in a limb in this particular model, so that changes in the details of the gradient determine changes in pattern. This is only one of the possible ways of generating patterns by field solutions. A somewhat different mechanism is proposed by Wilby & Ede (1976) and by Newman & Frisch (1979), who used diffusion–reaction equations for their fields and selected different solutions by means of changes in an effective diffusion constant. The similarity of the models is that both use solutions of the same field operator (the Laplacian) to give a prepattern defining where the skeletal elements will later appear in the limb. We could say of the present model that the proximo-distal wave of competence, together with the treatment of the process as an eigenvalue problem, thus giving quantised solutions across the antero-posterior axis, is what transforms a one-dimensional energy profile into a pattern which is periodic in two dimensions. (The third dimension of the limb is ignored in this treatment, since we are concerned only with the bone sequence. We assume that the periodicity of the solutions in the third dimension (dorso-ventral) is always one, making the elements in Fig. 1 all solids. Any treatment of muscle and cartilage pattern would of course have to consider three-dimensional field solutions, with boundary conditions partly determined by the cartilage condensation patterns.)

A final point about the model is that no attempt is made to describe the modifications following initial chondrogenesis, such as the differential growth of elements in different parts of the limb or fusions of basipodial elements. Thus our concern is solely with the variety of forms which is possible in the primary condensation pattern, and how this may be generated.

Invariance and variation

A simple modification of the form in Fig. 1 involves a reduction in the total number of elements as a result of both a shorter proximo-distal series and a reduced periodicity across the antero-posterior axis, as shown in Fig. 4. This pattern correlates with an altered 'energy' profile, as shown in Fig. 3 (dashed curve). The level generating field solutions of period 5 is not reached, so the limb is one with only four digits. This is the type of limb found in urodeles such as the axolotl (*Ambystoma*; forelimb) or the salamander, *Necturus*, which is shown in Fig. 5 in a drawing taken from Kent *Comparative anatomy of the vertebrates* (1969). This drawing shows alternative ways in which a five-digit limb could undergo

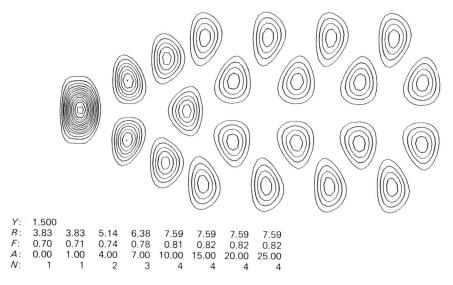

Y:	1.500							
R:	3.83	3.83	5.14	6.38	7.59	7.59	7.59	7.59
F:	0.70	0.71	0.74	0.78	0.81	0.82	0.82	0.82
A:	0.00	1.00	4.00	7.00	10.00	15.00	20.00	25.00
N:	1	1	2	3	4	4	4	4

Fig. 4. The limb pattern generated by the 'energy' profile of the dotted curve in Fig. 3. Y is an arbitrary scale factor. R, F, and A are as described in the Appendix, while N describes the Bessel function $\mathcal{J}_m$ which has been selected by the energy value in accordance with equation (A9). The first two Bessel functions corresponding to $N = 1$ at positions $A = 0$ and 1.0 fuse to give the first chondrogenic site, after which the functions are separated.

conversion to one with four digits, by 'loss' of a digit either preaxially or postaxially. However, our generative model suggests a much simpler and more plausible solution: in a transformation from one limb form to another, there is no conservation of elements so that there can be no question of one digit being lost and others conserved. Biologists have been misled by this type of question ever since Darwin adopted a view of evolution which stressed the adaptive detail of organismic morphology (i.e. the particular 'content' of organisms) rather than principles of form, and problems of inheritance rather than those of organisation. Great attention has been paid to the exact number of phalanges per digit, their relative length, and other features which could identify their individuality and so allow them to be recognised as units of inheritance when they turn up in *Eryops*, in the salamander, in the wing of the domestic hen, or in some other organism with an even greater 'loss' of digits. Thus the three digits of the hen's wing are identified as digits 2, 3, and 4 of the ancestral five-digit limb, the two of the antelope are

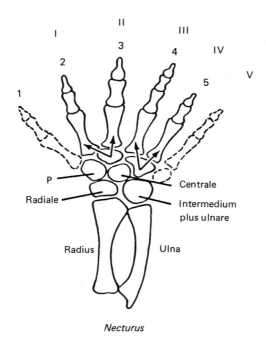

Necturus

Fig. 5. Hypothetical alternatives in the evolution of a four-digit limb from a five-digit 'common ancestor', according to the Darwinian concepts of homology and inheritance. The original pentadactyl limb has digits 1–5, or I–V, and either digit 1 or digit V is then selected against to give a four-digit limb, 2–5 or I–IV. P = pisiform. (After Kent, 1969.)

labelled 3 and 4, and the single toe remaining in the horse is identified as digit 3. Just how is such identification possible, especially when so much variation can occur in the details of the bone pattern within a single species, or even in an individual with multiple regenerates: a question to which we shall turn in a moment.

Figure 6 shows examples of the general categories of limb reduction to 3, 2 and 1 terminal digit, without any effort to reproduce the detailed pattern of any particular species. What emerges very clearly at this level of abstraction is just the point made by the pre-Darwinian rational morphologists: biological patterns must be seen and understood as wholes which undergo transformation, so that there is no identity of elements in different patterns. It is wrong even to suggest that the most proximal element of the limb is the 'same' in all tetrapods, even though the

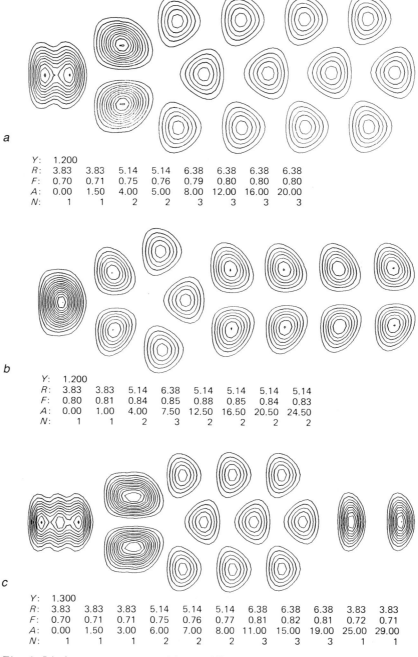

a

Y:	1.200							
R:	3.83	3.83	5.14	5.14	6.38	6.38	6.38	6.38
F:	0.70	0.71	0.75	0.76	0.79	0.80	0.80	0.80
A:	0.00	1.50	4.00	5.00	8.00	12.00	16.00	20.00
N:	1	1	2	2	3	3	3	3

b

Y:	1.200							
R:	3.83	3.83	5.14	6.38	5.14	5.14	5.14	5.14
F:	0.80	0.81	0.84	0.85	0.88	0.85	0.84	0.83
A:	0.00	1.00	4.00	7.50	12.50	16.50	20.50	24.50
N:	1	1	2	3	2	2	2	2

c

Y:	1.300										
R:	3.83	3.83	3.83	5.14	5.14	5.14	6.38	6.38	6.38	3.83	3.83
F:	0.70	0.71	0.71	0.75	0.76	0.77	0.81	0.82	0.81	0.72	0.71
A:	0.00	1.50	3.00	6.00	7.00	8.00	11.00	15.00	19.00	25.00	29.00
N:	1	1	1	2	2	2	3	3	3	1	1

Fig. 6. Limb patterns generated by modifications in the proximo-distal 'energy' profile of the limb bud, corresponding to limbs with (a) three, (b) two, and (c) one digit. Y, R, F, A and N are as described in the legend to Fig. 4 and in the Appendix.

element in this position is nearly always single. What is the same is the generating principle of the total pattern, so that there are invariants or universals which are recognised as common to the whole set of forms. These forms are thus equivalent under transformation (an idea embodied in Geoffroy St Hilaire's 'Principle of Connections' and later in Owen's concept of homology). Each element is a component of a total pattern, but has no inherited individuality in a genealogical sequence.

Embryological transformation

The conclusion reached above suggests that it may be possible to perturb limb morphogenesis in a single vertebrate species in such a way that the basic skeletal patterns of other vertebrates are produced (not, of course, the secondary modifications). This is because, if the only universal constraint is the generative mechanism of the presumptive chondrogenic pattern, then perturbations to the initial and the boundary conditions of the limb field should result in perturbations in the pattern even against a fixed genetic background, since in a field description genes cease to play the role of specific causes of form (cf. Webster & Goodwin, 1982). They contribute to the particular conditions which stabilise or select a specific field solution from the possible set, or a specific sequence of solutions which generate the field morphology, but their influence can be over-ridden by other particular conditions. Because of the absence of inherited individuality in the units of the limb, we should not expect to be able to identify elements in the perturbed limbs, but only general categories of tetrapod pattern. It is even possible that patterns will be observed which are not found in any tetrapod, since the expectation is that the set of possible forms which the generating mechanism can produce will be greater than the set which has been realised in evolution, i.e. the potential set is greater than the historically realised set, and individual species embody this potential in their fields, independently of genotype.

The investigation of limb regeneration in urodeles provides an opportunity for this type of observation, and we reproduce in Fig. 7 some patterns which have resulted from work with the axolotl, *Ambystoma mexicanum*, by M. Maden. A more complete study of this material is in progress, but we report here some observations of relevance to the present investigation. The sample shown in Fig. 7

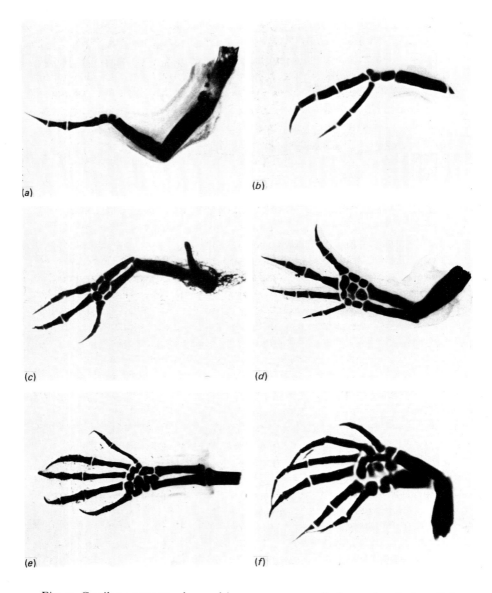

Fig. 7. Cartilage patterns observed in supernumerary limbs produced after 180°
ipsilateral rotations of blastemas in the axolotl, *Ambystoma mexicanum* (a) one,
(b) two, (c) three and (f) six-digits together with normal forelimb ((d) four-digit)
and hindlimb ((e) five-digit) patterns. (Courtesy of M. Maden.)

provides evidence supporting the general principles developed above, illustrating how plastic is the limb field in the patterns which can be generated and reinforcing the proposition that the field is the generative unit in development and evolution.

Transformations involving expansion

The limb pattern of Fig. 7*f* shows an expansion of elements across the proximo-distal axis, rising to six phalanges. Such a limb is usually interpreted to result from a fusion of two four-digit limbs, with the 'loss' of a digit in the process. However, there is no need to regard this type of pattern in this way if we are using a field description of limb morphogenesis, since it is perfectly possible for the generating process to involve 'energy' levels giving a periodicity of six (or more) across the antero-posterior axis. An example of such a limb is to be found in the fossil marine reptile *Ichthyosaurus*, shown in Fig. 8, where the number of elements spanning the antero-posterior axis rises to six and then falls off again. The digits are not separated in this case, since the limb was of paddle type, used for swimming. It is interesting to note that no attempt is made to identify the elements of such a limb in relation to the hypothetical 'common ancestor', except for the humerus and the radius and ulna, if the second set of elements in proximo-distal order constitute a pair (see Romer, 1956, Figure 129) – radius, ulna, and pisiform if these constitute a triplet (see Fig. 8 and Smith, 1960).

A computer simulation of this type of limb is shown in Fig. 9. In this particular example, the most proximal element arises from a 'fusion' of the maxima of functions with periods 1 and 2, giving a Y-shaped chondrogenic centre. Thus the number of elements goes from one to three without the presence of a separated solution of period 2, duplicating the pattern of Fig. 8. The type of variation seen in the bone sequence of *Ichthyosaurus* and the possibility that fusions may occur from the earliest stage of chondrogenesis further confuses the problem of identifying limb elements between species. Once again, we draw the obvious conclusion that the pattern of Fig. 8 is simply a transformation of the other limb forms conforming to the invariant generative principles of the limb field.

This conclusion is also reinforced by the absence of any transient archetypal limb pattern during early limb morphogenesis, as has been demonstrated so clearly by Hinchliffe & Griffiths (this

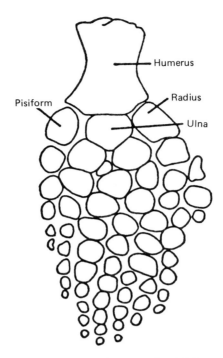

Fig. 8. Bone pattern in the limb of a fossil reptile, *Ichthyosaurus*. (After Smith, 1960.)

volume). However, the notion that there is inherited continuity of elements of pattern such as digits, as in Darwin's concept of continuity of descent from a common ancestor, or that there is a specific, isolatable material acting as a central organising agency of biological form (Webster & Goodwin, 1982), as in Weismann's concept of the germ plasm, dies hard. The extremely valuable studies of Nieuwkoop (this volume) show that the latter notion must also be abandoned as a general principle, since in urodeles there is no continuity of germ plasm from generation to generation, its appearance during embryogenesis being an epigenetic phenomenon. It seems likely that this will turn out to be the general case, with different species occupying different positions on the regulative-mosaic spectrum, so that the classical instances of apparent germ plasm continuity are resolvable in terms of early mosaicism in oogenesis, an earlier regulative phase nevertheless occurring. These observations all lead in the direction of an abandonment of

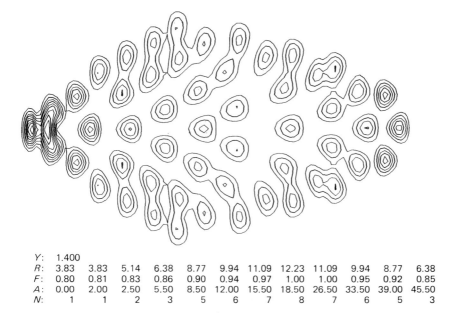

Y:	1.400										
R:	3.83	3.83	5.14	6.38	8.77	9.94	11.09	12.23	11.09	9.94	8.77	6.38
F:	0.80	0.81	0.83	0.86	0.90	0.94	0.97	1.00	1.00	0.95	0.92	0.85
A:	0.00	2.00	2.50	5.50	8.50	12.00	15.50	18.50	26.50	33.50	39.00	45.50
N:	1	1	2	3	5	6	7	8	7	6	5	3

Fig. 9. Contour plot of a pattern with higher harmonic solutions (eigensolutions) than that with a periodicity of 5, simulating the pattern of Fig. 8 and showing partial contour fusions which could result in variability of bone pattern. For variables, see Fig. 4 and Appendix.

the rather simplistic notions of the inheritance of sufficient causal elements of morphogenesis in an entity with the properties of a central organising agency, now called the 'genetic programme', and its replacement by an exact theory of morphogenesis in terms of universal generative principles expressed by field constraints, together with the particular influences operating during ontogenesis (including those arising internally from the genes) which result in an organism or specific form.

Discussion

We have made obvious the transformational equivalence of different tetrapod limb forms by concentrating on what we regard as the essential generative principles responsible for the patterns of chondrogenic sites. In so doing, we have ignored what we regard as the secondary influences in the limb field responsible for the systematic differences of bone structure observed across the antero-posterior axis, and the factors responsible for the differential growth of the chondrogenic sites in the proximo-distal axis. That these are considered as secondary, not primary, is one feature that distinguishes our treatment of limb morphogenesis sharply from a model such as that of Wolpert *et al.* (1975), in which the limb elements are derived from primary axial gradients by a process of 'interpretation' and there is no prepattern of elements in a field isomorphic with the manifest limb pattern. A major difference between these views is in the number of parameters involved in generating the pattern and in changing from one pattern to another. Applying Wolpert's chick limb model to the pentadactyl limb, and using the threshold mechanism of Lewis *et al.* (1977), about 24 parameters are needed to generate 5 discrete elements across the antero-posterior axis (a minimum of 4 parameters for each threshold value, of which there are 5, and the morphogen gradient mechanism, involving another 4 parameters or more). Each digit 'gained' or 'lost' involves changes in several parameters, for the threshold points must be properly positioned relative to the morphogen gradient to produce discrete, separate chrondrogenic sites. The proximo-distal pattern requires more threshold points in another monotonic gradient involving some 35 parameters or more. Clearly, genetic programmes of this type are expensive on particulars because there are no generative universals; and since, in such models, parameters are specified by genes, we would expect a large number of mutants affecting the details of limb morphology, for which there is little evidence. A polydactylous mutant with six digits requires either the introduction of a new threshold (four parameters from one gene) or a change in the gradient such that a 'hidden' threshold is revealed, in which case it might be concluded that the 'common ancestor' had six digits.

On the whole, the evidence seems strongly to favour a more robust pattern-forming mechanism involving fewer parameters, whose basic characteristic is the generation of spatially ordered arrays (whole patterns) of repeating elements which are primarily the same but differ by secondary variations superimposed on the primary pattern generator. This takes us back to the prepattern fields of Stern (1954) and Turing (1952), and away from positional information and interpretation, whose specific objective was to replace prepatterns by monotonic gradients and thresholds wherever there is evidence that the units produced are different from one another (difference in this context being called non-equivalence, so that equivalence comes to have the extremely restrictive connotation of identity rather than its more general meaning of similarity under some transformation, which is how we use it in relation to the concept of homology). The model proposed in this paper, involving a field theory which generates a morphogenetic prepattern of a type different from those based upon reaction and diffusion but with some fundamental characteristics in common, is economical on parameters, since the basic pattern varies in response to the proximo-distal 'energy' profile. This profile can be described by curves of three parameters, only one of which needs to be varied to generate the range of patterns shown in Fig. 1, 4 and 6. Genes are assumed to be involved in specifying the details of these curves, but the rest of the pattern-forming mechanism is assumed to be universal to the tetrapods. We must then assume a secondary mechanism (say another three-parameter gradient) across the antero-posterior axis to introduce variations in the intensities of the order parameter, u, leading to asymmetries in the units across this axis.

Tetrapod limbs, as a general category of vertebrate structure, are characterised in such an analysis by a high-level constraint relating to the generative particulars which distinguish limbs from fins and other appendages, presumably involving genes implicated in processes such as limb bud formation and chondrogenesis. However, as well as these particulars, the basis of evolutionary change, there will continue to be universal mechanisms involved which, in our model, are the morphogenetic fields themselves, with solutions characterised as eigensolutions in our formulation. Thus, at every level of a generative analysis, there is a combination of universal and particular factors at work.

The view of biological form advanced here is based upon the general scientific principle that diversity of structure is logically intelligible only if there is some unity or invariance which limits not only observed diversity (which is the sample that history happens to have given us), but possible or potential diversity. In the case of the typical vertebrate form which has come to be known as the pentadactyl limb (somewhat unfortunately, since there is nothing special about five digits), this invariance is to be found in the generative principles which are involved in primary limb morphogenesis. Our particular model involves constraints deriving primarily from the mechanism whereby a continuous 'energy' profile selects a series of eigen-function solutions of a field arising in a distally propagating domain of competence in the developing limb bud. In consequence, limb forms are defined by a series of discrete chondrogenic sites whose antero-posterior arrays differ in number of elements by a small integer (usually one) as one moves distally. If limbs commonly occurred with elements increasing distally in a sequence such as 1,2,4,8,16,32, . . . then different generative rules would be involved. Such a sequence describes the number of cells in holoblastic cleavage, which has been shown to be describable by a field treatment similar to that used here for vertebrate limb morphogenesis, but involving different selection (generative) rules on the field solutions (Goodwin & Trainor, 1980). Thus, we see that the general field approach to biological form gives a hierarchy of constraints within a global theory whereby different forms can be distinguished from one another while still united by general principles of spatial organisation as embodied in the field treatment. In this way both the unity and the diversity of biological forms may be grasped, and a generative basis provided for a rational taxonomy, as envisaged by the pre-Darwinian rational morphologists.

Another constraint arising from our field description of limb morphogenesis is very similar to that which Holder (this volume) has deduced in relation to the polar coordinate model of limb formation. His prediction is that limbs will not be found whose primary patterns have external digits larger than internal ones. Anomalies such as the lobster-claw syndrome are excluded because they arise from a secondary process which effectively destroys part of the primary pattern. A similar prediction can be made for the pattern-generating mechanism which we describe in this paper. This is that tetrapod limbs will not be found with more phalanges in the external digits than in the internal ones. The reason for this is that there cannot be an eigensolution which has blank spaces in the middle of an antero-posterior array. The converse can certainly occur, as illustrated in Fig. 1, where the interior three digits each have one more phalange than the external digits, because of a drop in the 'energy' profile distally. With antero-posterior asymmetry, there could be a very long external digit on one side of the limb, as in certain species of whale, but not on both.

The approach to morphogenesis described in this paper has the further consequence that the historical sequence of forms emerging during evolution is logically secondary to an understanding of the generative principles defining the potential set of these forms and their transformations; so that Darwinian taxonomy, which is genealogy, is secondary to rational taxonomy, and both are logically derivative from ontogenetic principles. The external selection principle on historically realised forms which is added by the functional or stability considerations embodied in the concept of fitness, while of obvious importance in many instances, is supplementary to those which operate within organisms in the generative process itself, and certainly can never be used as a causal substitute for the latter. Within the rationalist approach to biological morphology being proposed, the interesting possibility emerges that a stability constraint (such as 'fitness') on the set of possible morphological transformations may embody a general principle of directed or time-dependent transformation of the type which in physics gives such directional stability criteria as the second law of thermodynamics, imposing temporal asymmetry on the time-reversal symmetries of mechanics (Goodwin, 1982). It is evident that a much more highly elaborated theory of form and stability than that currently available is required before this type of theory can emerge, although it remains a part of the research programme which follows from the principles underlying the treatment of biological form and organisation advanced in this paper.

Grateful acknowledgement is made to Charles Trainor for substantial assistance with the computation involved in this work, to John Totafurno for the computer graphics of Fig. 2 and to Malcolm Maden for the photographs of limbs shown in Fig. 7. We are also indebted to the National Sciences and Engineering Research Council of Canada for financial support.

References

Ede, D. A. (1976). Cell interactions in vertebrate limb development. In *The cell surface in animal embryogenesis and development*, eds. G. Poste and G. L. Nicolson, pp. 495–543. Elsevier: Amsterdam.

French, V., Bryant, P. J. & Bryant, S. V. (1976). Pattern regulation in epimorphic fields. *Science* **193**, 969–81.

Goodwin, B. C. (1982). Development and evolution. *J. theor. Biol.* **97**, 43–55.

Goodwin, B. C. & Trainor, L. E. H. (1980). A field description of the cleavage process in embryogenesis. *J. theor. Biol.* **85**, 757–70.

Kent, G. C. (1969). *Comparative anatomy of the vertebrates*. C. V. Mosby Co.: St Louis.

Lawrence, P. A. (1970). Polarity and patterns in the postembryonic development of insects. *Adv. Insect Physiol.* **7**, 197–265.

Lewis, J. H., Slack, J. M. W. & Wolpert, L. (1977). Thresholds in development. *J. theor. Biol.* **65**, 579–90.

Newman, S. A. & Frisch, H. L. (1979). Dynamics of skeletal pattern formation in the developing chick limb. *Science* **205**, 662–8.

Romer, A. S. (1956). *The vertebrate body*. W. B. Saunders Co.: Philadelphia.

Smith, H. M. (1960). *Evolution of chordate structure*. Holt, Rinehart and Winston: New York.

Stern, C. (1954). Two or three bristles. *Am. Sci.* **42**, 213–21.

Turing, A. M. (1952). The chemical basis of morphogenesis. *Phil. Trans. R. Soc. B* **237**, 37–72.

Webster, G. C. & Goodwin, B. C. (1982). The origin of species: a structuralist approach. *J. Soc. biol. Struct.* **5**, 15–47

Wilby, O. K. & Ede, D. A. (1976). Computer simulation of vertebrate limb development. In *Automata, languages, development*, eds. B. Lindenmeyer and G. Rozenberg, pp. 143–61. North-Holland: Amsterdam.

Wolpert, L., Lewis, J. & Summerbell, D. (1975). Morphogenesis of the vertebrate limb. In *Cell patterning*, Ciba Foundation Sympsoium No. 29, pp. 96–119. Associated Scientific Publishers: Amsterdam.

Appendix: A generative model for the pentadactyl limb

We assume that the process of chrondrogenesis is 'staged' by a set of contingent factors which pick out, at each stage, a particular or characteristic solution from the set of all possible solutions of the morphogenetic field equations. The contingent factors (which may be related at the subcellular or cellular levels to metabolic activity, ultra-structural detail, membrane properties, etc.) define at the field level the domain of activity, the values of the field on the domain boundary, and the overall field 'energy'.

In our model, we regard the primary morphogenetic field $u(x,y)$ to be an 'order parameter', whose value at the point (x,y) in the limb bud describes the degree of order there; a high degree of order favours chondrogenesis.

Figure 2 in the text gives a three-dimensional perspective of the ordering field; peaks correspond to centres of chondrogenic activity. As described in the text, we concentrate on the proximo-distal aspect of the developing field with a two-fold simplication: first, we ignore dorso-ventral distinctions and treat the morphogenetic field as a two-dimensional field in the plane containing the proximo-distal and antero-posterior axes; secondly, we ignore antero-posterior asymmetries and consider only field solutions which are intensity symmetric about the proximo-distal axis.

At any particular stage in chondrogenic development we assume that the level of activity is quadratic in both the field amplitude and its gradient, an assumption which has numerous successful analogues in physical theory. We further assume that an optimality principle operates which requires investigation of the field energy (Hamiltonian)

$$E = \int Hr\,dr\,d\theta = \int \frac{1}{2}\left[\left(\frac{\partial u}{\partial r}\right) + \frac{1}{r^3}\left(\frac{\partial u}{\partial \theta}\right)^2 + \sigma u^2\right] r\,dr\,d\theta, \tag{A1}$$

subject to a normalisation of the field intensity

$$\int u^2\,r\,dr\,d\theta = 1. \tag{A2}$$

In these equations, σ is analogous to an elastic constant whose value establishes the relative importance of field strength as opposed to field smoothness (gradient). We use polar coordinates with $\theta = 0$ along the proximo-distal axis and $r = 0$ at the proximal centre of activity. The field domain is discussed below.

The equivalent minimisation problem is

$$\delta \int L\,dr\,d\theta = 0,$$

where

$$L = r\left(H - \frac{\alpha u^2}{2}\right),$$

and α is a Lagrange parameter. Defining $\beta = \sigma - \alpha$, and using the notation

$$u_r = \left(\frac{\partial u}{\partial r}\right), \qquad u_\theta = \left(\frac{\partial u}{\partial \theta}\right),$$

the Euler–Lagrange equation

$$\frac{\partial}{\partial r}\left(\frac{\partial L}{\partial u_r}\right) + \frac{\partial}{\partial \theta}\left(\frac{\partial L}{\partial u_\theta}\right) - \frac{\partial L}{\partial u} = 0$$

becomes

$$r^2\frac{\partial^2 u}{\partial r^2} + r\frac{\partial u}{\partial r} + \frac{\partial^2 u}{\partial \theta^2} - \beta r^2 u = 0. \tag{A3}$$

Separation of variables then yields the equations

$$\frac{d^2 g}{d\theta^2} + \lambda g = 0, \tag{A4}$$

$$\frac{d^2 f}{dr^2} + \frac{1}{r}\frac{df}{dr} - \left(\beta + \frac{\lambda}{r^2}\right)f = 0, \tag{A5}$$

where λ is a separation constant and

$$u(r,\theta) = f(r)g(\theta). \tag{A6}$$

If we take the domain of a field activity at any stage to be a semi-circular region of radius R with base perpendicular to the proximo-distal axis, as shown in Fig. 10, and require $u = 0$ on the boundaries, the following solutions are obtained:

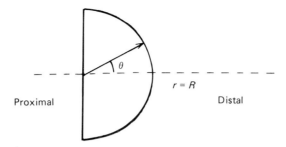

Fig. 10. Semi-circular domain of field activity or of competence in the developing limb bud, within which a field solution is set up whose form is determined by the boundary values on the domain and the energy level within it.

$$g = g_0 \frac{\sin m\theta}{\cos m\theta}, \text{ where } m = 0, \pm 1, \pm 2, \ldots,$$

and the λ values which are possible are restricted by the eigenvalue condition

$$\lambda = m^2 \geqslant 0.$$

Thus for $m = 0$, a null solution occurs,

for $m = \pm 1$, $g = \cos \theta$,

for $m = \pm 2$, $g = \pm \sin 2\theta$,

for $m = \pm 3$, $g = \cos 3\theta$,

for $m = \pm 4$, $g = \pm \sin 4\theta$.

(We have taken $g_0 = 1$ for convenience.)

The radial equation then becomes

$$\frac{d^2 f}{dr^2} + \frac{1}{r}\frac{df}{dr} - \left(\beta + \frac{m^2}{r^2} \right) f = 0, \tag{A7}$$

with solutions indexed by $|m| = 1, 2, 3, \ldots$, according to the possible values of m^2.

Writing $\beta = -k^2$ and defining $\varrho = kr$, this equation becomes Bessel's equation

$$\frac{d^2 f}{d\varrho^2} + \frac{1}{\varrho}\frac{df}{d\varrho} + \left(1 - \frac{m^2}{\varrho} \right) f = 0, \tag{A8}$$

with eigensolutions $f \longleftrightarrow \mathcal{J}_m(\varrho) = \mathcal{J}_m(kr)$, $m = 1, 2, 3, \ldots$ .. The corresponding $u(r, \theta)$ solutions are then (up to a normalisation constant)

$$u_1 = \mathcal{J}_1(kr) \cos \theta,$$
$$u_2 = \mathcal{J}_2(kr) \sin 2\theta,$$
$$u_3 = \mathcal{J}_3(kr) \cos 3\theta,$$
$$u_4 = \mathcal{J}_4(kr) \sin 4\theta \quad \text{etc.}$$

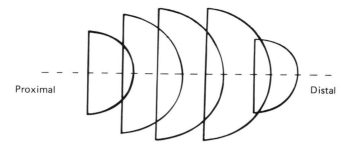

Fig. 11. Proximo-distal sequence of domains within which successive field solutions are established, generating the pattern of chondrogenic sites in the developing limb.

The total energy corresponding to expression (A1) for each solution u_m can then be calculated explicitly and simply (using the fact that u_m satisfies the differential equation (A3) and employing partial integration techniques) in a manner analogous to that in the Appendix of Goodwin & Trainor (1980), to yield

$$E_m = \frac{\sigma - \beta_m}{2} = \frac{\sigma + k_m^2}{2}. \qquad (A9)$$

The computer simulations used in this paper, and referred to in the main text, assume that successive stages of chondrogenesis take place in hemispherical domains of activity which shift distally and which may expand or contract depending upon contingent conditions (see Fig. 11). Although the order field $u(r, \theta)$ obeys a linear equation, the $u(r, \theta)$ solutions at successive stages develops at different times, so that earlier (more proximal) fields have generated fixed or determined presumptive chondrogenic patterns before later (more distal) fields develop. The result is that, rather than amplitudes, it is the intensities, which are proportional to u^2, that are additive.

At each stage, which Bessel solution will be selected depends on the metabolic energy available at that stage, since equation (A9) must be satisfied. The overall pattern obtained depends then on which succession of Bessel solutions is selected and in what size domains (R value), as well as on the domain locations along the proximo-distal axes. For example, Fig. 1 is a contour plot and Fig. 2 is a simulated landscape of u^2 intensity, corresponding to the legend values in Fig. 1: the main value (Y) indicates which Bessel function $\mathcal{J}_m$ has been selected according to equation (A9) by the (assumed) energy profile given by the solid lines in Fig. 3; the R values are the radial positions of the first zeros of the Bessel functions while the F values are the reciprocals of the scale factors k_n in the Bessel function, which are themselves determined according to (A9); finally the A values are the distal shifts in the origin of coordinates for successive stages, corresponding to a distal progression in the stages of early cartilage formation.

The process can be visualised as follows: the basic geometry together with minimisation condition leads to eigensolutions of Bessel's equation. Stable solutions develop only when the metabolising conditions satisfy (A9). At each stage of chondrogenesis a particular solution occurs and a pattern is laid down. As the metabolising level falls (or rises), successively higher (lower) solutions of Bessel's equation are selected and more (less) complicated patterns develop. The boundary conditions $u(r, \theta) = 0$, together with energy optimality, select out the first zero of the appropriate Bessel function to coincide with the domain of chondrogenic activity. From this perspective, successive stages of chondrogenic activity are defined by the successive time frames in which the metabolic activity reaches a level which satisfies some particular eigensolution of Bessel's eqution.

The model illustrates the importance of organising principles in developmental biology. While genetic factors play an important contingent role, for example in the determination of the energy profiles in Fig. 3, the process of patterning reflects properties of the morphogenetic field and is common to early limb development in many diverse species.

The prechondrogenic patterns in tetrapod limb development and their phylogenetic significance

J.R. HINCHLIFFE AND P.J. GRIFFITHS

The Zoology Department, The University College of Wales, Penglais,
Aberystwyth, Dyfed SY23 3 DA, UK

To avoid the taint of theory in morphology is impossible however much it may be wished.

T. H. Morgan

Limb homology and development: the archetype theory

The pattern of the pentadactyl limb which is such a striking feature of the skeleton of tetrapods constitutes *the* classic example of homology. As early as 1849, in his book *On the nature of limbs*, Richard Owen drew attention to the fundamental similarity of pattern in tetrapods, regardless of the use to which the limb was put, whether for flight, for swimming or for running. As an opponent of evolution, Owen had no explanation for this fundamental similarity – which he termed 'homology' – but to Darwin the pentadactyl pattern provided a splendid illustration of the principle of adaptive radiation. 'What can be more curious', asks Darwin (1872), 'than that the hand of man formed for grasping, that of a mole, for digging, the leg of a horse, the paddle of a porpoise and the wing of a bat, should all be constructed on the same pattern and should include similar bones and in the same relative positions'. To Darwin, the homology of the vertebrate limb, regardless of its various functional adaptations, indicated descent from a common ancestor.

In due course, as discoveries were made in fossils of the structure of the primitive amphibian limb skeleton (e.g. in *Eryops*) and of the Crossopterygian fish fin paddles (in rhipidistians such as *Sauripterus* and *Eusthenopteron,* regarded as possible tetrapod ancestors),

Abbreviations for all Figures:-
bc, basal commune; C, central carpals or tarsals; F, fibula; f, fibulare; fc, falciform; Fe, femur; H, humerus; i, intermedium; p, pisiform; R, radius; r, radiale; T, tibia; t, tibiale; U, ulna; u, ulnare; Y, element Y, according to Holmgren (1933), restricted to urodeles; 1–5, distal tarsals (dt) and carpals (dc); I–V, digits.

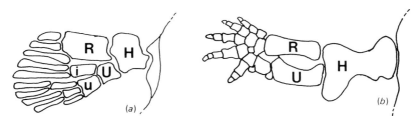

Fig. 1. Gregory's theory (Gregory & Raven, 1941) of the transformation of a Crossopterygian (*Sauripterus*) pectoral fin paddle skeleton (*a*) into the pentadactyl limb (*b*) of the primitive amphibian (*Eryops*).

Gregory and his colleagues (Gregory *et al.*, 1923; Gregory & Raven, 1941) were able to put forward a plausible scheme of evolutionary transition of fish fin into primitive tetrapod limb (Fig. 1). The homology between fin paddle and limb is more evident at the proximal end of the limb where amphibian stylopod (humerus, or femur) and zeugopod (radius, ulna or tibia, fibula) elements are clearly represented in the Crossopterygian fin paddle skeleton. More distally, from the wrist or ankle outwards, the correspondence has been more debatable, with little agreement between the different authorities as to which elements of the fish paddle survive, how they may be divided up, or whether the tetrapod digits represent newly evolved ('neomorph') structures.

   A number of attempts have been made to construct a fundamental pattern – or archetype – for the pentadactyl limb (Steiner, 1934; Holmgren, 1933, 1949, 1952, 1955) based on the skeletal pattern in the primitive amphibians (Fig. 2). The so-called canonical elements of the carpus according to Holmgren consist of four proximal elements (radiale, intermedium, ulnare, pisiform), four central elements (1–4), and five distal carpals, one at the base of each of the five digits. The precise composition of the pattern which makes up the pentadactyl limb is difficult to define, since digital number and phalangeal formula are very variable between different species, as in specialised forms such as adult bird wing and

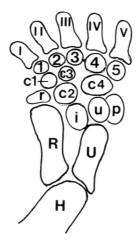

Fig. 2. 'Canonical' elements of the primitive archetypal carpus, according to Holmgren (1933, 1949, 1952).

horse foot (Hinchliffe & Johnson, 1980). There has been a long held (and pre-Darwinian) assumption that in development the general precedes the specialised in the formation of structure (Patterson, this volume), and it is still widely believed that, however specialised the adult limb may be, it passes through an archetypal pattern in its development. Holmgren thus considered that the canonical elements of the carpus could be identified in limb development throughout the tetrapods, with the exception of the urodeles, for reasons discussed later. In his wide-ranging surveys (Holmgren, 1933, 1952, 1955) of many species of anurans, reptiles, birds and mammals, the same archetype is described in the prechondrogenic pattern of the limb bud.

This approach is stated, in a general way, by Darwin, who wrote that 'The forelimbs, which once served as legs to a remote progenitor, may have become, through a long course of modification, adapted in one descendant to act as hands, in another as paddles, in another as wings: but . . . the forelimbs will not have been much modified in the embryo of these several forms; although in each form the forelimb will differ greatly in the adult stage'. Essentially this view is still echoed today in Jarvik (1980), who states that '. . . ontogenetic development of the extremities proceeds from a given number of metameric elements. According to a common opinion originating from Gegenbaur the extremity skeleton in tetrapods is constructed from a definite number of elements which are arranged, in the hand (manus) as well as the foot (pes) according to an original common ground plan'. Normally the limb skeleton in its development goes through a process in which a prechondrogenic condensation of closely packed cells forms in undifferentiated mesenchyme, the condensation gradually chondrifying through the secretion of cartilage matrix, and thus forming a cartilage 'model', which usually later hypertrophies and is replaced by bone. According to the archetype theory, the precartilage condensation pattern represents the pentadactyl limb archetype, which becomes gradually modified in later development into more specialised forms by processes of fusion or disappearance of individual elements, or by differential growth rates of different elements.

Descriptive accounts of the development of the limb skeleton have thus been seized on by palaeontologists and comparative anatomists in their search for reliable evidence of homology on which to base their phylogenies. Thus, identification of a single archetypal pattern of precartilaginous condensations will support the view of a monophyletic origin of the tetrapods. Another interpretation is that of Holmgren (1933, 1939) and Jarvik (1965, 1980) who argue that there is a fundamental difference in the condensation pattern in the urodele limb, from that in the 'eutetrapods' (anuran amphibians, birds, reptiles and mammals), which possess a distinctly different condensation pattern. These two authors argue that urodeles originate independently from the eutetrapods, and that the tetrapods are therefore diphyletic in origin. Jarvik (1980) sees the same fundamental seven-rayed pattern in the Crossopterygian *Eusthenopteron*, in the primitive amphibian limb, and in the prechondrogenic and chondrogenic pattern in the eutetrapod limb bud, for example in both the human embryonic hand and that of the lizard *Lacerta* (Fig. 3). Urodeles he regards as having a distinctly different pattern, with digits I and II being carried by a single basal carpal element, the intermedium ray thus being branched in urodeles, but carrying only a single digit in the eutetrapods.

Again, descriptive embryology has been used in the recent discussion as to the origin of birds and the significance of *Archaeopteryx*. Ostrom (1976) has claimed there is strict homology between the three digits of certain coelurosaurian dinosaurs and *Archaeop-*

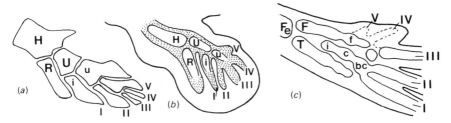

Fig. 3. According to Jarvik (1980) the fin paddle skeletal pattern of the Crossopterygian *Eusthenopteron* (*a*) is repeated in limb development in the pre- and chondrogenic pattern of the 'eutetrapod' group (anurans, reptiles, birds and mammals) as represented by the lizard, *Lacerta* (*b*). Jarvik claims the pattern of the intermedium ray and digits I and II is fundamentally different in urodeles, as represented by *Triturus* (*c*, after Sewertzoff, 1908).

teryx as part of his argument for a dinosaur origin for birds. These three digits are clearly anterior digits (I–III) in dinosaurs. The question then arises as to whether the three surviving digits in modern birds are anterior as would be shown if there were embryological evidence of the existence of two vestigial digits (IV, V) posteriorly. In this case the embryological evidence provided by the digits is not conclusive, since only one posterior vestigial digit can be found. A different viewpoint on bird evolution is provided by Walker (1972), who claims that birds and crocodiles are closely related, partly on the basis of the similarity of the developing pattern in the carpus and tarsus. This similarity is not, however, supported by more recent evidence of bird carpus and tarsus development (Hinchliffe, 1977). These three examples illustrate how the descriptive embryology of the limb may be used in the attempt at construction of phylogenetic relationships based on homology.

The use of embryological evidence in such a way is not straightforward, however. First, the evidence itself is often equivocal, since it is usually based on haematoxylin-stained sections of developing limbs, which do not give a clear picture of condensations and which therefore leave much scope for interpretation. The embryological studies quoted are usually of older papers, since little descriptive embryology of limb patterns, using the newer methods now available, has been published in the last forty years. The equivocal nature of the older evidence is apparent in reviewing the literature on the development of the pattern in the bird wing. Here, the number of precartilaginous carpal condensations is differently identified by different authors. Estimates range from 13 to 4 (e.g. 13 Montagna, 1945; 11 Holmgren, 1955; 7 Parker, 1889; 6 Lansdown, 1969; 4 Leighton, 1894). Since the number of cartilaginous elements in the wrist is small (4), authors such as Montagna & Holmgren invoke a process of fusion and regression to explain the reduction from the larger number of condensations. However, a recent investigation of the carpal condensation pattern, using autoradiography, revealed only four condensations, corresponding to the number of cartilaginous elements, and there appeared to be no process of fusion and regression (Hinchliffe, 1977). These findings appeared to question the existence of an archetypal condensation pattern in the wrist in birds, and thus a wider survey of the condensation pattern was attempted using autoradiography in a number of representative species: the axolotl (urodele), *Xenopus* (anuran) and the mouse. The previous findings in the chick limb (Hinchliffe, 1977) are first briefly reviewed; the new findings are presented and discussed in relation to the hypothesis of a limb archetype, and to control mechanisms of the development of limb skeletal pattern.

Improved autoradiographic methods for analysis of the prechondrogenic pattern

The condensations are ill-defined using standard histological methods, which also fail to distinguish prospective chondrogenic from tendon and myogenic condensations. A more specific method of identifying chondrogenic condensations is clearly needed. Using $^{35}SO_4$ the chondroitin sulphate component of cartilage intercellular material can be labelled. Searls (1965a) showed that if chick limb buds are labelled with $^{35}SO_4$ and processed for light microscopy, all the radioactivity left in the limb buds is bound to proteoglycan, mainly chondroitin-4 and chondroitin-6 sulphates. While some of this is initially localised in prospective soft-tissue and myogenic regions (Cioffi *et al.*, 1980), chondroitin sulphate synthesis eventually falls off in the non-chondrogenic regions and increases in the prechondrogenic condensations. The prechondrogenic pattern representing the future zeugopod elements can be detected by autoradiography at late stage 22 (Searls, 1965b), which is slightly before the increased packing of the condensations can be identified by histology (Hinchliffe, 1977). By application of $80\,\mu$Ci $^{35}SO_4$ to the chick embryo (4–7 days) vitelline circulation for 4 h (for details see Hinchliffe & Ede, 1973) the developing pattern of wing and leg prechondrogenic condensations and chondrogenic elements was mapped out. The resulting autoradiographs produced a more specific and clearly defined map of the condensations and their transition into cartilage elements than in the classical histological accounts.

We attempted to apply similar methods to *Xenopus* and axolotl (*Ambystoma*) larva (staged according to Nieuwkoop & Faber, 1967, or Rugh, 1960), but without initial success. Direct injection proved too difficult, while larvae immersed in various concentrations of $^{35}SO_4$ in freshwater failed to show much radioactivity in their limbs. Finally, we cut off the developing limbs and cultured them for 12 h in Holtfreter's solution to which $^{35}SO_4$ was added at an activity of $80\,\mu$Ci per ml. Details of fixation, use of stripping film and exposure were the same as for the chick limb buds. The results were more variable and the condensations less clearly defined than in the chick limb buds and there tended to be more background activity in the soft-tissue regions. Nonetheless, the patterns were more precise than those obtained by histology, though in some instances the number and form of elements is clearly debateable.

In the case of the mouse, pregnant mice of 13, 14 and 15 days were injected intraperitoneally for 12 h with $500\,\mu$Ci of $^{35}SO_4$, after which the embryos were obtained, and the limb buds then treated in the same way as chick limb buds. The day on which the copulation plug was found was designated day 1. The technical work was carried out by Dr M. Hicks. A similar study was made by Milaire (1978) using the $^{35}SO_4$ method, which he found to produce a pattern of chondrogenesis slightly in advance of histochemical methods (Masson's trichrome) involving staining for proteoglycans.

Where elements correspond they are labelled according to Holmgren's interpretation, but this does not mean that Holmgren's conclusions about homology are accepted.

Prechondrogenic condensation pattern in the chick wing and leg bud

This pattern has already been described (Hinchliffe, 1977) and here only the main features will be reviewed (Figs. 4, 5). Auto-

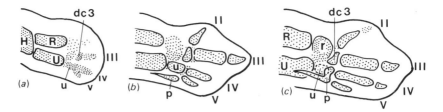

Fig. 4. Development of the pattern of prechondrogenic condensations and cartilaginous elements in the chick forelimb. *a* Stage 27 ($5\frac{1}{4}$ days); (*b*) stage 28 (5.5–6 days), and (*c*) stage 30 (7 days). Drawn from autoradiographs of the pattern of $^{35}SO_4$ incorporation.

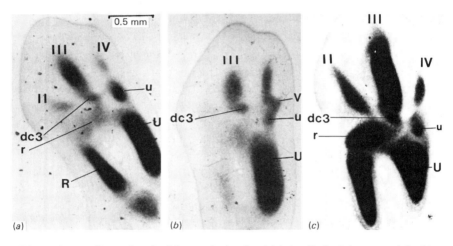

Fig. 5. Autoradiographs of $^{35}SO_4$ uptake by the chick forelimb. (*a*) stage 27/28; (*b*) stage 28; (*c*) stage 30.

radiography shows carpal condensations starting to appear at stage 27 (5 days) at the same time as the condensations of metacarpals 3–5. Four carpal elements are clearly defined by stage 28 (6 days); radiale, ulnare, pisiform and a single distal carpal at the base of metacarpal 3. After stage 30, the ulnare (by now chondrogenic) ceases to be active synthetically, and soon becomes pycnotic and disappears. It is the only chondrogenic element to disappear. Only condensations which later become chondrogenic are present, and no condensation fusion can be seen. Autoradiography does not reveal the 'missing' digit in the chick: the three main digits

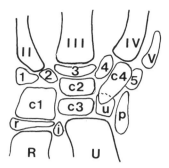

Fig. 6. Montagna's (1945) interpretation of the prechondrogenic pattern of the chick carpus at 6 days. This interpretation conforms with Holmgren's archetype.

(conventionally termed II–IV after Montagna) and a small posterior metacarpal (V) are present, but there is no sign of a fifth digit.

This account is very different from the classical description (Fig. 6) by Montagna (1945) which is summarised in such standard accounts of avian embryology as those by Hamilton (1952) and Romanoff (1960). All the previous accounts, though differing from each other in points of detail, describe a pattern of carpal condensations reasonably close to the ancestral amphibian carpus. Thus, at 6 days Montagna describes four proximal elements (radiale, intermedium, ulnare, pisiform), the four central elements (C_1–C_4) and finally five distal carpals (dc_1–dc_5) (Fig. 6). A process of fusion of some condensations and non-chondrification of others then accounts for the pattern of chondrogenic elements. The 13 condensations thus become 4 cartilages. In Holmgren's account (1955), 11 condensations are described, the identification of central carpals differing from that of Montagna. According to other accounts the number of condensations is considerably greater than the number of carpus cartilages: Sieglbauer (1911) describes 9 (duck), Steiner (1934) and Parker (1889) 7.

In our view, the condensation number is the same as the cartilage number, and there is no process of condensation fusion or of non-chondrification. The carpal condensation pattern is specialised and quite different from the archetype of primitive amphibians.

In the leg bud, autoradiography (Figs. 7, 8) shows that prechondrogenic tarsal elements first appear at stage 26 with the fibulare and a single distal tarsal (dt 2/3) at the base of metatarsals 2 and 3. Metatarsals 4 and 5 are also present. At stage 27, metatarsal 1 and

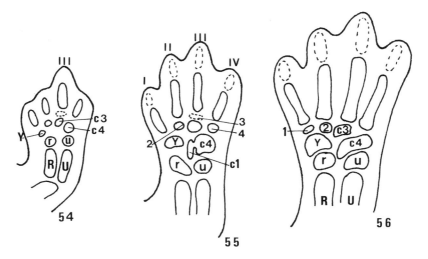

Fig. 10. Development of the pattern of prechondrogenic condensations and cartilaginous elements in the *Xenopus* forelimb. Drawn from autoradiographs of the pattern of $^{35}SO_4$ incorporation. Stage numbers given below.

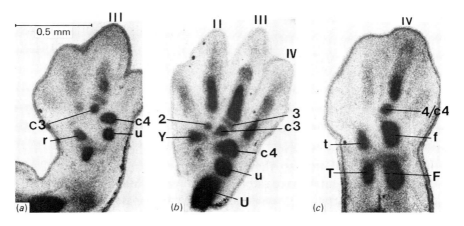

Fig. 11. Autoradiographs of $^{35}SO_4$ uptake by the developing limbs in *Xenopus*. (*a*), (*b*) forelimb stages 54 and 55; (*c*) hindlimb, stage 54.

cartilage element pattern, and not the archetypal pattern in primitive amphibians, since there are too few elements and the pattern in the proximal part of the carpus is quite different.

In the hindlimb, a description of the pattern is complicated by the appearance of the 10 tarsal condensations at different times

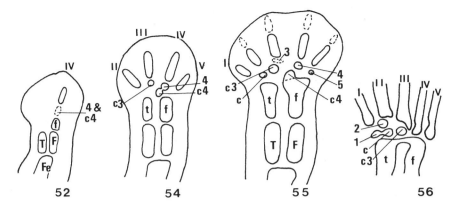

Fig. 12. Development of the pattern of prechondrogenic condensations and cartilaginous elements in the *Xenopus* hindlimb. Drawn from autoradiographs of the pattern of $^{35}SO_4$ incorporation. Stage numbers given below.

(Figs. 11, 12). The two proximal elements are the elongated tibiale and fibulare. Distally, each of the five metatarsals has an associated distal tarsal, and these appear in the order 4, 3, 2 and 5, 1. Of these, 3, 4 and 5 appear to fuse with their metatarsals, without becoming a separate chondrogenic element. There are three central elements, one each at the base of digits II and III, and a third which fuses with the distal end of the fibulare. Thus only six of the ten condensations form separate chondrogenic elements.

Holmgren's reconstruction (1933; Fig. 13) is comparable to our stage 54 with the exception of his distal 3, which appears later, at stage 55. Though interpretation is not easy on account of condensation and cartilage element fusion, the carpus pattern does not homologise readily with the archetype. There are too few elements, both proximally, where the intermedium is absent, and also centrally, where only three of the four archetype centrals can be found.

The prechondrogenic and chondrogenic pattern in the mouse forelimb

Descriptive accounts of the developing chondrogenic pattern in the mouse have been published by a number of authors (Holmgren, 1933, 1952; Dalgleish, 1964; Milaire, 1978). The chondrogenic

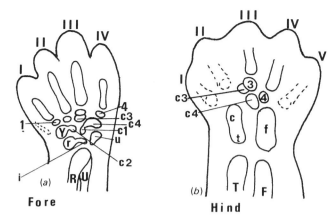

Fig. 13. Holmgren's (1933) interpretation of the prechondrogenic pattern of *Xenopus* carpus (*a*) and tarsus (*b*). This interpretation conforms with the archetype.

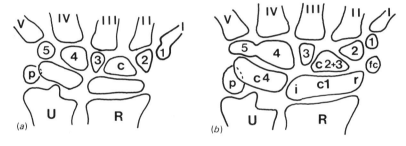

Fig. 14. Development of the pattern of (*a*) prechondrogenic condensations (13 days) and (*b*) chondrogenic elements (14 days) in the mouse forelimb. Drawn from autoradiographs of $^{35}SO_4$ incorporation partly from Milaire (1978) and partly from autoradiographs obtained by M. Hicks. The labelling in (*b*) follows Holmgren's (1952) interpretation which identifies two elements (C2 and 3; i, C1, r) as formed by fusions involving central condensations, thus enabling the pattern to conform with the archetype (see text).

elements of the carpus are clear by 14 days, when the pattern is essentially that of the adult (Fig. 14). If the falciform is excluded, there is a total of eight elements, three proximal (including pisiform), one central and four distal carpals, one for each digit except for IV and V which have a common distal carpal. Autoradiography reveals that, at 13 days, each of these chondrogenic elements is preceded by a condensation, except that at this stage there are separate distal carpals 4 and 5 (Fig. 14).

Holmgren (1933, 1952) identifies all these condensations, but in addition claims two chondrogenic elements are formed by fusion. These are (i) the element distal to the radius, claimed to be a fusion of radiale, central 1 and intermedium, and (ii) the central element, claimed to be a fusion of centrals 2 and 3. According to our autoradiographs, both these are single condensations. A careful study by Milaire (1978), also using $^{35}SO_4$ considers the first as a single, but the second as a double condensation (though his 12-day autoradiographs suggest continuity between C2 and C3). Dalgleish (1964), in a purely histological study of the mouse limb, also identifies all the eight carpal chondrogenic elements as single condensations at $12\frac{1}{2}$ days, the falciform appearing at 13 days. Dalgleish concludes that '. . . no real evidence is found for the previously reported four centrals other than the single definitive centrale of the mouse hand. Several structures . . . have been mistakenly considered such vestiges. Lack of such evidence casts doubt upon the conclusions of others that, phylogenetically, the mouse limbs are amongst the oldest tetrapod appendages.'

The adult mouse carpus is relatively unspecialised and its form is very similar to the standard pattern in mammals with unspecialised limbs and in primitive reptiles (Romer & Parsons, 1977). Even so, the condensation pattern of the mouse carpus more closely resembles that of the resulting cartilages than it does the archetypal pattern represented by the primitive amphibian carpus. To give one example: at eight, the number of mouse carpal condensations is too low.

The prechondrogenic and chondrogenic pattern in the axolotl forelimb and hindlimb

In the stage 43 (stages of Rugh, 1960) forelimb, radius and ulna and digits I and II are clearly defined (Figs. 15 & 16). In the carpus, only the basal commune (bc) is well defined, but there are three areas of increased synthesis, (i) distal to the radius, (ii) proximal to the bc, (iii) distal to the ulna. Gradually these three areas resolve themselves as (i) radiale and 'Y', by stage 46, (ii) intermedium and central, by stage 45 and (iii) ulnare (stage 46), distal carpal 3 (stage 45) and 4 (stage 46). The additional digits III and IV become clear by stage 46. The metacarpal of digit I develops a distinct articulation towards bc by stage 44. By stage 47, the chondrogenic pattern of

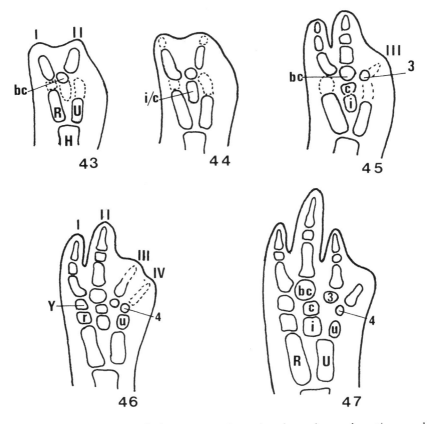

Fig. 15. Development of the pattern of prechondrogenic condensations and cartilaginous elements in the axolotl forelimb. Drawn from autoradiographs of the pattern of $^{35}SO_4$ incorporation. Stage numbers given below.

the larval limb (essentially that of the adult also) has been completed. Autoradiography does not clearly indicate that digit 1 is part of a single ray including the radius, on the other hand it does not support the view (Holmgren, 1933; Jarvik, 1964, 1965, 1980) that digits I and II and the intermedium ray form initially as a single Y-shaped bifurcation, with the b.c. at the base of the digits. The head on the metacarpal 1 which articulates with b.c. is secondary, a conclusion which Thomson (1968) also reaches from the descriptions of other urodeles by Sewertzoff (1908) on *Triturus* and Schmalhausen (1910) on *Hynobius* (= 'Salamandrella'). The condensation pattern in the axolotl is very similar to the final chondrogenic pattern (although the intermedium and central sepa-

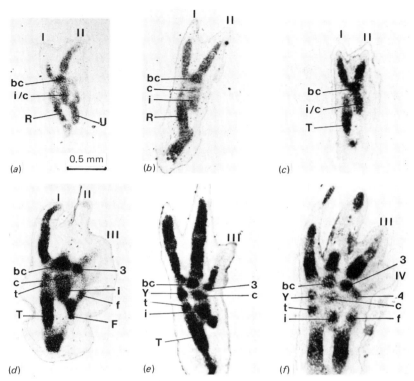

Fig. 16. Autoradiographs of $^{35}SO_4$ uptake by the developing limbs in the axolotl. (a), (b) Forelimb stages 44 and 45. (c)–(f), successive stages of hind limbs.

rate from a single initial condensation), and there is no fusion or loss of elements.

The hindlimb in the axolotl follows an almost exactly similar pattern of chondrogenesis to the forelimb. The eight carpal elements all have precise tarsal equivalents (Fig. 16), although in addition there is a fifth digit which has an associated distal tarsal element.

The developing chondrogenic pattern in urodeles has been used by Holmgren (1933, 1949) and Jarvik (1964, 1980) as important evidence in their separation from the 'eutetrapod' group. Jarvik (1964) states that in urodeles 'the intermedium ray is branched and carries two digits (I & II) in urodeles, whereas it . . . is unbranched and carries one digit (I) in other tetrapods . . . . There is always a basal commune in the urodelian limb'. The autoradiographic pattern does not support this view, and thus provides no ontogenetic evidence for the phylogenetic separation of urodeles from eutetrapods. But since the urodele pattern is very different from that of the species chosen as representative of anurans (which lack an intermedium), birds and mammals; and since each of these patterns is specialised, prechondrogenic pattern cannot be used as evidence of homology in deciding the possible monophyletic origin of tetrapods.

Possible mechanisms of pattern change in evolution: mutant genes changing condensation pattern

The archetypal condensation pattern theory must assume as a logical consequence that the specialisation of the tetrapod limb occurring during its adaptive radiation involves developmental mechanisms which affect only the postcondensation processes, such as growth and chondrification. Thus if evolution favours, as it frequently does, a large tibia in relation to the fibula, then there must be differential growth during chondrogenesis of tibia relative to fibula from initially rather similar-sized condensations (Hinchliffe & Johnson, 1980, 1982). Alternatively elements may fuse or disappear, as happens with several chondrogenic elements in *Xenopus* carpus and tarsus and the ulnare in the chick wing (Hinchliffe, 1977). These processes seem to be the consequence of a growth programme – presumably genetically controlled – inherent in the condensation. If isolated, and grown by culture *in vitro* , such elements grow to their characteristic size, whether large or small. Thus Hicks (1982) isolated early chondrogenic rudiments of chick tibia and fibula and found that, when grown *in vitro*, the growth rate of the tibia was much greater than the fibula (for review see Hinchliffe & Johnson, 1982), as *in vivo*.

While differential growth of cartilage elements clearly exists, it is quite clear that genes control, directly or indirectly, the initial pattern of condensations. There are thus no good grounds for making the archetype in some way sacrosanct from modification by natural selection acting on genetically controlled variation. It is only in a few 'laboratory' species such as the mouse that we know much about the effects of gene mutations on the process of skeletogenesis, but it is clear from the studies of Grüneberg (1963) that the initial condensation pattern is susceptible to modification by gene action. In *Oligosyndactylism* the footplate is narrower than

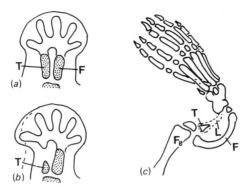

Fig. 17. Limb development in the luxate mutants in the mouse. Mesenchyme is effectively shifted from making tibia to making enlarged or additional preaxial digits. (*a*) Normal and (*b*) luxate leg bud at $12\frac{1}{2}$ days. A typical resultant skeleton, in this case Strong's luxoid, is illustrated in (*c*), in which the tibia is reduced and there is an additional preaxial digit. L = ligament. (After Hinchliffe & Johnson, 1980; Grüneberg, 1963.)

Fig. 18. Limb development in the brachypod mutant in the mouse. Mesenchyme is misallocated between metacarpal and phalangeal condensations. Normal (n) and brachypod (b) positions of metacarpalphalangeal future joint in 13-day leg bud. The brachypod metacarpals are abnormally small. (After Hinchliffe & Johnson, 1980.)

normal, so that the digital condensations are either squashed closer together, or digit II may be missing completely. In the luxate group (review Grüneberg, 1963; Hinchliffe & Johnson, 1980), some limb bud mesenchyme is in effect frequently shifted from making one proximal condensation to making another more distally. Thus in the hindlimb bud, the tibial condensation is much reduced in size, while in the footplate there is an anterior excess of mesenchyme which goes to form an enlarged anterior (first) digit and frequently also an additional preaxial digit (Fig. 17). One of the effects of the *brachypod* mutation is to misallocate mesenchyme between digital condensations: the mutant metatarsal condensation is smaller than normal, the additional mesenchyme being allocated to the adjacent phalangeal condensation (Fig. 18). These few examples cited from

a mass of available evidence make it clear that there exists genetically controlled variation of the normal condensation pattern. There is
thus no reason why the pattern of condensations – including the
archetype – should not have been repeatedly remodelled during the
adaptive radiation of the pentadactyl limb.

The prechondrogenic pattern and phylogeny

Though based on a very small number of species, the evidence
discussed above, from axolotl, *Xenopus*, chick and mouse, suggests
that, far from passing through an archetypal phase (or two phases,
if urodeles are separated from eutetrapods) the limb prechondrogenic pattern in these species is already very specialised –
almost as specialised in fact as are their adult limb skeletons. This
suggests that there is no general pattern of condensation, common
to all tetrapod groups. In consequence, there is no justification for
constructing a eutetrapod group on the basis of similarity of
condensation pattern in *Xenopus*, chick and mouse. From the avian
evidence it is very clear that the pattern of condensation in ankle and
wrist regions is much more closely related to the later avian skeletal
pattern of cartilage elements than to condensation pattern in
Xenopus, or the mouse, or for that matter, to the adult limb or fin
paddle skeletal pattern in primitive amphibian or Crossopterygian
fish ancestor. In view of these differences, the argument whether in
urodele limb ontogeny from the condensation stage digits I and II
have a common base on the basal commune or whether this is
secondary ceases to have important phylogenetic implications, since
there is no homologous carpus or tarsus pattern throughout the
eutetrapod group from which the urodele pattern can be distinguished. This conclusion rules out the condensation pattern as
useful evidence in constructing overall schemes of tetrapod evolution though it may still provide evidence of homology in relation to
particular problems within the tetrapods, such as the relationships
of birds and crocodiles. In our view, the identification of a
condensation archetype has arisen because investigators have superimposed their preconceptions on histological evidence which was
unclear.

As attention currently focuses on models which generate limb
patterns, this may seem rather a negative conclusion. But such

models must in the end aim to generate actual limb patterns, and it is perhaps worth emphasising the rather specialised and complicated way in which the final pattern of tarsus and carpus emerges. At present most of the models simply omit carpus and tarsus. Finally, it is worth pointing to evidence in the mouse of genetic modification of condensation pattern, which provides a possible mechanism for the modification in evolution of limb skeletal structure through selection pressure.

References

Cioffi, M., Searls, R. L. & Hilfer, S. R. (1980). Sulphated proteoglycan accumulation during development of the embryonic chick-limb bud studied by electron microscopic autoradiography. *J. Embryol. exp. Morph.* **55**, 195–209.

Dalgleish, C. (1964). Development of the limbs of the mouse. Ph.D. thesis, Stanford University, USA.

Darwin, C. (1872). *On the origin of species*, 6th edn. John Murray: London.

Gregory, W. K., Miner, R. W. & Noble, G. K. (1923). The carpus of *Eryops* and the structure of the primitive Chiropterygium. *Bull. Am. Mus. nat. Hist.* **48**, 279–88.

Gregory, W. K. Raven, H. C. (1941). Studies on the origin and early evolution of paired fins and limbs. *Annls N. Y. Acad. Sci.* **42**, 273–360.

Grüneberg, H. (1963). *The pathology of development.* Blackwells: Oxford.

Hamilton, H. L. (1952). *Lillie's development of the chick.* Holt, Rinehart & Winston: New York.

Hicks, M. (1982). An analysis of the differential growth of chondrogenic elements in the chick embryo limb. Ph.D. thesis, University College of Wales, Aberystwyth.

Hinchliffe, J. R. (1977). The chondrogenic pattern in chick limb morphogenesis: a problem of development and evolution. In *Vertebrate limb and somite morphogenesis*, eds. D. A. Ede, J. R. Hinchliffe and M. Balls, pp. 293–309. Cambridge University Press: Cambridge.

Hinchliffe, J. R. & Ede, D. A. (1973). Cell death and the development of limb form and skeletal pattern in normal and *wingless* (*ws*) chick embryos. *J. Embryol. exp. Morph.* **30**, 753–72.

Hinchliffe, J. R. & Johnson, D. R. (1980). *The development of the vertebrate limb.* Oxford University Press: Oxford.

Hinchliffe, J. R. & Johnson, D. R. (1982). The growth of cartilage. In *Cartilage*, vol. 2, ed. B. K. Hall pp. 255–95. Academic Press: New York.

Holmgren, N. (1933). On the origin of the tetrapod limb. *Acta zool.* **14**, 184–295.

Holmgren, N. (1939). Contribution to the question of the tetrapod limb. *Acta zool.* **20**, 89–124.

Holmgren, N. (1949). On the tetrapod limb problem – again. *Acta zool.* **30**, 485–508.

Holmgren, N. (1952). An embryological analysis of the mammalian carpus and its bearing upon the question of the origin of the tetrapod limb. *Acta zool.* **33,** 1–115.

Holmgren, N. (1955). Studies on the phylogeny of birds. *Acta zool.* **36,** 243–328.

Jarvik, E. (1964). Specialisations in early vertebrates. *Annls Soc. r. zool. Belg.* **94,** 11–95.

Jarvik, E. (1965). On the origin of girdles and paired fins. *Israel J. Zool.* **14,** 141–72.

Jarvik, E. (1980). *Basic structure and evolution of vertebrates,* vol 2. Academic Press: London.

Lansdown, A. B. G. (1969). An investigation of the development of the wing skeleton in the quail (*Coturnix coturnix japonica*). *J. Anat.* **105,** 103–14.

Leighton, L. (1894). The development of the wing of *Sterna wilsonii. Am. Nat.* **28,** 671–774.

Lutz, H. (1942). Beiträge zur Stammesgeschichte der Ratiten. *Rev. suisse Zool.* **49,** 299–399.

Milaire, J. (1978). Etude morphologique, histochimique et autoradiographique du développement du squelette des membres chez l'embyon de Souris. 1. Membres antérieurs. *Arch. Biol.* **89,** 169–216.

Montagna, W. (1945). A re-investigation of the development of the wing of the fowl. *J. Morph.* **76,** 87–113.

Nieuwkoop, P. D. & Faber, J. (1967). *Normal table of* Xenopus laevis. North-Holland: Amsterdam.

Ostrom, J. H. (1976). Archaeopteryx and the origin of birds. *Biol. J. Linn. Soc.* **8,** 91–182.

Owen, R. (1849). *On the nature of limbs.* J. Van Voorst: London.

Parker, W. K. (1889). On the structure and development of the wing in the common fowl. *Phil. Trans. R. Soc. Lond. Ser. B,* **179,** 385–98.

Romanoff, A. L. (1960). *The avian embryo.* Macmillan: New York.

Romer, A. S. & Parsons, T. S. (1977). *The vertebrate body,* 5th edn. W. B. Saunders Co: Philadelphia.

Rugh, R. (1960). *Experimental embryology.* Burgess Publishing Co.: Minneapolis, Minn.

Schmalhausen, I. I. (1910). Die Entwicklung des Extremitätenskelettes von *Salamandrella keyserlingii. Anat. Anz.* **37,** 431–66.

Searls, R. L. (1965a). An autoradiographic study of the uptake of S-35-sulphate during the differentiation of limb bud cartilage. *Devl. Biol.* **11,** 155–68.

Searls, R. L. (1965b). Isolation of mucopolysaccharide from the precartilaginous embryonic chick limb bud. *Proc. Soc. exp. Biol. Med.* **118,** 1172–6.

Sewertzoff, A. N. (1908). Studien über die Entwicklung der Muskeln, Nerven, und der Skeletts der Extremitaten der hiederen Tetrapoda. Beiträge zu einer Theorie des pentadactylen Extremität der Wirbeltiere. *Bull. Soc. im. Natural. Moscou* (N. S.) **21,** 1–432.

Sieglbauer, F. (1911). Zur Entwicklung der Vogelextremität. *Z. wiss. Zool.* **97,** 262–313.

Steiner, H. (1934). Ueber die embryonale Hand- und Fuss-Skelett-Anlage bei den Crocodiliern, sowie über ihre Beziehungen zur Vogel-Flügel anlage und zur ursprunlichen Tetrapoden-Extremitat. *Rev. suisse Zool.* **41,** 383–96.

Thomson, K. S. (1968). A critical view of the diphyletic theory of rhipidistian–amphibian relationships. In *Current problems of lower vertebrate phylogeny*, ed. T. Ørvig, pp. 285–306. Almquist & Wiksel: Stockholm.

Walker, A. D. (1972). New light on the origin of birds and crocodiles. *Nature* **237,** 257–63.

Some problems in the development and evolution of the chordates

P.D. NIEUWKOOP* AND L.A. SUTASURYA†

*Hubrecht Laboratory, Utrecht, the Netherlands, and †Department of Biology, ITB, Bandung, Indonesia

#0431

Introduction

During recent years, an important shift in the general approach to the problem of evolution has become manifest: a shift away from the idea that the genome is the sole target for the process of evolution and towards the presently more widely advocated view that the developing organism as a substrate for the expression of the genome also plays an important role in evolution. Embryonic development may, in fact, direct the evolutionary process by interfering with the expression of certain mutational changes and by facilitating the expression of others, thus influencing the general process of natural selection (see Lewin, 1981). As embryologists, we have never questioned the importance of embryonic development in the process of evolution. We therefore prefer to think in terms of *directional* rather than *random* evolution.

What deductions can we make about the relationship between embryonic development and evolution? It is a well-known fact that closely related species show a much greater similarity in their embryonic development than less closely related ones. The similarity holds particularly for early development, which may be virtually indistinguishable in related species and may still show pronounced similarities in less closely related ones. This conclusion may also be formulated in the opposite sense: 'the less closely species are related, the earlier in their development do differences usually begin to manifest themselves'. This is a generalisation which may hold for the majority of selected criteria but not necessarily for all criteria. I therefore want to add the following qualification: the more significant and fundamental are the observed differences in

the embryonic development of different species, and the earlier
they become manifest, the less closely related are the species or the
more ancient was the deviation of their phylogenetic history. I do
realise that the terms 'significant' and 'fundamental' are subjective,
but I am afraid we cannot yet do without this subjectivity. When we
set value upon this form of reasoning, the study of embryonic
development can contribute significantly to our insight into the
process of evolution, and particularly into macroevolution.

I have selected three problems in the development and evolution
of the chordates: the role of head formation, which played such an
important part in comparative embryology in the second half of the
nineteenth and the beginning of the twentieth centuries; the origin
of the germ cells in anuran and urodele amphibia as an important
argument for a diphyletic origin of the tetrapods; and the so-called
'land egg' of reptiles and birds.

Head formation in the chordates

Although the fossil record of the chordates shows the usual gaps for
the actual transition from one group into the other – the so-called
missing links – palaeontology has nevertheless furnished important
arguments in favour of the descent theory of the chordates; a theory
in which birds and mammals have descended from different groups
of reptiles, which in turn arose from amphibian ancestors. The
tetrapod amphibia have descended from osteichthyan fishes, which
themselves may have originated from agnathan fishes; these in turn
may have arisen from cephalochordate ancestors. Finally, the
cephalochordates may be phylogenetically related to the urochordates.

The nervous system of the urochordates is only a larval structure,
which is, for the greater part, of spinal cord character and is
underlain by a larval notochord. It also contains a small cerebral
vesicle which cannot, however, be compared to any part of the
vertebrate brain. In the cephalochordate *Branchiostoma*, the nervous system resembles the rhombencephalon and spinal cord of the
vertebrates, a prosencephalon being absent. The nervous system is
entirely underlain by the notochord, which extends even beyond its
anterior tip. The cyclostomes have a very small prosencephalon,
consisting of a single telencephalon with a single olfactory organ
and a small diencephalon with two eyes. The latter is connected

with a small mesencephalon, followed by a well-developed rhomb-encephalon and spinal cord. The latter three parts of the nervous system are underlain by a notochord, flanked by two rows of somites, and the prosencephalon by a small and ill-defined area of prechordal mesoderm. In the rather highly developed and very mobile selachian and teleostomial fishes, the prosencephalon is well developed and underlain by a similarly well-developed prechordal plate. The prosencephalon and the underlying prechordal mesoderm seem slightly smaller in the less mobile amphibians, but this difference may be secondary in nature and related to growth processes in the developing embryo (see also below). In all amniotes, the prosencephalon is more highly developed than in the Anamniota and is underlain by a well-developed prechordal mesoderm. Depending on the mobility of the animal and the predominance of certain sensory functions (smell, sight or hearing), as well as on the extent of the integrative functions of the central nervous system, particular portions of the brain are more strongly developed in different groups of reptiles, birds and mammals, but these differences are mainly secondary in nature. It can therefore be stated that the evolution of the chordates is characterised by two phenomena: a progressive development of the prosencephalon and a parallel development of the prechordal mesoderm.

What can experimental embryology contribute to a better understanding of this 'cephalisation' process in the evolution of the chordates? Defect, grafting and recombination experiments have demonstrated during the past half-century that, in all the various groups of the chordates, the central nervous system develops epigenetically, i.e. by a process of induction occurring in the competent dorsal ectoderm under the influence of the underlying invaginating mesodermal archenteron roof, particularly the notochordal anlage and, where present, also the more anterior prechordal mesoderm, which is responsible for the induction of the prosencephalon.

In the 1940s, the Japanese school of Okada, with Takaya, Hama and Kato as collaborators, found that important changes in differentiation tendencies and inductive capacity occur in the presumptive prechordal mesoderm of the amphibian embryo during gastrulation, a phenomenon which they studied extensively in the succeeding decades (see Okada & Takaya, 1942*a, b*; Okada &

Hama, 1943, 1944, 1945; Hama, 1949; Kato, 1963). The presumptive prechordal endo- and meso-derm, when cut out of the uninvaginated dorsal blastoporal lip of the early gastrula and combined with competent gastrula ectoderm, differentiates only into notochord and somites and induces rhombencephalon and spinal cord, but no prosencephalon. The same material, when isolated directly after its invagination around the dorsal blastoporal lip and combined with competent gastrula ectoderm, differentiates into prechordal endoderm and mesoderm and induces prosencephalon instead of rhombencephalon and spinal cord. Other experiments have shown that the presumptive prechordal mesoderm gradually loses its notochord-forming capacity during the invagination process, an observation confirmed by Miss Hoessels (1957).

What do we know about this transformation process? Actually not very much! It can also be achieved by simple cultivation of the presumptive prechordal endo- and meso-derm in Holtfreter solution for about 10 h, the time normally required for its invagination. It can likewise be evoked by combining the material with non-competent neurula ectoderm (Kato & Okada, 1956). The transformation can be prevented by treating the presumptive prechordal mesoderm with the vegetalising agent, LiCl, or by combining it with neuralised ectoderm (Masui, 1960). It is evident that the properties of the ectoderm with which it interacts play an important role in the expression of the transformation. Unfortunately, nothing is known about the mechanisms involved.

Extensive investigations on the regional determination of the central nervous system (Nieuwkoop *et al.*, 1952; Saxén & Toivonen, 1962; and reviews by Nieuwkoop, 1973, and Toivonen, 1978, in the amphibians and by Hara, 1978, in the birds) have made it very likely that the induction of the central nervous system occurs in two steps: (1) a general neuralisation of the ectoderm ('activation') by the invaginating prechordal plate and anterior notochordal region of the archenteron roof which determines the spatial extension of the neural anlage; and (2) by a superimposed caudalising ('transforming') action of the chorda-mesoderm which shows a cranio-caudal gradient in intensity and transforms the more caudal neurectoderm into mes- and rhomb-encephalon, spinal cord and tail somites, respectively. It is not known whether the activation and transforming actions are related in character; if this were so, the cephalisation process would represent a gradual shift in the nature of the inductive action from 'transformation' to 'activation'.

Although this cephalisation process has only been studied in the amphibians, we would like to propose the following working hypothesis: The evolution of the chordates is characterised by a more and more pronounced shift in the differentiation tendencies, as well as the inductive capacity of the most anterior part of the invaginating archenteron roof, a process occurring during gastrulation. This process has apparently not yet started in the cephalochordates, is in its initial phase in the primitive cyclostome fishes, and encompasses more and more caudal levels of the archenteron roof in the gnathostome fishes, the amphibia, reptiles, birds and mammals.

The origin of the primordial germ cells in the vertebrates

The primordial germ cells (PGCs) of a given animal or plant species do not merely represent the precursors of an arbitrary cell type of the organism, but are the forerunners of the next generation of individuals. They are therefore the elements in which the evolution process has found its expression for billions of years. The different mechanisms by which they are formed during embryonic development may be of special interest for an evaluation of certain steps in the evolution process.

A thorough investigation of the PGCs has only been carried out in the amphibians, although the situation in birds has also been well studied. In the amphibians two basically different mechanisms have been demonstrated in the two main groups, the anuran and urodele amphibia. In the anurans, the PGCs develop from specific vegetal (endodermal) blastomeres which are characterised by the presence of a so-called 'germinal plasm' containing specific cytoplasmic organelles, the germinal granules; germ cell formation therefore shows a strongly preformistic character (Bounoure, 1939; Blackler, 1970; Williams & Smith, 1971) (Fig. 1a–c). Germinal granules are already present in the oocyte (Czołowska, 1969, 1972). In the urodeles, the PGCs develop epigenetically; they arise from arbitrary, common cells of the animal (ectodermal) moiety of the blastula, being induced along with the mesoderm by the vegetal yolk mass, particularly its ventral portion, which induces blood

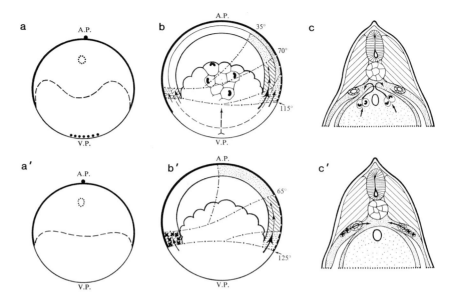

Fig. 1. Preformistic origin of the primordial germ cells (PGCs), characterised by the presence of germ plasm (in black) and their subsequent migration in the anuran amphibians. (*a*) Uncleaved egg, (*b*) late blastula with indication of endoderm and mesoderm induction and (*c*) early larva with migration route (arrows) of PGCs.

(*a'–c'*) Purely epigenetic origin of the PGCs (crosses) in the urodele Amphibia during mesoderm induction (*b'*) and subsequent migration route of the PGCs (*c'*). A.P. = animal pole; V.P. = vegetal pole.

islands and lateral plate mesoderm (Sutasurya & Nieuwkoop, 1974) (Fig. 1*a'–c'*). These (and other) facts are strongly at variance with Weismann's 'Keimplasma-Theorie' (1885, 1892): there is no question of a specific distribution of (nuclear) determinants (= genes) over the different cell types, but only of differential functioning of the common genome under the influence of different cytoplasmic factors; in the case of the PGCs these factors are either preformed during oogenesis or arise epigenetically as a result of the interaction of adjoining cell groups.

It is evident that germ cell formation is fundamentally different in the two groups of Amphibia. Moreover, germ cell formation occurs very early in development, i.e. as early as oogenesis in the anurans and during blastulation in the urodeles. There are also pronounced differences in mesoderm formation in the two groups:

predominantly or completely internal in the Anura (internal marginal zone) (Nieuwkoop & Florschütz, 1950; Keller, 1975, 1976)
and essentially external in the Urodela (external marginal zone)
(Vogt, 1929; Nieuwkoop, 1947). In our opinion, all these arguments plead strongly in favour of a very ancient bifurcation in the
phylogenetic history of the two groups of Amphibia and even
suggest a diphyletic origin of the tetrapods from different ancestral
fishes (Nieuwkoop & Sutasurya, 1976).

The proposed diphyletic origin of the tetrapods leads to two new
questions: (1) from which groups of ancient fishes have the two
groups of tetrapods descended? and (2) to which groups of the
amniotes do each of the two lines possibly lead? Jarvik (1968), who
also advocates a diphyletic origin of the tetrapod amphibia, suggests
on the basis of comparative anatomical and palaeontological data
that the anuran and urodele amphibia have evolved respectively
from the osteolepiform and porolepiform osteichthyan fishes in the
Devonian or Silurian, 400 to 450 million years ago. Jarvik believes
that all the amniotes are derived from anuran ancestors, the
urodeles forming a separate branch. Embryological evidence has
led us (Nieuwkoop & Sutasurya, 1976) to suggest, however, that
the amniotes may actually represent two separate, parallel lines of
evolution, the anuran line leading to the birds and the urodele line
to the mammals. This suggestion is based on the origin of the PGCs
in the different groups. Experimental evidence pleads in favour of
an epigenetic origin of the PGCs from the totipotent, ectodermal
moiety of the embryonic anlage in the mammals (Gardner &
Rossant, 1976), whereas investigations on the origin of the PGCs in
the birds plead in favour of an early origin of the PGCs from the
primary hypoblast (Clawson & Domm, 1969). However, chimaeric
recombinations of primary hypoblast and totipotent epiblast of
chick and quail early blastoderms, recently performed by Eyal-
Giladi, suggest that, in the birds, the PGCs originate from the
epiblast (Eyal-Giladi *et al.*, 1981). The origin of the PGCs in birds
may therefore not be principally different from that in mammals.
We are at present investigating the origin of the PGCs in the
reptiles; we now work on the marine green turtle, *Chelonia mydas*,
and hope later to extend this work to the Sauropsida. These
investigations may furnish evidence for or against our working
hypothesis of two separate evolutionary lines from the osteichthyan
fishes to the birds and mammals, respectively (Nieuwkoop &

Sutasurya, 1979), since the hypothesis implies the existence of two different groups of reptiles. Unfortunately we cannot yet say anything definite about the origin of the PGCs in the turtles, so that the problem is still open.

The essential properties of the vertebrate 'land egg'

Although our present investigations on the properties of the vertebrate 'land egg' are descriptive rather than explanatory in nature, the problem is, in our opinion, so interesting from an embryological as well as from an evolutionary point of view that we would like to include a discussion of the 'land egg' problem in this symposium.

For those who are not familiar with the 'land egg' problem we want to recall that all Anamniota (fishes and amphibians) develop either permanently (fishes) or at least up to metamorphosis (amphibians) in an aquatic habitat, either sea, brackish or fresh water. Though the Amphibia are predominantly land animals, they return to an aquatic habitat for reproduction. The reptiles are the first real land animals; they not only live permanently on land, but their eggs also develop in a terrestrial habitat. This makes special demands upon the so-called 'land eggs'.

The reptiles are at present represented by the lizards, the related snakes, the crocodiles and the turtles. Palaeontological evidence indicates that the turtles are an old and conservative group of reptiles. It may therefore be assumed that turtle eggs still resemble in their essential features the first successful land eggs of the tetrapods. The ancestors of the present marine turtles were land animals which secondarily adapted themselves to an aquatic habitat. Like all other turtles they lay their eggs in a terrestrial habitat, and these may therefore just as well be taken as typical land eggs as those of land turtles.

The study of the normal development of the marine green turtle, *Chelonia mydas*, and in particular the elaboration of adequate operation techniques for green turtle embryos (Sutasurya & Nieuwkoop, unpublished) have confronted us with certain special properties of the marine turtle egg which are, in our opinion, essential for the development of the 'land egg'.

The marine turtle has spherical eggs with a diameter of 40 to 50 mm. They are deposited in a flask-shaped hole made by the

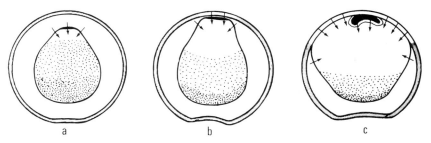

Fig. 2. Sagittal sections through eggs of *Chelonia mydas*: (*a*) at the time of deposition – subgerminal cavity formation underneath the embryonic anlage by transport of fluid (indicated by arrows) from albumen layer to yolk sac; (*b*) after about 24 h of cultivation – enlargement of subgerminal cavity and attachment of embryo to shell membrane; (*c*) after 2 to 3 days of cultivation – marked extension of subgerminal cavity and of attachment area of embryo and adjacent yolk sac to shell membrane.

female in the beach above the high water line in sand of the proper humidity. At the time of deposition the embryo has already reached an early to middle gastrula stage. At that stage the yolk is no longer spherical but slightly pear shaped (see Fig. 2*a*), as a result of the local elevation of its blastodermal portion caused by the uptake of fluid by the blastoderm from the overlying albumen layer. During the next 24 hours, the rapid uptake of fluid by the blastoderm and the surrounding region of the yolk sac reduces the overlying albumen layer to a very thin, transparent film situated between embryo and shell. Subsequent water uptake by the embryo leads to a withdrawal of fluid from the overlying shell membrane and calcareous shell. In the latter, the water is replaced by air, causing the appearance of a white spot in the egg shell directly overlying the blastoderm (see Fig. 2*b*). During further development, the white area extends rapidly, covering about one-fourth of the egg circumference after three days at *c*. 30 °C and about half of the circumference after two to three more days of cultivation (see Fig. 2*c*). Because of the strong absorption of fluid from the albumen layer, the blastoderm becomes firmly attached to the albumen film and shell membrane. At the same time the yolk sac changes into a thin-walled vesicle partially filled with a clear fluid. The yolk sac is very fragile, so that turning of the egg after attachment of the blastoderm to the shell easily leads to rupture of the yolk sac and subsequent death of the embryo (see also Holder & Bellairs, 1962; Yntema, 1964).

The following characteristics seem to represent the essential features of this typical 'land egg':

(1) The shell of the turtle egg is quite permeable to water. In a dry environment, the *Chelonia* egg shrinks rapidly and may even partially collapse. In a very humid environment the egg markedly swells, often becomes oval shaped, and develops a high internal pressure, which may even lead to rupture of the shell. In the normal environment as selected by the female, the eggs neither shrink nor swell, so that the egg itself must contain a sufficient amount of water for the full development of the embryo. The water storage of the egg is mainly in the albumen layer which has a water content of more than 95%.

(2) The embryo is without external food supply during its entire development, so that it must contain a sufficient store of reserve food for its development into a complete miniature adult and even for the initial postnatal period, including hatching, emergence from the sand, rapid migration to the sea, and the first adaptation to marine life. This store is in the form of a large amount of yolk present in the very large egg, which has a diameter of 25 to 35 mm. Such a large egg can only undergo a meroblastic type of cleavage, leading to the formation of a very small, flat, embryonic anlage on top of the large, dense, non-cellular yolk mass; the latter tends to stay in a downward position, lifting the embryonic anlage upwards.

(3) The main problem for the development of such a large egg is the accessibility of the huge store of reserve food and water for the small embryonic anlage. In the 'land egg' this problem is solved by the creation of an internal 'culture medium' for the embryonic anlage. This is achieved by a massive transfer of fluid from the overlying albumen layer into the rapidly expanding subgerminal cavity inside the yolk sac, as well as by a gradual liquefaction of the yolk, so that the embryonic anlage actually becomes suspended in its own nutritive culture medium.

(4) In order to facilitate access of the rapidly developing embryo to nutrients and water an extra-embryonic vascular system develops precociously in the wall of the yolk sac, its blood circulation being driven by the large embryonic heart.

(5) The position of the embryonic anlage directly adjacent to the shell facilitates the necessary gas exchange between the embryo and the environment.

(6) The embryonic anlage later becomes separated from the egg

membrane by the formation of the amniotic cavity, inside which the embryo can freely perform its morphogenesis.

(7) An outgrowth of the embryonic hindgut, the allantoic sac, expands into the exocoelic cavity and fuses with the overlying chorion, forming the chorioallantoic membrane; the allantois first serves as a depot for waste products during the rapid growth of the embryo, while the chorioallantoic membrane later functions as the main extra-embryonic respiratory organ through its extensive vascular network.

Summarising, it may be concluded that in evolution nature has found a very simple solution for the development of an egg in a non-aqueous, gaseous environment: (1) by supplying it with a large store of water and reserve food, and (2) by creating an internal 'tissue culture medium' for the developing embryo.

The warm-blooded birds are clearly more highly specialised than the reptiles and their eggs represent a more sophisticated form of the 'land egg'. The bird egg is less dependent on a humid environment, but requires an elevated temperature which birds maintain by brooding, simultaneously preventing the egg from excessive drying. The bird egg has a much thicker outer calcareous shell than the reptilian egg, giving it a high rigidity and preventing evaporation more efficiently, but allowing a proper gas exchange. The bird egg has an internal air chamber usually situated at the blunt end of the egg, while the so-called chalazae loosely attach the yolk to the egg membrane at both ends of the oval egg.

In the examples given we have used embryological data for the support of the descent theory of the chordates and interpreted certain specific egg properties in the framework of this theory. We have also modified this theory on the basis of other embryological data. The question is now: is it really permissible to do so?

References

Blackler, A. W. (1970). The integrity of the reproductive cell line in the Amphibia. *Curr. Top. devl Biol.* **5**, 71–87.

Bounoure, L. (1939). *L'origine des cellules reproductrices et le problème de la lignée germinale.* Gauthiers-Villars: Paris.

Clawson, R. C. & Domm, L. V. (1969). Origin and early migration of primordial germ cells in the chick: a study of the stages definitive primitive streak through 8 somites. *Am. J. Anat.* **125**, 87–112.

Czołowska, R. (1969). Observations on the origin of the 'germinal cytoplasm' in *Xenopus laevis. J. Embryol. exp. Morph.* **22**, 229–51.

Czołowska, R. (1972). The fine structure of the 'germinal cytoplasm' in the egg of *Xenopus laevis. Wilhelm Roux Arch. EntwMech. Org.* **169**, 335–44.

Eyal-Giladi, H., Ginsburg, M. & Farborov, A. (1981). Avian primordial germ cells are of epiblastic origin. *J. Embryol. exp. Morph.* **65**, 139–47.

Gardner, R. L. & Rossant, J. (1976). Determination during embryogenesis. In *Embryogenesis in Mammals*, eds. K. Elliott and M. O'Connor, Ciba Foundation Symposium No. 40, pp. 5–25. Elsevier: Amsterdam.

Hama, T. (1949). Explantation of the urodelean organizer and the process of morphological differentiation attendant upon invagination. *Proc. Jap. Acad.* **25**, 4–11.

Hara, K. (1978). 'Spemann's organizer' in birds. In *Organizer – A milestone of a half-century from Spemann*, eds. O. Nakamura and S. Toivonen, pp. 221–65. Elsevier/North-Holland Biomedical Press: Amsterdam.

Hoessels, E. L. M. J. (1957). Evolution de la plaque préchordale d'*Ambystoma mexicanum*: sa différenciation propre et sa puissance inductrice pendant la gastrulation. Ph.D. Thesis, University of Utrecht, the Netherlands.

Holder, L. A. & Bellairs, d'A. (1962). The use of reptiles in experimental embryology. *Br. J. Herp.* **3**, 54–61.

Jarvik, E. (1968). Aspects of vertebrate phylogeny. In *Current problems of lower vertebrate phylogeny*, ed. T. Ørvig, pp. 497–527. Almquist & Wiksell: Stockholm; and Wiley Interscience Publ.: New York.

Kato, K. (1963). Neuro-notochordal relationship in the development of the explanted pieces taken from the dorsal lip of *Triturus* gastrula. *Mem. Coll. Sci. Kyoto. Univ. Ser. B* **30**, 29–39.

Kato, K. I. & Okada, T. S. (1956). A mutual relationship between the explanted piece of dorsal blastoporal lip and the enveloping ectoderm in *Triturus* gastrula. *Mem. Coll. Sci. Kyoto. Univ. Ser. B* **23**, 1–9.

Keller, R. E. (1975). Vital dye mapping of the gastrula and neurula of *Xenopus laevis*. I. Prospective areas and morphogenetic movements of the superficial layer. *Devl. Biol.* **42**, 222–41.

Keller, R. E. (1976). Vital dye mapping of the gastrula and neurula of *Xenopus laevis*. II. Prospective areas and morphogenetic movements of the deeper layer. *Devl. Biol.* **51**, 118–37.

Lewin, R. (1981). Seeds of change in embryonic development. *Science* **214**, 42–4.

Masui, Y. (1960). Differentiation of the prechordal tissue under influence of lithium chloride. *Mem. Konan Univ.* **4**, 79–102.

Nieuwkoop, P. D. (1947). Experimental investigations on the origin and determination of the germ cells, and on the development of the lateral plates and germ ridges in the urodeles. *Arch. néerl. Zool.*, **8**, 1–205.

Nieuwkoop, P. D. (1973). The 'organization centre' of the amphibian embryo: its origin, spatial organization and morphogenetic action. *Adv. Morphogen.* **10**, 1–39.

Nieuwkoop, P. D., Boterenbrood, E. C., Kremer, A., Bloemsma, F. F. S. N., Hoessels, E. L. M. J., Meyer, G. & Verhegen, F. J. (1952). Activation and organization of the central nervous system. I. Induction and activation. II.

Differentiation and organization. III. Synthesis of a new working hypothesis. *J. exp. Zool.* **120,** 1–108.

Nieuwkoop, P. D. & Florschütz, P. A. (1950). Quelques caractères spéciaux de la gastrulation et de la neurulation de l'oeuf de *Xenopus laevis* Daud. et de quelques autres anoures. I. Etude descriptive. *Archs. Biol.* **61,** 113–50.

Nieuwkoop, P. D. & Sutasurya, L. A. (1976). Embryological evidence for a possible polyphyletic origin of the recent amphibians. *J. Embryol. exp. Morph.* **35,** 159–67.

Nieuwkoop, P. D. & Sutasurya, L. A. (1979). *Primordial germ cells in the chordates. Embryogenesis and phylogenesis.* Cambridge University Press: Cambridge.

Okada, Y. K. & Hama, T. (1943). Examination of regional differences in the inductive activity of the organizer by means of transplantation into ectodermal vesicles. *Proc. Imp. Acad. Tokyo* **19,** 48–53.

Okada, Y. K. & Hama, T. (1944). On the different effects of the amphibian organizer following culture, transplantation and heat treatment. *Proc. Imp. Acad. Tokyo* **20,** 36–40.

Okada, Y. K. & Hama, T. (1945). Prospective fate and inductive capacity of the dorsal lip of the blastopore of the *Triturus* gastrula. *Proc. Jap. Acad.,* **21,** 3242–8.

Okada, Y. K. & Takaya, H. (1942a). Experimental investigation of regional differences in the inductive capacity of the organizer. *Proc. Imp. Acad. Tokyo* **18,** 505–13.

Okada, Y. K. & Takaya, H. (1942b). Further studies on the regional differentiation of the inductive capacity of the organizer. *Proc. Imp. Acad. Tokyo* **18,** 514–19.

Saxén, L. & Toivonen, S. (1962). *Primary embryonic induction.* Logos Press: London.

Sutasurya, L. A. & Nieuwkoop, P. D. (1974). The induction of the primordial germ cells in the Urodeles. *Wilhelm Roux Arch. EntwMech. Org.* **175,** 199–220.

Toivonen, S. (1978). Regionalization of the embryo. In *Organizer – A milestone of a half-century from Spemann,* eds. O. Nakamura and S. Toivonen, pp. 119–56. Elsevier/North-Holland Biomedical Press: Amsterdam.

Vogt, W. (1929). Gestaltungsanalyse am Amphibienkeim mit örtlicher Vital-Färbung. II. Gastrulation und Mesodermbildung bei Urodelen und Anuren. *Wilhelm Roux Arch. EntwMech. Org.* **20,** 384–706.

Weismann, A. (1885). *Die Continuität des Keimplasmas als Grundlage einer Theorie der Vererbung.* Fischer: Jena.

Weismann, A. (1892). *Das Keimplasma. Eine Theorie der Vererbung.* Fischer: Jena.

Williams, M. A. & Smith, L. D. (1971). Ultrastructure of the 'germinal plasm' during maturation and early cleavage in *Rana pipiens. Devl Biol.* **25,** 568–80.

Yntema, C. L. (1964). Procurement and use of turtle embryos for experimental procedures. *Anat. Rec.* **149,** 577–86.

The evolution of patterning mechanisms: gleanings from insect embryogenesis and spermatogenesis

KLAUS SANDER

Institut für Biologie I (Zoologie), Albertstrasse 21a, D-7800 Freiburg i Br.,
West Germany

Introduction

The study of insect embryogenesis, like embryology in general, was much stimulated by early Darwinism. In the face of a century's writings, both expressly and implicitly, on the links between ontogenesis and evolution, it seems impossible to provide new insights in this field except on rather remote specialized levels. Much that a developmental physiologist on the basis of his special knowledge could say on evolution has been said before on the basis of reasoning from descriptive data, albeit in different terminology (see e.g. Riedl, 1979). If, nonetheless, part of this essay deals with general aspects and this in a fairly reductionist way, it is because views which are held commonplace among the experts of developmental biology may be far from familiar to those of evolutionary biology (and vice versa), and therefore exposing some principles may be profitable to both.

Ontogenetic networks provide restraints and opportunities for evolution

A basic fact to bear in mind when speculating on ontogenesis and evolution is outlined in Fig. 1. Ontogenesis is a network of more or less interdependent functions, and changes in one function are likely to affect many others to different degrees. Concepts like induction, inhibition, resource sharing, and pleiotropism testify to the network character of ontogenesis; relevant examples from insect embryogenesis are, for instance, the inductive effects of the ectoderm on the mesoderm (Bock, 1941) and several mutations

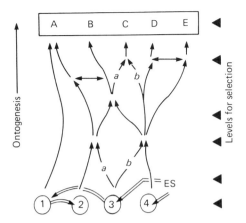

Fig. 1. Generalized diagram representing a small portion of the ontogenetic network leading from genotype (*bottom*, genes encircled) to phenotype (*top*, traits A–E in box). Double-headed arrows mark direct regulatory effects between genes and from extrakaryotic signals (ES) on genes. Pathways marked *a* and *b* could represent biochemical pathways, inductive interactions, morphogenetic movements and so on, depending on the levels of organization considered. Arrowheads at the right are intended to show that selection does not only act on the larval or adult phenotype and its functions, but also (and sometimes differently) on all steps in its generation.

which are pleiotropic in the sense that they cause identical anomalies in most or all segments of the developing body (Lewis, 1978; Nüsslein-Volhard & Wieschaus, 1980; Struhl, 1981). Consequently, natural selection must be viewed as an optimization process between all functions interlaced in a living system, including those required for ontogenesis, and the single property for which all these functions are optimized is maximum reproductive success through the chain of generations. However, the individual functions which can be discerned in this network may be subjected to quite specific selective pressures. Ontogenetic patterning mechanisms, for instance, might be under considerable pressure with respect to reliability.

A second important point is that networks tend to protect their individual components against gross change or straightforward elimination. The mutative reduction of one particular component will frequently be detrimental to some interconnected functions in the system, and thus should fail to spread in a population, even if the reduction as such would seem of advantage. On the other hand, a property which by itself might not seem of adaptive value could be strengthened within the system if it contributed to overall optimization; with further change in other components of the system, it might even provide the basis for some major evolutionary innovation. This would then, with hindsight, represent a case of 'preadaptation', while the former type of protection provides a selectionist explanation for 'recapitulation'. Both these terms refer to well-established phenomena but their use is fraught with controversy to this day, because of inacceptable earlier explanations and associations. In our context, they will be considered as (*a*) network-dependent opportunities for evolutionary innovation and (*b*) network-dependent restraints effecting evolutionary conservation.

The network character of ontogenesis has a consequence not always recognized. Assume two organisms which were derived from a common ancestor in the not too distant past, and therefore are still depending on comparable ontogenetic networks. Let these share a character (function and/or structure), for instance C in Fig. 1, considered of high adaptive value. The differences between their respective modes of living should require that one organism invests more than the other in function B, while the other's strategy involves strengthening function D. Assuming some sharing of metabolic pathways between these three functions, the optimal way of reaching function C (or generating the structures required for it) might lead through pathway *a* in one organism and through pathway *b* in the other. We would then find that a structure or function derived without apparent change from a common ancestor could be generated using different ontogenetic pathways. That such partial differences indeed exist was shown for instance by comparative studies on the induction of Meckel's cartilage (Hall, this volume) and by the different origins of regenerating eye lenses in various amphibians (Wolff, 1895; Freeman, 1963); in the latter case apparently different mechanisms are even co-existing within the ontogenetic repertoire of a single species (Spemann, 1915).

The point of these considerations is that selective pressures conserving a given character need not at the same time conserve the ontogenetic pathways by which that character is generated. Some functions may be served by partially differing means when, because of changes in other components of the network, a new pathway contributes more to overall optimization than its predecessor. We shall discuss this using the generation of the 'phylotypic' stage in insect ontogenesis, the germ band.

The insect germ band – a conserved pattern arising by modified pathways

All insects during ontogenesis pass a stage known as the germ band. It consists of head lobes (procephalic anlage with antennal buds), 3 segments bearing the buds of the mouth parts, 3 thoracic and 8–11 abdominal segments. Internally, the germ band is characterized by segmental subdivision of the mesoderm and by some other common features. This stage represents a generalization called 'Körpergrundgestalt' by Seidel (1960), which term may be translated as 'basic body pattern' (Sander, 1976*a*). A comparable stage has long been recognized in some other phyla. It can be defined as the first stage that reveals the general characters shared by all members of that phylum, and therefore was called the 'phyletic' stage by Cohen (1977). I suggest the term 'phylotypic' instead, because phyletic refers to phylogenesis rather than to characters typical of individual phyla. Incidentally it is the stage separating 'primitive development' from 'definitive development' in the terminology of the classical embryologist (e.g. Schleip, 1929). Different members of a phylum embark on ontogenesis from very different starting conditions – compare for example the egg cells and early development of trout, frog, chick and mouse – and thence converge on the phylotypic stage (marked for instance in vertebrates by chorda, neural tube, somites, gill clefts, tubular heart, etc.). Thereafter, development diverges towards the differently specialized post-embryonic stages, as noted already by von Baer (1828) and Haeckel (1866). Generally speaking, the phylotypic stage is the stage of greatest similarity between forms which, during evolution, have differently specialized both in their modes of adult life and with respect to the earliest stages of ontogenesis which are strongly influenced by special modes of reproduction (e.g. ovipary v. vivipary). It is also the earliest stage which on a general scale permits establishing homologies (Spemann, 1915). The divergent specialization following the phylotypic stage can be demonstrated in insects, using for instance their mouth parts, which, despite their incredible variation in larvae and adults, develop from nearly identical rudiments in the germ band stage (see p. 144).

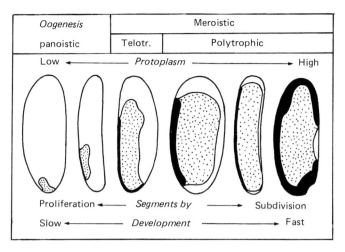

Oogenesis	Meroistic	
panoistic	Telotr.	Polytrophic

Low ◄——————— *Protoplasm* ———————► High

Proliferation ◄——— *Segments by* ———► Subdivision

Slow ◄——————— *Development* ———————► Fast

Fig. 2. Diagram summarizing data on Krause's (1939) egg types from the short germ type (*left*) to the long germ type (*right*). Species shown (*from the left*): *Tachycines* (stone cricket), *Acheta* (cricket), *Notonecta* (water bug), *Forticula* (earwig), *Apis* (honey bee), *Drosophila* (fruit fly). Dotted area shows extent of the germ anlage (embryonic rudiment, unsegmented) seen from the left-hand side; the relative thickness of the germ anlage is indicated in solid black. After polytrophic oogenesis (eggs shown at the right) embryogenesis is much rapid than after panoistic oogenesis (eggs shown at the left). Telotr. = telotrophic. (Modified from Bier, 1970.)

I will now discuss the primitive development of insects in order to document that the highly conserved spatial pattern of the germ band arises by evidently different ontogenetic pathways and mechanisms in various groups. These different pathways are at the roots of the different 'egg types' of insects established by Krause (1939) (Fig. 2). In 'short germ' development, the germ anlage consists essentially of the head anlage and a small bud at its posterior rim which produces the germ band segments one after the other in antero-posterior succession. Well-known examples of this type are found among the apterygote bristle tails (Thysanura, commonly considered close to the basic stock of winged insects) (Larink, 1969) and in some primitive hemimetabolans (Krause, 1939). The other extreme, called the long germ type, must have evolved independently in several orders, notably the hymenopterans and dipterans. In the best known example, the fruit fly *Drosophila*, the anlagen for all individual germ band segments are already of equal size when the germ anlage forms (Lohs-Schardin *et al.*, 1979). The head lobes, which in short germ development dominate the germ anlage, are here confined to a fraction of the germ anlage which approximately corresponds to their share of the germ band. Stated differently, the germ band segments in extreme short germ development arise by proliferation, like the body segments from the trochophora larva in annelids and the nauplius larva in crustaceans, while in long germ development the segments arise by the fairly simultaneous subdivision of a pre-established stretch of embryonic blastema (Sander, 1981). The latter process is less time consuming, which may have been its main evolutionary advantage, but at the same time obviously requires major modifications in the egg cell and the mechanisms of patterning. In passing, it should be mentioned that most insect eggs fall between these extremes, in that some segments either at the anterior or in the middle of the germ band form by subdivision of the germ anlage, as in the long germ type, and only the remaining segments (mainly the abdominal) arise sequentially by proliferation (see Sander 1976*a*).

The differences between these egg types are clearly reflected by their different reaction to comparable experiments. This finding is at the basis of Seidel's (1936) distinction between structurally 'indeterminate' and 'determinate' development. In a less preferable terminology, these differences could be characterized as resulting from a low or high degree of embryonic mosaicism: the determinate type evidently is marked by a larger number of determinants prelocalized in different regions of the egg cell. The best recognized of these local signals are the germ cell determinants (Illmensee & Mahowald, 1976) and the anterior determinants (Kalthoff, 1979) discussed further on p. 148. Thinking of existing models (Sander, 1981), indeterminate development might be more aptly characterized by the successive generation and handing out of positional values, as implied for instance in the progress zone model of chick limb patterning (Summerbell *et al.*, 1973), while long germ development lends itself better to global systems of positional information based on localized reference points (Wolpert, 1969) or gradient sources (Sander, 1960; Crick, 1970; Meinhardt, 1978).

The most striking differences in the mode of generating the germ band are known from the hymenopterans. In this group, which is mainly characterized by long germ development (e.g. Schnetter, 1935; Bull, 1982), members of several families have turned to endoparasitism, and this is sometimes connected with polyembryony

(Ivanova-Kasas, 1972). These forms lay their yolkless eggs in the eggs or early larval stages of other insects. In some chalcidid wasps the minute egg starts development by forming a trophamnion (p. 147). Thereafter, the central portion of the egg cell containing the zygote nucleus undergoes total cleavage. The daughter cleavage cells, up to some hundreds in number, then separate and each gives rise to a germ band and larva. These germ bands cannot possibly be shaped by the patterning mechanisms employed by long germ hymenopterans, which in all likelihood involve prelocalization of different centres, determinants or reference points in separate regions of the original egg cell (Schnetter, 1934; Jung *et al.*, 1977). As pointed out by Ivanova-Kasas (1972), the secondary acquisition of total cleavage must have been accompanied by profound changes in developmental mechanisms including those responsible for embryonic patterning. Unfortunately, knowledge of germ band development in these forms is very scanty. Therefore, the question remains open which modifications occurred, and whether the secondary mechanisms differ from those acting in insects with primary total cleavage (some apterygotans), as would be expected from Dollo's law claiming irreversibility of the evolutionary process.

A limitation to the concept of homology

Homology is a basic concept of biology which (under different names) was established long before the advent of evolutionary and developmental biology (Remane, 1960). Perhaps because of its all-pervading importance, it was at times employed beyond useful limits, and these limits always were and still are a matter of debate. Eminent developmental biologists like Spemann (1915) and Baltzer (1950) have struggled to define homology or its equivalent in relation to developmental physiology. Their thoughts cannot be reviewed here, but one limitation concerning homology and developmental biology should be obvious from the preceding section, as from earlier examples (p. 139): the ontogenetic development of homologous structures may take fairly different courses in different species even if the structures thus formed differ little in shape and function.

While all stages of evolutionary transition must certainly have been fully functional, there is neither proof nor need for the widespread belief that homology of anatomical structures (based on common descent) implies generation of these structures by identical ontogenetic mechanisms, whether at the level of genes or of complex interactions. As outlined above (p. 139) selective pressures may to some degree act differently on developmental pathways and their end-products. Therefore, both may evolve in partial divergence. A useful analogy, though subject to obvious restrictions, is provided by the evolution of the motor car and of the methods by which it is manufactured. Just consider the continuous evolution of wheels, engine, steering, doors, front grille etc. from the 'Tin Lizzy' to the latest models, and the differences in 'ontogeny' exemplified by the transition from the workman to the robot. The end-product was selected for style, function, reliability, and safety, its ontogeny mainly for economy.

What evolutionary restraints could have conserved the germ band stage?

A look at the fossil record of adult insects (Hennig, 1981) suggests that the germ band must have persisted since the palaeozoic. What are the factors that lead to the conservation of this uniform pattern through more than 200 million years? The general answer is that it must represent an 'interphene' (Riedl, 1979), i.e. it must serve hidden but vital functions in ontogenesis.

In order fully to realize the extent of conservation obtaining in the germ band stage, we must attend to details like for instance the three head segments which form the mouthparts. In most insects, these segments look very much like all other segments in the germ band, and the appendage buds forming on them are comparable throughout the insect kingdom; for instance, the third segment always carries a pair of buds which later on fuse to form the labium. In contrast to this uniformity, when looking at the adult forms we find that an incredible diversification of mouth parts has taken place and, even more surprisingly, that hardly any external traits of the individual segments are preserved in the head capsule which carries the mouth parts. What is the good of making gnathal segments or paired labial buds, and then fusing and distorting them beyond recognition?

The answer can only rest in restraints resulting from the network character of ontogenesis. Insects are apparently constructed piecemeal (Lawrence & Morata, 1976), and segments are building pieces which obey some identical rules, for instance in constructing their share of the central nervous system, and in behaving as compartments when generating the pattern of their integument (García-Bellido *et al.*, 1979). Changing the rules in some segments but not in others might endanger reliability of development in the latter so much that overall optimization favoured retaining the underlying general mechanisms rather than introducing short cuts towards specialized structures. Another or an additional possibility is that combinatorial determination of developmental pathways might be more economical in terms of coding capacity required (see Sander, 1982).

To put it another way, uniformity may be imposed on the germ band stage by the fact that developmental mechanisms must be foolproof. This requirement possibly also restricts the number of formal principles employed for spatial patterning, as well as the chances for evolutionary change once a reliable mechanism has been arrived at.

Germ band extension viewed as an interphene

The assumption that a foolproof ontogenetic mechanism should be worth quite some expenditure of energy might explain yet another highly conserved trait of the germ band stage, namely germ band extension. In most long germ eggs, the germ band extends considerably before segmentation but soon afterwards contracts to its original length (Fig. 3). In many instances, the extended germ band exceeds its final length by more than 50%, so that the tip of the tail, which initially (and ultimately) occupies the posterior egg pole, moves up the dorsal egg side until nearly touching the head lobes (e.g. *Bruchidius*: Jung, 1966). This exercise, which is reminiscent of anatrepsis in lower insects (Sander, 1976*b*), no doubt consumes a lot of energy and moreover involves additional risks of failure. If, nonetheless, it has persisted through the ages, it must serve some vital function as an interphene (Riedl, 1979). I suggested earlier (Sander, 1975) that this function is to increase the space between segmental reference points or compartment borders, so that a system of intrasegmental positional information can be set up. Conceivably such a system, based for instance on gradients, might require a certain minimum of space, or might increase in

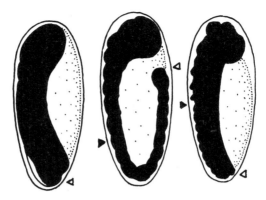

Fig. 3. Extension and contraction of the germ band in the beetle *Bruchidius obtectus*. Eggs seen from the left-hand side. Germ anlage (left egg) and germ band shown in black, yolk system dotted. Anterior end at the top, posterior end marked by open arrowheads, first abdominal segment marked by solid arrowheads. Age at 20 °C: *left* 41 h, *middle* 72 h, *right* 120 h. (Modified from Jung, 1966.)

reliability with increasing length. Once the cells in the system have acquired their positional values; (see French, this volume) they may be returned to their earlier crowded arrangement. Evolution might thus have weighed the energy expenditure connected with germ band extension against the risks of modifying the mechanisms of segmental patterning, and, if so, the latter apparently were so large as to protect germ band extension against evolutionary decay.

However, many insects apparently do without germ band extension. In short germ development and related intermediary types (Sander, 1976a), the growing germ band does not overshoot its final length, which, however, is mostly greater than in long germ development, thus directly providing space for intrasegmental patterning. Among long germ eggs, some hymenopterans are notable for lacking germ band extension (see for example Bull, 1982). However, here the segment borders (which appear in sequence) may be spaced further apart at the beginning than later on (Schnetter, 1935), and this might serve the same purpose.

Opportunistic evolution of extrakaryotic patterning signals

Entomology abounds with examples of evolutionary opportunism, i.e. instances where evolution, so to speak, seized upon some structure and entrusted it with new functions in addition to, or instead of, those functions apparently responsible in the first place for its selection. A clear-cut instance from insect development are the polar bodies. They owe their existence no doubt to a compromise between two requirements: storage of provisions for the embryo, and reduction of chromosome number via meiotic division(s). Once evolved, they were adapted in different groups for different functions (for review see Tremblay & Caltagirone, 1973). These functions include serving as a substitute for the sperm cell (in autogamy), generating a large part of the embryonic body (e.g. in a bag moth), harbouring symbiotic microorganisms (in some scale insects), and sequestering nutrients for the embryo in parasitic hymenopterans. In the latter instance (Ivanova-Kasas, 1972), the outer layer of the egg cell separates and is populated by descendants of two polar body nuclei; the resulting envelope is known as the trophamnion, which in turn may set free some daughter cells which further enhance its functions (see Jackson, 1928).

Embryonic patterning in most insects depends on spatial cues laid down during oogenesis (reviewed in Sander, 1976*a*). In the panoistic ovariole of lower insects, the main cue may be polarity. The oocyte must somehow become polarized by influences from the surrounding maternal tissues, represented primarily by the follicular epithelium. The main effect of oocyte polarity may be to guide embryonic cells towards the posterior region where they aggregate to form the germ anlage or embryonic rudiment while the more anterior regions of the egg cell are without any importance for embryonic patterning. This is quite different in the most highly evolved insect groups, characterized by long germ development and absence of blastoderm cell aggregation. Here, removal of sizeable anterior parts of the egg is rarely tolerated, and local damage at the anterior pole can lead to global changes in embryonic patterning and polarity, for instance replacing the most anterior germ band structures by the most posterior in reverse polarity (Kalthoff, 1979).

These highly evolved groups are also characterized by a special type of oogenesis (Fig. 2), and this invites some speculations on the opportunistic evolution of spatial cues for long germ patterning. Long germ development usually results from meroistic-polytrophic oogenesis, where the oocyte remains connected with some sister cells which furnish it with large amounts of RNA, cell organelles etc. (Telfer, 1975). The nurse cells are located anterior to the oocyte in the follicle, and consequently their products enter the oocyte anteriorly and in a polarized fashion; in *Drosophila* the streaming connected with this import is quite dramatic when viewed in time-lapse films (Gutzeit & Koppa, 1982). The polarized arrangement must depend on ovariole polarity and it could have evolved because electrical polarity may serve for transporting charged molecules from the nurse cells to the oocyte (Woodruff & Telfer, 1980). Recent evidence indicates that oocyte polarity in *Drosophila* does not derive directly from the follicular epithelium but rather is mediated via the nurse cells. In the mutant *dicephalic*, the nurse cells tend to split into two groups located at either pole of the growing oocyte (Lohs-Schardin, 1982). This condition invariantly triggers, at both egg poles, the formation of the micropyle, a structure normally marking the anterior pole only (Fig. 4). Few eggs of this abnormal type produce larvae, but in all observed cases the body pattern was abnormal; frequently it is marked by anterior structures at both ends, and by a switch of antero-posterior polarity about halfway between poles (double-head type). Taken together, these and other observations (discussed in Gutzeit & Sander, 1983) indicate that in *dicephalic* follicles part of the nurse cells are pushed to the posterior pole mainly for reasons of geometry, but once located there they signal 'anterior' to the adjacent follicle cells which shape the vitelline membrane. In some unknown and less reliable way, they also signal 'anterior' to the oocyte; the anterior determinants in the ooplasm of chironomids (Kalthoff, 1979) may represent a link in the chain transmitting such a signal to the embryonic patterning mechanisms. The nurse cells, whose paramount function doubtless is to nourish the oocyte, are thus seen to have acquired simultaneously or subsequently another function, namely that of establishing a crucial and perhaps multifunctional 'reference point' for embryonic patterning.

Another but more speculative example for the opportunistic acquisition of a signalling function is provided by the eggs of stick insects (phasmids). Like many other insects eggs, these carry a channel (micropyle) in their shells which permits the sperm to reach the oocyte. The germ anlage forms between the micropyle region and the posterior egg pole. Some phasmids, among them the well-known walking stick *Carausius morosus*, can produce abnormal eggs carrying two micropylar regions instead of one (Pijnacker,

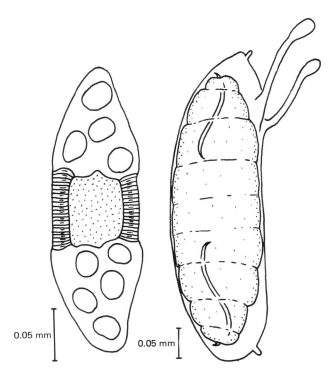

Fig. 4. The *dicephalic* mutant of *Drosophila* (see Lohs-Schardin, 1982). *Left* figure represents *dic* follicle with nurse cell group split in two and nurse cells located at either pole (only nuclei shown). *Right* figure is a *dic* egg with a micropylar cone at either pole and a 'double-head' embryo inside showing mouth parts at either end. (From Sander & Nübler-Jung, 1981.)

1971; Mazzini & Scali, 1980). Cappe de Baillon (1940) found that a germ analage can form underneath each micropylar region (Fig. 5), provided the micropylar channel is not reduced. He interpreted this finding to mean that the micropyle somehow provides the signal which attracts the blastoderm cells forming the germ anlage. If the micropyle indeed provides such a cue, and there is some additional evidence supporting this interpretation, then a whole chain of opportunistic alterations of function may have occured during evolution. To begin with, during evolution of the egg shell, a small area must have remained free from obstacles to sperm access. The sperm entering the egg cell through that area may then have

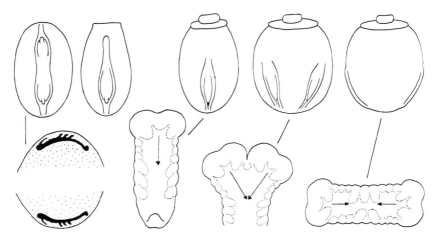

Fig. 5. Abnormal eggs from stick insects (phasmids) demonstrating the correlation between micropyle and formation of the germ anlage. *Left*, an abnormal and a normal egg of *Clonopsis gallica* are shown. The abnormal egg carries posterior traits at the anterior pole (*top*) and produces two embryos (shown in black underneath the enlarged polar caps). *Right*: the group of figures shows one normal egg and two abnormal eggs of *Carausius morosus*. The abnormal eggs carry micropyles on two meridians instead of one only. Under each micropyle a germ anlage forms, and the outgrowing germ bands fuse on meeting at the posterior pole. They may continue in a common thorax and abdomen (middle) or form a 'double-head' monster comprising opposite axial polarities (arrows). (Based on Cappe de Baillon (1940) and earlier publications.)

attracted the oocyte nucleus (as seen in many present-day insect species, Sander 1982*b*), and this in turn could have left some trace attracting the blastoderm cells. However, *Carausius* belongs to those phasmids which lack functional males and propagate exclusively by parthenogenesis. Therefore, if the original signal was provided by sperm entry, the signalling function must subsequently have been taken over by some property of the micropyle as such, possibly by the improved oxygen supply (Wigglesworth & Beament, 1950) or some such stimulus. Alternatively this property may of course have served as the signal from the beginning, or both micropyle and sperm had acquired signal function and one signal persisted while the other was lost; such double safeguarding is to be expected in network ontogenesis and may be more frequent than usually thought of.

Origins, levels and functions of patterning mechanisms in insects

Assuming that short germ development is the most ancient type in insects, budding of segments ('epimorphic' or successive patterning, Sander, 1981) should have been the primary mode of generating the basic body pattern. We shall discuss two questions based on this assumption:

(1) How did the subdividing or 'morphallactic' patterning mechanism(s) (Sander, 1981) of the long germ type eggs evolve from short germ budding?

(2) How does epimorphic patterning as inferred from experiments on insect larvae compare to the primary budding mechanism?

From a formal point of view it may be stated that long germ development is achieved by transferring some functions from early embryogenesis back to oogenesis. Krause (1981) suggested that the process of growth which in short germ development accompanies the transformation from germ anlage to germ band might have been shifted back into oogenesis in the long germ type. There, oogenesis is frequently characterized by extensive growth of those regions of the oocyte which are located behind the oocyte nucleus. Whether this coincidence is more than formal remains open. But doubtless such a transfer has occurred with respect to the biosynthetic machinery for embryogenesis. The long germ oocyte is furnished by the nurse cells with large amounts of 'semi-finished products' like RNA, ribosomes, and even mitochondria, while the short germ oocyte contains much less of these provisions (Bier, 1970). Consequently, in short germ development much of the biosynthetic machinery has to be assembled from 'raw materials' like protein and nucleotide precursors in the yolk system during early embryogenesis, and accordingly embryonic development takes much longer (6 weeks in the stone cricket *Tachycines* at about 24 °C as compared to 1 day in *Drosophila*) (see Fig. 2). Patterning during rapid development may require a number of prelocalized determinants, and indeed some of these are probably supplied by the very nurse cells which make rapid development possible (p. 147). Again using an analogy, we might think of the pattern of prelocalized determinants as a spatial blueprint, and anyone experienced in building a model plane or tailoring a dress will appreciate that these 'ontogenetic' tasks would take much more time if, lacking a blueprint, one were to rely merely on written directions (comparable to genetic information in the zygote nucleus).

The role of ooplasmic determinants in long germ patterning has frequently been linked to morphogenetic gradients (review see Sander, 1976a). However, some experimental results yield better to other models, and an ultimate decision is not yet possible. For believers in the evolutionary conservation of mechanisms, intercalation of positional values (Sander, 1981) would be a more attractive idea, especially in view of the double head and double abdomen malformations which will be discussed below. But as yet no satisfactory model has been evolved for this type of patterning, whereas the possibilities of generating morphogenetic gradients have been well studied and stimulated (Meinhardt, 1977, 1978).

Patterning by intercalation is best documented for post-embryonic stages and therefore will be discussed starting from these. As shown especially in experiments by Bohn (1970) and Nübler-Jung (1977), postembryonic cells possess a positional value (Wolpert, 1969) based on their position within the confines of the respective patterning system. When cells of strongly differing positional values are juxtaposed, mitosis will occur and provide new cells which acquire the positional values lacking between the originally confronted cells (Nübler-Jung, 1977). It is attractive to consider this model also in connection with germ band patterning in short germ embryogenesis. Some arguments have been collected previously in support of this view (Sander, 1981), but another should be stressed here in addition. Intercalation is not dependent on polarity (Sander & Nübler-Jung, 1981), that is, a given cell may border cells of slightly more posterior positional values posteriorly *and* anteriorly, and yet fail to initiate intercalation. Such a situation is found in the double-head monsters described by Cappe de Baillon (1940) from *Carausius* (Fig. 5). Mirror-type duplications in embryos of long germ insects (Kalthoff, 1979; Nüsslein-Volhard, 1979; Lohs-Schardin, 1982) might be explained by symmetrically sloping morphogenetic gradients built up in the mutant ovary or in the experimentally altered eggs. However, this explanation cannot hold for *Carausius* because all evidence indicates that there the double head originates from two separate short germ anlagen which bud towards each other. If the description of

Cappe de Baillon is correct, the interesting point should be that budding does not continue once the two blastemas meet (if it did, it might result in a *duplicitas cruciata* – Krause, 1981) but that apparently the blastemas stop budding and form a stable configuration with a mirroring plane of polarity. This might be interpreted to mean that budding in short germ development is triggered by something comparable to a discrepancy in positional values between the head lobes and the cells at their posterior rim. Intercalation would continue until this discrepancy was filled in by a series of cells differing but little in positional values from their neighbours. In normal development, this stable situation would be reached when all the germ band segments have been intercalated between head and endpiece. In the double-head monsters of *Carausius*, it would be reached when positional values from either side are continuous up to the level of mirroring symmetry. On the lines of this model, one might envisage that, in long germ eggs, intercalation of positional values (without new cells) perfects mirror image symmetry during double monster formation.

In the larval segment, intercalation is well established, and in *Drosophila* it might take place already at the germ band stage, because some segmentation mutants (Nüsslein-Volhard & Wieschaus, 1980) develop patterns reminiscent of polarity reversal in regenerating larval segments (Lawrence & Wright, 1981). This might then reflect a shift of functions from larval into embryonic development, like the one assumed by Krause (1981) to occur from embryonic development into oogenesis.

It is thus possible to propose that the same principle is used at successive levels of patterning in ontogenesis and perhaps in evolution, first at the level of the entire germ anlage, which thereby is converted in the segmented germ band, and thereafter again within each segment of that germ band. This possibility has been suggested earlier using gradients as an example (Sander, 1975), but it may as well be extended to intercalation as a mode of patterning, especially in view of the fact that in evolutionary terms this should be the more primitive mode. A third level on which the principle might be employed is the one revealed by regenerative patterning, which is the expression of general form control (Sander, 1975, 1981) rather than a mechanism specifically evolved for regeneration. Even leg regeneration must be the evolutionary descendant of general form control because its initial (non-functional) evolutionary stages should have been a waste of material rather than an adaptive asset to the individuals involved.

Insect spermatozoa – models for studying the molecular evolution of patterning mechanisms?

However interesting embryonic patterning may appear under evolutionary aspects, it has some formidable disadvantages. To mention only one, embryogenesis is so complex that one can never hope to analyse the whole process down to the molecular level, but only some crucial steps at best. Moreover, experiments mimicking major steps of evolution seem impossible because of the complex network character which spoils prospects for 'experimental pattern evolution' – the approach certainly most tempting to the developmental physiologist. I would like, therefore, to draw attention to a potential model system for studying the molecular evolution of patterning mechanisms: namely, insect spermatogenesis. Spermatozoa represent clear-cut cell shapes and intracellular patterns (for reviews see Baccetti, 1970; Phillips, 1970, Dallai, 1979), and the number of genes involved in their construction is almost certainly much lower than the number required for embryogenesis.

Insect spermatozoa display a wide range of evolutionary phenomena, for instance a remarkable radiation of structure within different groups and convergent evolution of traits like gigantism or immotility (the latter of course in 'co-evolution' with compensatory changes in the genital ducts which must take over sperm transport) (Dallai, 1979). Insect spermatozoa might even rely on kin selection – many insects produce several sperm types some of which might be 'helpers' promoting the 'reproductive success' of the others!

In the context of this essay, the apparently low complexity of the network for sperm patterning is of special interest. The ultrastructural pattern of insect spermatozoa is based on a few elements like chromatin, microtubules and their monomeres, biomembranes and mitochondria, but these are modified in structure and arrangement by influences from a number of gene loci. It is these influences that mainly account for diversity. Several mutations modifying individual elements (Kemphues *et al.*, 1979) and/or affecting their aggregation into patterns are known (Shoup, 1967; Hess & Meyer, 1968; Romrell *et al.*, 1972; Wilkinson *et al.*, 1974).

The best-studied pattern mutations are those affecting the different lamp-brush-type loops forming on the Y-chromosome in *Drosophila hydei* during early meiosis (Hess & Meyer, 1968). Deficiencies for each of these loops are followed by a characteristic syndrome of anomalies disrupting the intracellular spermatid pattern; duplication of the loci is followed by a considerable increase in sperm length (which even in the wild type amounts to an impressive 6 mm). Efforts to characterize the normal functions of these loci at the molecular level are under way, and could potentially lead to the complete description of intracellular patterning. This aim could be furthered by using physical or chemical agents which cause 'phenocopies' of the mutant effects when applied during appropriate stages of spermatogenesis. Even more interestingly, such treatment may occasionally lead to quite new intracellular patterns (Meyer, 1970) which in one respect or the other might resemble patterns found in other insect species. These 'hopeful monsters' among sperm are as yet chance findings but, if studied systematically with the methods available today, spermiogenesis might provide new insights into molecular patterning mechanisms and their relations to evolutionary events.

The author is indebted to his colleague Günter Osche for advice on literature discussing evolution, to Mrs Margrit Scherer for typing the manuscript, and to the Deutsche Forschungsgemeinschaft for supporting his research on patterning in insects (SFB 46).

References

Baccetti, B. (1970). The spermatozoon of arthropoda. IX. The sperm cell as an index of arthropod phylogenesis. In *Comparative spermatology*, ed. B. Baccetti, pp. 169–82. Academic Press: London.

Baer, C. E. von (1828). *Über Entwickelungsgeschichte der Thiere. Beobachtung und Reflexion*. Bornträger: Königsberg.

Baltzer, F. (1950). Entwicklungsphysiologische Betrachtungen über Probleme der Homologie und Evolution. *Rev. suisse Zool.* **57**, 451–77.

Bier, K. H. (1970). Oogenesetypen bei Insekten und Vertebraten, ihre Bedeutung für die Embryogenese und Phylogenese. *Zool. Anz. Suppl.* **33**, 7–29.

Bock, E. (1941). Wechselbeziehungen zwischen den Keimblättern bei der Organbildung von *Chrysopa perla* L. *Wilhelm Roux Arch. EntwMech. Org.* **141**, 159–247.

Bohn, H. (1970). Interkalare Regeneration und segmentale Gradienten bei den Extremitäten von *Leucophaea*-Larven (Blattaria). I. Femur und Tibia. *Wilhelm Roux Arch. EntwMech. Org.* **165**, 303–41.

Bull, A. L. (1982). Stages of living embryos in the jewel wasp *Mormoniella* (*Nasonia*) *vitripennis* (Walker) (Hymenoptera: Pteromalidae). *Int. J. Morphol. Embryol.* **11**, 1–23.

Cappe de Baillon, P. (1940). L'embryogenie des monstres doubles de Phasmes. *Bull. biol. Fr. Belg.* **74**, 197–248.

Cohen, J. (1977). *Reproduction*. London: Butterworth.

Crick, F. (1970). Diffusion in embryogenesis. *Nature* **225**, 420–2.

Dallai, R. (1979). An overview of atypical spermatozoa in insects. In *The spermatozoon*, eds. D. W. Fawcett and J. M. Bedford, pp. 253–65. Urban & Schwarzenberg: Baltimore.

Freeman, G. (1963). Lens regeneration from the cornea in *Xenopus laevis. J. exp. Zool.* **154**, 39–65.

García-Bellido, A., Lawrence, P. A. & Morata, G. (1979). Compartments in animal development. *Scient. Am.* **241**, 102–10.

Gutzeit, H. O. & Koppa, R. (1982). Time-lapse film analysis of cytoplasmic streaming during late oogenesis of *Drosophila. J. Embryol. exp. Morph.* **67**, 101–11.

Gutzeit, H. O. & Sander, K. (1982). Establishment of polarity in the insect egg. In *Fertilization*, 2nd edn, eds. Ch. Metz and A. Monroy (in press).

Haeckel, E. (1866). *Generelle Morphologie*, 2 vols. G. Reimer: Berlin.

Hennig, W. (1981). *Insect phylogeny*. John Wiley & Sons: New York.

Hess, O. & Meyer, F. (1968). Genetic activities of the Y chromosome in *Drosophila* during spermatogenesis. *Adv. Genet.* **14**, 171–223.

Illmensee, K. & Mahowald, A. P. (1976). The autonomous function of germ plasm in a somatic region of the *Drosophila* egg. *Exp. Cell Res.* **97**, 127–40.

Ivanova-Kasas, O. M. (1972). Polyembryony in insects. In *Developmental systems: Insects*, vol. 1, eds. S. J. Counce and C. H. Waddington, pp. 243–71. Academic Press: New York.

Jackson, D. J. (1928). The biology of *Dinocampus* (*Perilitus*) *rutilus* Nees, a braconid parasite of *Sitona lineata* L. – Part I. *Proc. Zool. Soc. Lond.* 597–630.

Jung, E. (1966). Untersuchungen am Ei des Speisebohnenkäfers *Bruchidius obtectus* Say (Coleoptera). 1. Mitteilung: Entwicklungsgeschichtliche Ergebnisse zur Kennzeichnung des Eitypus. *Z. Morph. Ökol. Tiere* **56**, 444–80.

Jung, E., Nuss, E. & Wolf, R. (1977). Geschnürte *Pimpla*-Eier zeigen nur im hinteren Teilembryo Segmentausfall: Sind abgeänderte Ooplasmaströmungen die Ursache? *Verhandlungen der Deutschen Zoologischen Gesellschaft* 1977, p. 397. Fischer-Verlag: Stuttgart.

Kalthoff, K. (1979). Analysis of a morphogenetic determinant in an insect embryo (*Smittia* spec., Chironomidae, Diptera). In *Determinants of spatial organization*, eds. St. Subtelny and J. Konigsberg, pp. 97–126. Academic Press: New York.

Kemphues, K. J., Raff, R. A., Kaufmann, T. C. & Raff, E. C. (1979). Mutation in a structural gene for a β-tubulin specific to testis in *Drosophila melanogaster*. *Proc. nat. Acad. Sci., USA* **76**, 3991–5.

Krause, G. (1939). Die Eitypen der Insekten. *Biol. Zbl.* **59,** 495–536.

Krause, G. (1981). Homology studies on insect egg systems. In *Progress in developmental biology*, ed. H. W. Sauer, Fortschritte der Zoologie, vol. 26, pp. 307–33. Gustav Fischer Verlag: Stuttgart.

Larink, O. (1969). Zur Entwicklungsgeschichte von *Petrobius brevistylis* (Thysanura, Insecta). *Helgoländer wiss. Meeresuntersuch.* **19,** 111–55.

Lawrence, P. A. & Morata, G. (1976). The compartment hypothesis. In *Insect development*, ed. P. A. Lawrence, pp. 132–49. Blackwell Scientific Publications: Oxford.

Lawrence, P. A. & Wright, D. A. (1981). The regeneration of segment boundaries. *Phil. Trans. R. Soc. Ser. B* **295,** 595–9.

Lewis, E. B. (1978). A gene complex controlling segmentation in *Drosophila*. *Nature* **276,** 565–670.

Lohs-Schardin, M. (1982). *Dicephalic* – a *Drosophila* mutant affecting polarity in follicle organization and embryonic patterning. *Wilhelm Roux Arch. EntwMech. Org.* **191,** 28–36.

Lohs-Schardin, M., Cremer, Ch. & Nüsslein-Volhard, C. (1979). A fate map for the larval epidermis of *Drosophila melanogaster*: localized cuticle defects following irradiation of the blastoderm with an ultraviolet laser microbeam. *Devl. Biol.* **73,** 239–55.

Mazzini, M. & Scali, V. (1980). Ultrastructure and amino acid analysis of the eggs of the stick insects, *Lonchodes pterodactylus* Gray and *Carausius morosus* Br. (Phasmatodea: Heteronemiidae). *Int. J. Insect Morph. Embryol.* **9,** 369–82.

Meinhardt, H. (1977). A model of pattern formation in insect embryogenesis. *J. Cell Sci.* **23,** 117–39.

Meinhardt, H. (1978). Space-dependent cell determination under the control of a morphogen gradient. *J. theor. Biol.* **74,** 307–21.

Meyer, G. F. (1970). Phenocopies of Y deficiencies and extraordinary differentiation types in spermiogenesis of *Drosophila*. In *Comparative spermatology*, ed. B. Baccetti, pp. 347–54. Academic Press: New York.

Nübler-Jung, K. (1977). Pattern stability in the insect segment. I. Pattern reconstitution by intercalary regeneration and cell sorting in *Dysdercus intermedius* Dist. *Wilhelm Roux Arch. EntwMech. Org.* **183,** 17–70.

Nüsslein-Volhard, C. (1979). Maternal effect mutations that alter the spatial coordinates of the embryo of *Drosophila melanogaster*. In *Determinants of spatial organization*, ed. S. Subtelny and J. Konigsberg, pp. 185–211. Academic Press: New York.

Nüsslein-Volhard, C. & Wieschaus, E. (1980). Segmentation in *Drosophila*: mutations affecting segment number and polarity. *Nature*, **287,** 795–801.

Phillips, D. M. (1970). Insect flagellar tubule patterns: theme and variations. In *Comparative spermatology*, ed. B. Baccetti, pp. 263–73. Academic Press: New York.

Pijnacker, L. P. (1971). The origin of abnormal micropyle apparatus of eggs of *Carausius morosus* Br. (Cheleutoptera, Phasmidae). *Netherlands J. Zool.* **21,** 366–72.

Remane, A. (1960). Die Beziehungen zwischen Phylogenie und Ontogenie. *Zool. Anz.* **164,** 306–37.

Riedl, R. (1979). *Order in living organisms: a systems analysis of evolution.* Wiley: New York.

Romrell, L. J., Stanley, H. P. & Bowman, J. T. (1972). Genetic control of spermiogenesis in *Drosophila melanogaster*: an autosomal mutant (ms(2)10R) demonstrating disruption of the axonemal complex. *J. Ultrastruct. Res.* **38**, 578–90.

Sander, K. (1960). Analyse des ooplasmatischen Reaktionssystems von *Euscelis plebejus* Fall. (Cicadina) durch Isolieren und Kombinieren von Keimteilen II. *Wilhelm Roux Arch. EntwMech. Org.* **151**, 660–707.

Sander, K. (1975). Bildung und Kontrolle räumlicher Muster bei Metazoen. *Verh. dt. zool. Ges.* **67**, 58–70.

Sander, K. (1976a). Specification of the basic body pattern in insect embryogenesis. *Adv. Insect Physiol.* **12**, 125–238.

Sander, K. (1976b). Morphogenetic movements in insect embryogenesis. In *Insect development*, ed. P. A. Lawrence, pp. 35–52. Blackwell Scientific Publications: Oxford.

Sander, K. (1981). Pattern generation and pattern conservation in insect ontogenesis – problems, data and models. In *Progress in developmental biology*, ed. H. W. Sauer, Fortschritte der Zoologie, vol. 26, pp. 101–19. Gustav Fischer Verlag: Stuttgart.

Sander, K. (1982). Rekapitulation aus der Sicht des Entwicklungsphysiologen: Die konservierende Rolle funktioneller Verknüpfungen in der Ontogenese. In 23rd Phylogenetisches Symposium, ed. O. Kraus, *Verhandlungen des Naturwissenschaftlichen Vereins Hamburg*, **25**, (in press).

Sander, K. (1983). Fertilization in insects. In *Biology of fertilization*, eds. Ch. Metz and A. Monroy (in press).

Sander, K. & Nübler-Jung, K. (1981). Polarity and gradients in insect development. In *International cell biology 1980–1981*, ed. H. G. Schweiger, pp. 497–506. Springer Verlag: Berlin.

Schleip, W. (1929). *Die Determination der Primitiventwicklung.* Akademische Verlags-Gesellschaft: Leipzig.

Schnetter, M. (1934). Physiologische Untersuchungen über das Differenzierungszentrum in der Embryonalentwicklung der Honigbiene. *Wilhelm Roux Arch. EntwMech. Org.* **131**, 285–323.

Schnetter, M. (1935). Morphologische Untersuchungen über das Differenzierungszentrum der Honigbiene. *Z. Morph. Ökol. Tiere* **29**, 114–95.

Seidel, F. (1936). Entwicklungsphysiologie des Insekten-Keims. *Verh. dt. zool. Ges.* **38**, 291–336.

Seidel, F. (1960). Körpergrundgestalt und Keimstruktur. Eine Erörterung über die Grundlagen der vergleichenden und experimentellen Embryologie und deren Gültigkeit bei phylogenetischen Überlegungen. *Zool. Anz.* **164**, 245–305.

Shoup, J. R. (1967). Spermiogenesis in wild type and in a male sterility mutant of *Drosophila melanogaster*. *J. Cell Biol.* **32**, 663–75.

Spemann, H. (1915). Zur Geschichte und Kritik des Begriffs der Homologie. In *Die Kultur der Gegenwart*, ed, P. Hinneberg, pp. Teil 3, Abt. 4, 63–86. Verlag von B. G. Teubner: Leipzig.

Struhl, G. (1981). A gene product required for the correct initiation of segmental determination in *Drosophila*. *Nature* **293**, 36–41.

Summerbell, D., Lewis, J. H. & Wolpert, L. (1973). Positional information in chick limb morphogenesis. *Nature* **244,** 492–6.

Telfer, W. H. (1975). Development and physiology of the oocyte–nurse cell syncytium. *Adv. Insect Physiol.* **11,** 223–319.

Tremblay, E. & Caltagirone, L. E. (1973). Fate of polar bodies in insects. *A. Rev. Ent.* **18,** 421–44.

Wigglesworth, V. B. & Beament, J. W. L. (1950). The respiratory mechanisms of some insect eggs. *Q. Jl. microsc. Sci.* **91.** 429–52.

Wilkinson, R. F., Stanley, H. P. & Bowman, J. T. (1974). Genetic control of spermiogenesis in *Drosophila melanogaster*: The effects of abnormal cytoplasmic microtubule populations in the mutant ms(3)10R and its colcemid-induced phenocopy. *J. Ultrastruct. Res.* **48,** 242–58.

Wolff, G. (1895). Entwicklungsphysiologische Studien. I. Die Regeneration der Urodelenlinse. *Arch. EntwMech. Org.* **1,** 380–90.

Wolpert, L. (1969). Positional information and the spatial pattern of cellular differentiation. *J. theor. Biol.* **25,** 1–47.

Woodruff, R. I. & Telfer, W. H. (1980). Electrophoresis of proteins in intercellular bridges. *Nature* **286,** 84–6.

Note added in proof
In an earlier publication, Cappe de Baillon (1927) claims that the double monster diagrammatically shown at the right in Fig. 5 contains a single abdomen compressed in the yolk. This would contradict the speculation on p. 153.

Cappe de Baillon, P. (1927). Recherches sur la tératologie des Insectes. In *Encyclopédie entomologique*, vol. 8, pp. 1–291. Lechevalier: Paris.

Development and evolution of the insect segment

VERNON FRENCH

Zoology Department, University of Edinburgh, Edinburgh EH9 3JT, Scotland

Introduction

Insects exhibit an enormous range of larval and adult morphologies. Within the class there are great differences in the ways in which the embryo is formed within the egg (Sander, 1976, and this volume) and subsequently grows and forms structures during postembryonic life. In recent years the development of a large number of insects has been examined, using both the techniques of traditional embryology and those of developmental genetics. Insect evolution has also been widely studied, to some extent by the examination of fossils, but mainly through the comparative morphology of extant insects, other arthropods and annelid worms. The changes in morphology which arise during evolution come directly from changes in the generative processes of development. An examination of some aspects of the development of the different insect segments and of different insects gives us some insight into the ways in which developmental mechanisms may have been modified during insect evolution.

Insect evolution

As a result of extensive studies of the comparative morphology of annelids, onychophorans, myriapods and insects (e.g. Snodgrass, 1935; Manton, 1977), insects are considered to have evolved gradually from annelid-like ancestors consisting of a number of morphologically identical simple segments plus non-segmental anterior and posterior extremities (Fig. 1a). The segments each developed a pair of simple appendages, while the anterior extremity

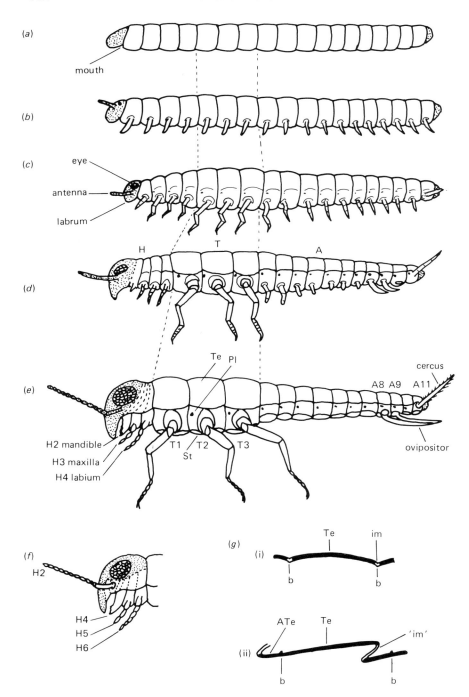

(a) mouth

(b)

(c) eye
antenna
labrum

(d) H T A

(e) Te Pl
cercus
A8 A9 A11
H2 mandible
H3 maxilla
H4 labium
T1 T2 T3
St
ovipositor

(f)
H2
H4
H5
H6

(g) (i) Te im
b b

(ii) ATe Te 'im'
b b

formed some sensory structures. Regions of the body segments secreted thickened cuticle, forming hard plates or sclerites separated by flexible regions and, similarly, the appendages became jointed. Groups of adjacent segments became similarly modified to form the distinctive body regions: the head (with appendages lost or modified into mouthparts), the thorax (with walking legs), and the abdomen (with most appendages lost but some remaining and adapted for sensory or reproductive functions).

The typical simple body segment consisted of dorsal tergite, lateral pleura bearing spiracle and appendage (if any), and ventral sternite. Figure 1e (adapted from Snodgrass, 1935) shows the primitive apterygote with an abdomen of 11 segments (plus posterior end) and a head of 4 segments (plus non-segmental anterior end, bearing eyes and antennae). However, careful observation of early insect embryos (Anderson, 1973) suggests that the head results from fusion of 6 anterior segments and that the antennae are segmental in origin (Manton, 1977), as shown in Fig. 1f. The morphology of extant pterygote insects has become much more complex. The pleura has developed sclerotised plates (the

Fig. 1. Suggested course of insect evolution (*a–e, g* are adapted from Snodgrass, 1935).

(*a*) Theoretical worm-like ancestor consisting of a series of identical segments with non-segmental anterior and posterior ends (stippled).

(*b, c, d*) Stages representing the evolution of simple legs and antennae, of jointed legs and body sclerites, and of adaptations of body segments to form distinct head (H), thorax (T) and abdomen (A) regions. The eye, antenna and labrum are formed from the non-segmented region anterior to the mouth.

(*e*) Basic apterygote insect with body segments consisting of tergite (Te), pleura (Pl) and sternite (St). Appendages form walking legs on the thoracic segments (T1, T2, T3), form the mandibles on the second head segment (H2), the maxillae on H3 and the labium on H4, while the abdominal appendages are mostly lost but form reproductive structures (such as the ovipositor) and the cerci on A11.

(*f*) Alternative version of the insect head, agreeing with modern descriptive and experimental evidence. The first head segment contributes to the labrum, the second (H2) forms the antennae, and mouthparts are formed by the fourth to sixth segments (H4–H6).

(g) Body segmentation. In primary segmentation (i), the simple tergite (Te) of one segment is separated from that of the next by a groove or intersegmental membrane (im) which corresponds to the segmental border (b). In secondary segmentation (ii), the tergite (Te) of one segment is fused to another sclerite, the aerotergite (ATe) of the more anterior segment, so the flexible region ('im') does *not* correspond to the segment border (b).

pleurites), possibly by incorporation of the basal levels of the leg, and there are complex patterns of fusion between pleurites and sternites. Meso- and meta-thoracic wings have evolved, either from the edge of the tergite (Snodgrass, 1935) or from the pleura (Kuklova-Peck, 1978). In most insects, secondary segmentation (Snodgrass, 1935) means that the area between successive sclerites, the 'intersegmental membrane' does not correspond to the border between embryonic segments (Fig. 1*g*).

An obvious feature of insect morphology is segmentation. In some sense, the segment is the basic body unit (especially in hypothetical ancestral forms), although the different segments often differ markedly in size and structure. Segmentation is also a fundamental feature of insect development.

Insect development

Embryonic development

The insect embryo forms from the ventral part of the layer of cells, the blastoderm, which surrounds the yolk. The segments first become visible at the germ band stage (Sander, this volume) and then develop more or less independently to form the larval structures. It is often difficult to determine the positions of the segment borders by simple observation (Fig. 1*g*), but in some insects the technique of *clonal analysis* has precisely defined the segment borders (Lawrence, 1981) and shown that the segments become specified at very early embryonic stages. An example is provided by the work of Lawrence (1973) on the abdominal segments of *Oncopeltus*.

X-irradiation of *Oncopeltus* embryos will occasionally alter a single nucleus so that the clone of epidermal cells descended from it will have abnormal pigmentation, visible beneath the larval insect's transparent cuticle. When embryos are irradiated at very early cleavage stages, the resulting clones are very large, usually extend into several larval abdominal segments and may include sternite and tergite (Fig. 2*a*). However, clones resulting from a later, germ band stage irradiation are confined to a single abdominal segment (Fig. 2*b*), respecting a precise intersegmental border and also respecting a line between tergite and sternite (Lawrence, 1973; Wright & Lawrence, 1981). These results suggest that, prior to

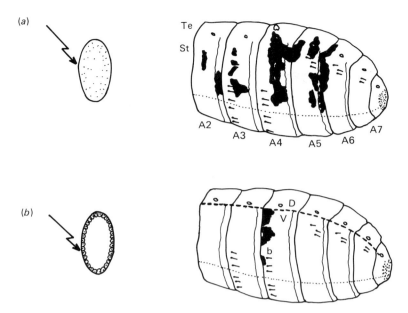

Fig. 2. Clonal analysis of the abdomen of *Oncopeltus* (adapted from Lawrence, 1981, and Wright & Lawrence, 1981). X-irradiation of embryos (*left*) generates clones of cells with altered pigmentation on the larval abdomen (*right* – shown in ventral/lateral view).

(*a*) After irradiation at cleavage stage, a single clone (black) will extend into several abdominal segments and can cross from sternite (St) to tergite (Te) although it will usually not cross the ventral mid-line (dotted line).

(*b*) After irradiation at blastoderm stage, most clones are restricted to one segment and often one edge runs precisely along the groove of the segment border (b) – *Oncopeltus* has primary segmentation. Clones also respect a precise dorso-ventral restriction line (D/V – heavy dashed line). All clones generated at germ band stage respect these restriction lines.

germ band stage, a small group of cells (about 100 – Wright & Lawrence, 1981) becomes determined to form a precise region, the segment, and that this commitment remains stable through embryonic and larval development.

Many experimental results indicate that the segments of the *Drosophila* embryo are determined at or before blastoderm stage. Clones induced at this stage are restricted to one segment (Steiner, 1976; Wieschaus & Gehring, 1976; Lawrence *et al.*, 1978). Ligature of the egg at earlier stages alters the pattern of segments, whereas fragments of the egg ligatured at blastoderm stage form their normal segments (Schubiger & Wood, 1977). A blastoderm stage fate map of the embryo (Fig. 3*a*) has been constructed from the result of ligaturing, and of damaging precise areas and scoring defects in structures of the larva (Lohs-Schardin *et al.*, 1979*a*; Underwood *et al.*, 1980) and the adult (Lohs-Schardin *et al.*, 1979*b*).

Postembryonic development

Postembryonic development consists of a number of cycles of growth, secretion of a new cuticle and moulting to shed the old cuticle. This process can produce the adult insect in several different ways.

In hemimetabolous insects (such as cockroaches or bugs), the first instar larva hatches from the egg in a form very similar to that of the adult. The epidermis of the segments and their appendages makes the larval cuticular structures and, finally, those of the adult. There may be many minor differences in pattern between larva and adult but the only major difference is that some regions of the meso- and meta-thorax (the wing pads) grow considerably and form the adult wing structures prior to the final moult.

In holometabolous insects there is an immobile pupal stage, and at least some of the adult structures are formed by cells which did not make larval parts. In insects such as beetles or moths, most of the adult structures (e.g. abdominal segments, legs) are derived from the corresponding larval parts, although they undergo great changes in growth and structure in passing from larva to pupa to adult. Other adult structures (such as the wings) are formed from a part of the embryonic segment which invaginated in the embryo or larva, and grew as an internal group of cells: an *imaginal disc.* In other holometabolous insects such as *Drosophila* and other flies, there is an extreme separation between larval and imaginal cells.

Within an embryonic segment of *Drosophila* (Fig. 3*a*) different populations of cells develop in different ways. The legless larva (Fig. 3*b*) consists of *larval* cells which do not divide (but become very large and polyploid as the larva grows and moults twice) and *imaginal* cells which remain diploid. The cells of the invaginated head, thoracic and genital *imaginal discs* divide but do not secrete cuticle through larval life. Just after pupariation, the discs evert and join as the intervening head and thoracic larval epidermis dies. The disc cells secrete the pupal cuticle and then the final adult cuticle. Larval abdominal segments 1–7 each contain eight nests of *imaginal histoblast* cells which do not divide but lie within the larval epidermis, secreting patches of larval and pupal cuticle. In the pupa, the histoblasts divide rapidly, spread out to form the epidermis of the adult abdomen as the intervening larval cells die, and then secrete the adult cuticle (Fig. 3*c*).

During the evolution of holometabolous insects from their hemimetabolous ancestors, different populations of cells within the embryonic segment have acquired different responses to the embryonic and postembryonic sequence of hormonal conditions (Fig. 3*d*). In the extreme case of the *Drosophila* head and thoracic segments, two sets of cells now have quite different responses and form totally different sets of structures. A segment of the adult fly is formed from a small number of the cells of the embryonic segment, which is estimated to consist of approximately 180 presumptive epidermal cells when determined at blastoderm stage. Hence, in the mesothorax, about 9 cells form each of the wing discs and about 10 cells each of the leg discs (see Wieschaus, 1978; Lawrence, 1981). Similarly, in an abdominal segment, 10–12 cells may form the histoblast nets on each side of the embryo (Madhavan & Madhavan, 1980).

Many different kinds of experiments performed on insects and other animals have demonstrated that the cells of the early embryo or larva gradually acquire their different developmental fates through interactions with their neighbours. They will develop differently if they are prevented from interacting or made to interact with cells that are not normally their neighbours. The most influential theoretical approach to this process of pattern formation has been the idea of positional information (Wolpert, 1969, 1971), which suggests that spatial patterns of developmental fate arise in two separate steps. A *map* of positional values is generated over a region of cells, perhaps as a result of special boundary properties of cells at the edge. For example, a simple one-dimensional map could be a concentration gradient between maintained maximum and minimum levels. The cells then *interpret* their positional value (their map coordinates) in accordance with their genotype and their developmental history, becoming committed and eventually differentiating to form particular structures. As pointed out by Wolpert (1971) and others, morphology may change during evolu-

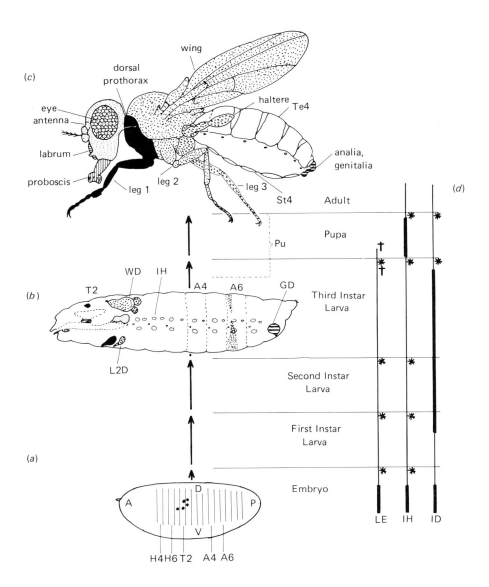

tion because of changes in the map of positional values or changes in the ways in which these values are interpreted.

If the ancestors of insects had a fixed number of identical segmental units (Fig. 1a), then what has changed about the mechanisms of development, enabling insects to evolve such a varied and rich morphology?

The number of insect segments

Ancestral insects are thought to have possessed 6 head, 3 thoracic and 11 abdominal segments (Fig. 1). This number of segments or segment rudiments is often discernible in the early embryo (Anderson, 1973), but in modern adult insects it is not clear which embryonic segments contribute to the head, and the abdomen usually has fewer than 11 obvious segments. The number of

Fig. 3. Development of *Drosophila* from the embryo to the first, second and third instar larvae, the last of which forms a puparium (*Pu*) within which the pupa develops, and finally the adult fly.

(*a*) At blastoderm stage the egg contains a surface layer of cells (the blastoderm) surrounding the central yolk. The thin lines show the estimated extent of the epidermis of the larval segments (which are determined at this stage) in the anterior–posterior (AP) and dorsal–ventral (DV) axes. The abdominal (e.g. A4) and thoracic (e.g. T2) segments are estimated from the nature of late embryonic and larval defects following localised damage to the blastoderm, while the position of head labial (H6), maxillary (H5) and mandibular (H4) segments are derived only from observation of damaged embryos. The spots indicate the estimated position of precursors of the three leg discs, the haltere disc and wing disc, and are based on the location of adult defects following blastoderm damage. Note that the discs map within the corresponding larval segments.

(*b*) Third instar larva showing invaginated head (dotted), thoracic segments (e.g. T2) and abdominal segments with nests (thin circles) of imaginal histoblasts (IH). On the sixth abdominal segment (A6) only, the pattern of cuticular hairs is shown. Imaginal discs such as the wing disc (WD), second leg disc (L2D) and genital disc (GD) are shown with symbols matching those of the corresponding parts of the adult fly.

(*c*) Adult fly showing structures of the head, thorax and abdomen. Te4, St4: tergite and sternite of fourth abdominal segment.

(*d*) Diagram in register with developmental stages (*a–c*), showing the behaviour of the larval epidermis (LE), the imaginal histoblasts (IH) and the imaginal discs (ID). Periods of cuticle secretion prior to an ecdysis (thin horizontal line) are shown by asterisks, and periods of cell division by thick vertical lines. Death of larval epidermis in the last larval instar (head and thoracic epidermis) and in the pupa (abdominal epidermis) is shown by a cross.

embryonic head and abdominal segments and their subsequent paths of development have been extensively studied in *Drosophila*.

Development of the *Drosophila* head

Drosophila has extreme holometabolous development (Fig. 3) and the larval and adult heads are made by different cells. The larval head segments are almost completely invaginated inside the thorax but scanning electron microscope studies of the embryonic stages preceding head involution (Turner & Mähowald, 1979) indicate that larval head structures probably derive from five embryonic segments (Fig. 4a). The head of the adult fly develops from three pairs of imaginal discs, and it has long been debated from which embryonic segments it is derived. This has recently been investigated by gynandromorph fate mapping. This technique (see Janning, 1978) measures the frequency with which the male/female border falls between two structures on the gynandromorph adult, to produce an estimate of the separation of the two sets of precursor cells on the surface of the blastoderm. Gynandromorph mapping of the adult thorax and anterior abdomen broadly agrees with other types of fate map (Fig. 3a) and suggests that segments develop from more-or-less equal-sized bands on the blastoderm. The gynandromorph map of the adult head (Struhl, 1981b) suggests that the head discs originate from the first, second and sixth segments of the embryo (Fig. 4b, c). The presence of three intervening segments is deduced because the eye–antenna and labial primordia are separated by approximately the blastoderm distance between abdominal segments 1 and 5.

Fig. 4. Development of the *Drosophila* head.

(a) The head segments of the embryo before head involution (redrawn from scanning electron micrographs of Turner & Mahowäld, 1979). Visible structures are the clypeo-labrum (labr), the large procephalic lobe (pl) with optic lobe (op) and antennal sense organ (ant), and the mandibular (man), maxillary (max) and labial (lab) segments. (ser) serosa; (T1) prothorax; (T2) mesothorax.

(b) Diagram of the segment primordia of the head (H1–H6) and thorax (T1, T2) at the cellular blastoderm stage, based on the gynandromorph fate map of Struhl (1981b). Dotted arrows indicate the presumed development of the larval cells to form embryonic head segments, and heavy arrows the development of imaginal cells to form the adult.

(c) The head and anterior thorax of the adult. Note that the eye, antenna and most of the head capsule apparently derives from segment H2 which has rotated so that the anterior compartment (A) is behind the posterior (P) one.

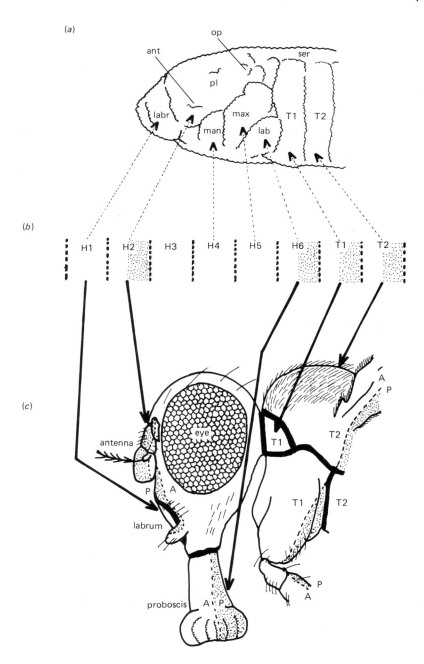

It seems that at least one embryonic segment does not form larval structures and three do not develop into the adult, but all six segments are initially formed as equal-sized regions on the blastoderm.

Development of the *Drosophila* abdomen

In *Drosophila,* the larval and the adult abdomen both have a reduced number of segments. The larva has eight clear segments (Fig. 3*b*), but scanning electron microscopy shows that the embryo starts with at least 10 abdominal segments (Fig. 5*ci*). Segments

Fig. 5. Evolution of the posterior abdomen of adult Diptera. Comparisons between the development of the posterior abdominal segments in the hypothetical ancestral dipteran (*a*), *Calliphora* and *Musca* (*b*) and *Drosophila* (*c*) show that female genitalia form from segment A8, male genitalia from A9 and anal plates from A10/11. There is clearly a tendency for the imaginal cells to fuse into a single genital disc, and also a tendency for non-genital parts to fail to develop.

(*a*) Hypothetical situation in the primitive dipteran. Primordia of the abdominal segments (e.g. A6–A11) are formed on the embryonic blastoderm (i). The situation in the larva is unknown but, in the adult male (ii), segment A8 forms tergites and sternites; A9 forms tergite (Te9), slightly reduced sternite (St9) and the male genitalia (apart from the gonad); and segments A10/11 form the hindgut and analia (An). In the female (iii), however, A8 forms tergite (Te8), slightly reduced sternite (St8) and female genitalia, while A9 just forms tergite and sternite.

(*b*) Development of genitalia in *Musca* and *Calliphora*. The blastoderm fate map is unknown. The larva (i) has (at least) two pairs of histoblast nests (IH) in abdominal segments A1–A7. In the terminal larval segment (A8) there is an anterior pair of lateral genital discs (LGD) which form reduced sternite (st8) and tergites (te8) in the male (ii), and form these plus the genitalia in the female (iii). The larva also has a single median genital disc (MGD) which forms just analia in the female (iii) but also forms the reduced ninth tergite (te9) and genitalia in the male.

(*c*) Development of genitalia in *Drosophila*. Gynandromorph fate mapping indicates that the segment primordia are arranged more or less as shown in *a*(i), and the early embryo (i – redrawn from scanning electron micrographs of Turner & Mähowald, 1979) has 10 or 11 abdominal segments. Segments fuse, however, so that the larva (ii) has seven abdominal segments with histoblasts (IH) plus a terminal segment (A8) with a single genital disc (GD). The disc consists of a segment A10/11 part which forms hind gut and analia; a segment A9 part which grows slowly and does not form adult parts in a female (iv), but forms genitalia and a rudimentary tergite (te9) in the male (iii); and also a segment A8 part which does not develop adult structures in the male but forms a reduced tergite (te8) and genitalia in the female. Note that the histoblasts (IH) of segment A7 form an adult segment in the female but not in the male. ser = serosa.

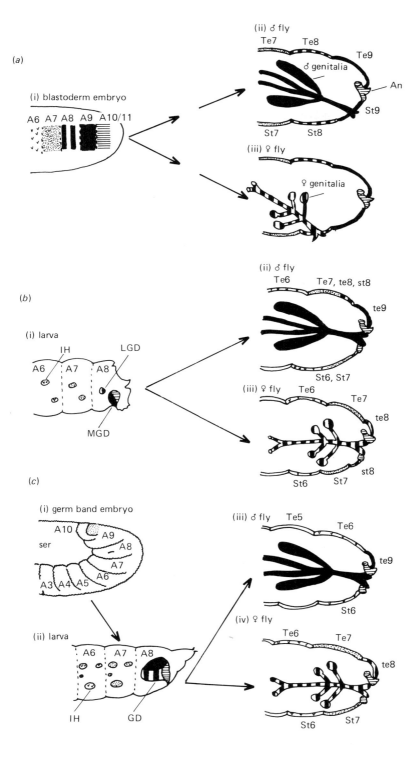

(a)

(i) blastoderm embryo

A6 A7 A8 A9 A10/11

(ii) ♂ fly
Te7   Te8
Te9
♂ genitalia
An
St9
St7   St8

(iii) ♀ fly
♀ genitalia

(b)

(i) larva
IH   LGD
A6   A7   A8
MGD

(ii) ♂ fly
Te6   Te7, te8, st8
te9
St6, St7

(iii) ♀ fly   Te6   Te7
te8
st8
St6   St7

(c)

(i) germ band embryo
A10   A9
ser   A8
A7
A6
A3 A4 A5

(ii) larva
A6   A7   A8
IH   GD

(iii) ♂ fly   Te5   Te6
te9
St6

(iv) ♀ fly
Te6   Te7
te8
St6   St7

behind the seventh fuse and change relative positions, so that the segment 8 spiracles come to lie at the extreme posterior end (Turner & Mähowald, 1979). The larva has nests of imaginal histoblast cells in the first seven abdominal segments and a single invaginated genital disc in the fused eighth segment. The histoblasts produce the adult anterior abdomen consisting of six segments in the male and seven in the female. The genital disc forms the adult *analia* consisting of the hindgut and the sexually dimorphic anal plates, the *genitalia* consisting of external structures plus the ducts leading to the gonads, and also the rudimentary eighth tergite (in the female) and ninth tergite (in the male).

Using a modified gynandromorph mapping technique, Schupbach *et al.* (1978) produced blastoderm fate maps of female, male and gynandromorph embryos. There is an anterior–posterior sequence in the position of the seventh abdominal segment, the female genitalia (plus eighth tergite), the male genitalia (plus ninth tergite) and the analia. Hence, it is clear that the female genitalia, male genitalia and analia are the separate products of embryonic segments 8, 9 and 10/11, respectively (Fig. 5*a*i). The larval genital disc has an *anal* (segment 10/11) portion (which produces either male or female analia); a *male* (segment 9) portion (which produces genitalia if male but, if female, grows slowly and forms no adult structures *in situ* or after transplantation); and a *female* (segment 8) portion producing no structures when male, but forming the genitalia in a female fly (see Fig. 5*c*; Epper, 1980; Wieschaus & Nöthiger, 1982). In a gynandromorph, the segment 8 primordium can be female (XX) while segment 9 is male (XO), and both will develop, giving a fly with full sets of both male and female genital structures (Nöthiger *et al.*, 1977). A similar result occurs in the mutant, *Doublesex-dominant* (dsx^D) where both primordia develop within female (XX) flies (Epper, 1981). The mutation, *transformer* (*tra*) converts XX flies to phenotypic males, in some way repressing the female and activating the male primordium (Epper, 1980; Wieschaus & Nöthiger, 1982).

It seems that all the ancestral abdominal primordia are formed in the early *Drosophila* embryo, that some do not form mature larval structures, and that only some segments posterior to the sixth abdominal will develop into adult structures, depending on their sex chromosomes (and on the activity of autosomal genes such as tra^+ and dsx^D +). The posterior abdominal primordia do not seem fully equivalent to the other embryonic segments, however, since they appear smaller on the blastoderm fate map (Schupbach *et al.*, 1978), and there is evidence that rapidly growing clones can cross between eigth and ninth abdominal segments (Wieschaus & Nöthiger, 1982).

There has probably been gradual evolution of the posterior abdomen (Epper, 1980) from an ancestral dipteran where all posterior body segments but only the appropriate genitalia were formed (Fig. 5*a*), to the state in *Musca* (Dubendorfer, 1971) and *Calliphora* (Emmert, 1972) where genital and anal primordia are partially fused and posterior abdominal segments are considerably reduced (Fig. 5*b*), to the position in *Drosophila* with one fused disc forming only the analia and the appropriate set of genitalia plus tiny tergite (Fig. 5*c*).

Development of different insect segments

The embryonic segments, which appear very similar at early germ band stages, develop into the varied segments of the larval and adult insect. The basis of the differences in the development of the individual segments has been investigated in several insects.

Larval abdominal segments

A clone induced in the *Oncopeltus* abdomen after germ band stage is restricted to one segment and may run precisely along the intersegmental or dorso-ventral border (Fig. 2*b*), but its shape is indeterminate in other respects. This shows that the pattern of structures formed within the segment does not arise from a fixed mechanism: cells make cuticular structures appropriate to the *position* which they occupy (Lawrence, 1973). Many grafting experiments on the abdominal segments of *Oncopeltus* and other insects indicate that position is specified by the level of a gradient running down the segment. This is well illustrated by the work of Stumpf (1966) on the tergites of the wax moth, *Galleria*.

Abdominal segment 6 of the *Galleria* pupa bears a cuticular pattern with a prominent transverse ridge (Fig. 6*a*). When a square of larval epidermis is rotated 180°, the resulting pupa has an altered pattern with a deflected main ridge and a separate ring of ridge cuticle (Fig. 6*b*). This can be readily understood if segment 6

contains an antero-posterior gradient which can smooth, experimentally created discontinuities, and the level of which leads
cells to make anterior, ridge or posterior type cuticle (Fig. 6c).
Stumpf (1968) then investigated segment 4, which is covered by
uniform cuticle. An epidermal graft from segment 6 proceeded to
make its normal cuticular pattern when grafted to an equivalent
position in segment 4 (Fig. 6d) but formed other segment 6
structures when put into another site. A posterior graft made more
anterior structures (including ridge) when put into anterior segment 4 (Fig. 6e), while a graft from anterior segment 6 formed
more posterior cuticle (including ridge) when grafted to posterior
segment 4. These results show segment 4 contains the *same*
antero-posterior gradient as segment 6, but forms a different
pattern because segment 4 cells interpret gradient levels in a
different way.

It seems that at least the abdominal segments of an insect embryo
are determined at early stages and each segment sets up the same
spatial map. Morphological differences arise because the different
states of segmental determination lead the cells to interpret their
maps in different ways.

Fig. 6. Grafting experiments on the abdominal tergites of the wax moth, *Galleria*.
   (*a*) Cuticular structures of (i) larval abdominal segment 6, (ii) pupal segment 6
and (iii) pupal segment 4. In pupal segment 6 the tergite between the intersegmental membranes (im) bears transverse stripes of anterior-type (a), ridge (r) and
posterior-type (p) cuticle, while pupal segment 4 has only posterior-type cuticle. A
dashed line marks the position of the presumptive ridge on the larval segment 6.
   (*b*) (i) 180° rotation of a region of larval segment 6 epidermis including the
presumptive ridge characteristically disrupts the pupal pattern (ii).
   (*c*) Schematic anterior–posterior (AP) section through the graft shown in (*b*),
demonstrating how the new pattern could arise after interaction of gradient levels
of graft and host. Cells interpret gradient levels according to thresholds a′, r′, p′
and hence form the AP sequence of cuticle types shown below the gradient.
   (*d*) (i) Larval segment 6 epidermis grafted to a corresponding position on larval
segment 4 forms its normal cuticular pattern in the pupal stage (ii).
   (*e*) (i) Presumptive posterior segment 6 epidermis grafted to an anterior site on
larval segment 4 forms some anterior cuticle (a) and a ridge (r) in addition to
posterior-type cuticle (p).
   (*f*) Schematic AP section through the graft shown in *e*, demonstrating how the
new pattern could result if host segment 4 (A4) and graft segment 6 (A6) both
contain the same gradient. Host cells will always form posterior-type cuticle, but
the grafted cells form a, r or p cuticle depending on their gradient level.

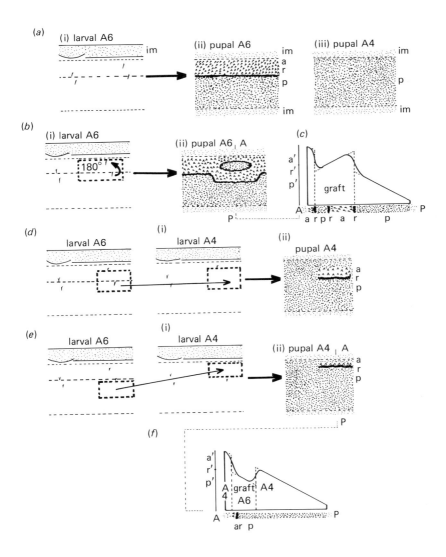

Larval segmental appendages

Much of the morphological diversity of insects derives from the variety of structures formed by the segmental appendages. Hence, the head appendages form antennae and a wide variety of feeding structures, thoracic appendages form legs of various kinds, and abdominal appendages form cerci or complex reproductive structures or may fail to develop (although they are often visible early in embryogenesis).

As in other parts of the insect, it seems that the structures made by cells of an appendage depend on the positions they occupy. Most studies of the cellular interactions which convey 'positional information' have been performed on legs, particularly those of cockroaches. If different proximal–distal levels of a segment of the cockroach leg are grafted together (Fig. 7b), growth is stimulated at the junction and the intervening mid-segment region is produced by *intercalary regeneration* (Bohn, 1970; Bullière, 1971). Similarly, when strips of epidermis are moved around the leg circumference and grafted to a different position (Fig. 7c), the intervening section of the circumference is formed by intercalary regeneration

Fig. 7. Grafting and regeneration in the larval cockroach leg.

(a) *Camera lucida* drawings of the left prothoracic (i) and metathoracic (ii) leg of fifth instar *Blabera craniifer*, showing the differences in size and cuticular patterns on the coxa (co), femur (fe), tibia (ti) and tarsus (ta). A, P, L, M: anterior, posterior, lateral and medial faces of the leg. (a ii) also shows five proximal-distal levels (a–e) marked on the tibia and 12 positions (1–12) marked around the circumference of a schematic transverse section through the femur.

(b) Grafting together proximal (a) and distal (e) levels of the metathoracic tibia stimulates local growth and the formation of an intercalary regenerate (shaded) consisting of the intermediate levels (b, c, d).

(c) Schematic cross-sections of the metathoracic femur with positions labelled as in (a ii). Grafting medial face of the left femur into the anterior face of the host left femur (i) confronts cells from different circumferential positions (ii) and results in local growth and the formation of an intercalary regenerate (shaded) consisting of tissue normally separating the confronted positions (iii), by the shortest route around the circumference (e.g. between positions 4 and 8 a new medial face – positions 5, 6, 7 – has formed).

(d) Diagram of graft between left and right metathoracic tibiae, reversing the medial–lateral (ML) axis of the graft. A supernumerary set of distal parts (shaded) with host orientation has regenerated at the points of maximum discontinuity (medial and lateral) between host and graft.

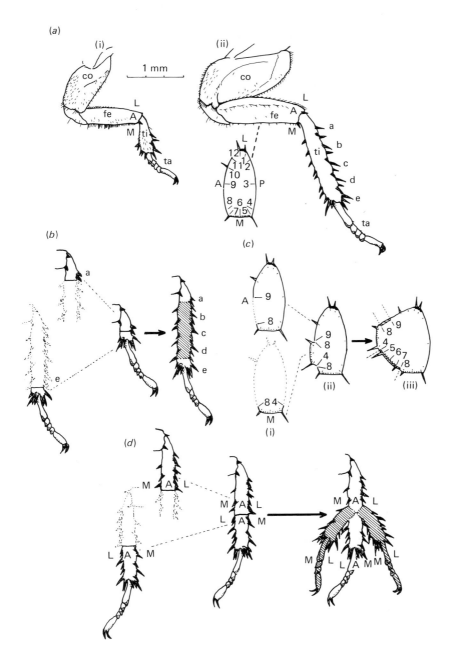

(French, 1978). When a leg is amputated and grafted onto the contralateral stump, one transverse axis of the graft is reversed relative to the host. After reversal of, say, the medial–lateral axis (Fig. 7*d*), supernumerary legs of host handedness and orientation are regenerated from the junction at medial and lateral positions (Bohn, 1965, 1972; Bullière, 1970). These results contribute to a model of pattern regulation in insect legs (French *et al.*, 1976; French, 1981), but their importance in the present context lies in grafts done *between* pro- and meta-thoracic legs. Grafts between corresponding proximal–distal levels (Bohn, 1970), or circumferential positions (French, 1980), simply heal, while grafts between different levels or positions provoke intercalary regeneration. Similarly, axial grafts between the different legs simply heal if orientation is conserved, but produce two supernumeraries if one graft axis is reversed (Bullière, 1970; Bohn, 1972; French, 1982).

This work suggests that the epidermis of the pro- and meta-thoracic legs contains the same map of positional values, so the legs differ in size and details of cuticular structure (Fig. 7*a*) because their cells have different segmental determination and hence interpret the map differently.

Some grafting experiments have been done on other appendages. Cricket abdominal cerci form intercalary regenerates between different proximal–distal levels (M. Schubiger, personal communication) and supernumerary regenerates after contralateral grafts (Palka & Schubiger, 1975). Similarly, grafted cockroach antennae simply heal if orientation is conserved, but produce two supernumerary regenerates after reversal of one transverse axis (French & Domican, 1982). These results, of course, only indicate that the different appendages may be spatially organised in a similar way, and grafts between them are needed to test whether they contain the same spatial maps. Unfortunately, grafts between leg and antenna (French & Domican, 1982) and between leg and cercus (M. Schubiger, personal communication; V. French, unpublished) are frequently rejected and results are inconclusive. It is a common finding that epidermis grafted from a different body region will reduce contact with host epidermis, isolating the graft from the haemolymph and leading to graft rejection. This presumably results from differences in the adhesive properties of the cells (Nardi, 1977), and there is some evidence from grafts between the abdominal segments of *Galleria* (Barbier, 1966) and *Oncopeltus*

(D. Wright, personal communication) that segmental adhesive properties change gradually down the length of the body.

If epidermal grafts healed and interacted satisfactorily, all the body segments and appendages could be explored, and the similarities and differences between their positional maps could be deduced from the patterns of intercalary regeneration resulting from the various graft combinations. In general, this has not proved possible, although it seems that these sorts of grafts can be done between legs and abdominal segments in larval *Tenebrio* (V. French & D. Wright, unpublished). Similarly, if it were possible to graft between very different species, the basis of major morphological differences could be tested. However, in general, grafts only heal satisfactorily between related species (e.g. cockroaches – Bohn, 1972; crickets – French, 1982; beetles – V. French, unpublished) with very similar cuticular patterns.

Drosophila imaginal discs

As shown in Figs, 3, 4 and 5, the different segments of the *Drosophila* adult bear very different structures and are formed from specific groups of imaginal cells (discs and histoblast nests) originating from the segments of the embryonic blastoderm.

Clones induced at blastoderm stage are restricted (at least in most head and thoracic segments) to anterior or posterior *compartments* (García-Bellido *et al.*, 1973) within the segment. Clones in the thorax subsequently become restricted to dorsal or ventral structures (e.g. to the wing disc or second leg disc (Wieschaus & Gehring, 1976)), and within the developing wing disc other compartments arise in larval life (García-Bellido *et al.*, 1973). Apart from compartmental restrictions, the shape of clones is indeterminate: cells develop according to their position in the disc. If fragments of a mature disc are implanted into mature larval hosts, they metamorphose with the host and form a specific set of the disc structures. This enables a fate map to be made of the disc (e.g. Bryant, 1975). If a fragment from the edge of the wing disc is given time to grow before metamorphosis, however, it forms a mirror-image duplication, while the remaining large fragment regenerates to form a complete pattern. If fragments from the opposite dorsal and lateral edges of the disc are mixed, they will interact and form an intercalary regenerate, as will mixed fragments from anterior and posterior edges (Haynie & Bryant, 1976). These results have given rise to a detailed model of the arrangement of positional values and rules for cellular interaction in imaginal discs (French *et al.*, 1976) but, in the context of evolution, the interest lies in experiments where specific fragments of *different* discs have been mixed together. Adler (1979) studied wing and haltere discs and found that the dorsal fragment of the wing disc and lateral fragment of haltere disc formed an intercalary regenerate (as did lateral wing disc and dorsal haltere disc) while no intercalation occurred between dorsal wing and dorsal haltere (or between the lateral fragments). These results suggest that the two discs contain, at least approximately, the same array of positional values but interpret them differently.

This approach has not been widely used to analyse the positional homology between different discs (although there are intriguing hints of homologies between wing and leg discs – Karlsson, 1979). Cells of different discs tend to sort out because of their different adhesive properties (García-Bellido, 1966), rather than healing and interacting. A different approach to the relationship between discs involves analysis of homoeotic mutants.

Homoeotic mutations cause the structures of one imaginal disc to be replaced by those normally formed by another disc. A complete transformation is not very informative about the organisation within the discs. However, partial transformations may occur because the mutation is 'leaky' and only affects some cells, or because the disc has been made a genetic mosaic of wild type and mutant cells. Postlethwaite & Schneiderman (1971) studied $Antp^R$ where parts of the antenna are replaced by ectopic mesothoracic leg structures. The nature of the leg structures produced depended on the proximal–distal and circumferential position of the transformed cells, so that a detailed correspondence map can be made between antenna and leg (Fig. 8). This result is easily understood if the leg and antenna discs normally contain the *same* map of positional values. Those mutant antenna cells which transform to 'leg' determination will interpret their unaltered spatial information to make structures normally found at that site in a leg disc. The resulting appendage will be an integrated chimera of parts of the antenna and complementary parts of the leg.

A study of the chimeric structures formed by a number of homoeotic mutants which partially transform between proboscis,

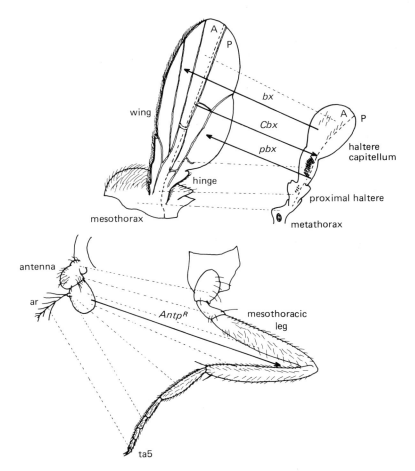

Fig. 8. Homoeotic transformation and positional homology in *Drosophila* (adapted from Postlethwaite & Schneiderman, 1971, and Morata, 1975). *Antennapaedia* (*Antp^R*) transforms part of the antenna into parts of the mesothoracic leg so that a chimeric appendage is formed. The transformation is variable in extent and position-specific (e.g. the arista (ar) transforms only to last tarsal segment (ta5)) so that a correspondence map can be drawn, indicating that the mesothoracic leg disc and the antenna part of the eye–antenna disc have the same positional information.

A similar positional homology between the wing disc and haltere disc is shown by patterns of partial transformation. *Bithorax* (*bx*) transforms the anterior compartment (A) of the haltere capitellum, the proximal haltere and the dorsal metathorax into, respectively, the wing, hinge and dorsal mesothorax. *Postbithorax* (*pbx*) causes similar transformation between posterior compartments (P), and *Contrabithorax* (*Cbx*) causes position-specific partial transformation of wing disc to haltere disc.

antenna, first, second and third legs (Struhl, 1981a) indicates that at least the distal parts of the anterior and posterior compartments of all these ventral appendages have the same positional values.

The other detailed study of homoeotics concerns the wing and haltere (Morata & García-Bellido, 1976; Morata, 1975). The *bithorax* mutation transforms anterior haltere to anterior wing, and *bithorax* clones induced in a normal haltere disc form position-specific wing structures, resulting in the formation of a chimeric appendage. Similar chimeras can be produced by partial transformations caused by leaky alleles of *bithorax*, by *postbithorax* (posterior haltere to posterior wing) and by *Contrabithorax* (wing to haltere). The detailed correspondence map which can be drawn between wing and haltere (Fig. 8) again suggests that these very different structures are alternative interpretations of the same map of positional values.

There are many other homoeotic mutations transforming between segments (e.g. *bithoraxoid* appears to transform dorsal and ventral histoblast nests of the first abdominal segment to the anterior haltere and third leg respectively – Kerridge & Sang, 1981) but precise positional homologies have not been demonstrated.

Thus, only from the wing and haltere discs and from the antenna and leg discs is there good evidence that different segmental structures are formed by different interpretations of the same segmental map. Rather weaker evidence that *all* segments may have a common map comes from studies of lethal mutations causing extreme transformations of the larval segment pattern.

Drosophila larval segments

There are many homoeotic mutations which cause extreme transformations of the segmental pattern of the *Drosophila* larva (Fig. 9a) but are lethal long before adult stage.

A few maternal-effect mutations change the nature, polarity and number of segments of the embryo. For example, *bicaudal* mothers can produce embryos consisting of two posterior ends in mirror symmetry (Nüsslein-Volhard, 1977), while *dicephalic* mothers can produce double anterior embryos (Lohs-Schardin, 1982). These mutations presumably alter conditions during oogenesis so that the positional map formed prior to blastoderm stage is abnormal and

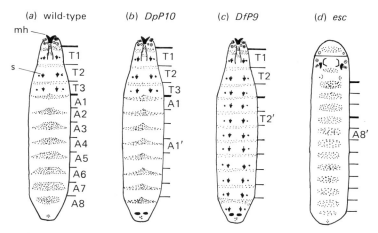

Fig. 9. Pattern of larval segments formed by wild-type (*a*) and mutant (*b–d*) *Drosophila* (adapted from Struhl, 1981*d*). Diagrams are all schematic ventral views of the late embryo or first instar larva (except that the extreme anterior part of the embryo is shown in dorsal view in *d*).

(*a*) Pattern of the wild-type larva. The head is largely invaginated but some characteristic structures such as mouthhooks (*mh*) can be seen. The thoracic segments (T1, T2, T3) bear characteristic sensory organs (*s*), and the thoracic and abdominal segments (A1–A8) bear characteristic patterns of ventral hairs (stippling).

(*b*) Phenotype of the deletion *DpP10*. All segments posterior to the first abdominal (A1) resemble first abdominal segments in their pattern of hairs (A1′).

(*c*) Phenotype of the deletion *DfP9*. All segments posterior to the mesothorax (T2) resemble metathorax in their pattern of hairs and in having sensory organs (T2′).

(*d*) Phenotype of the mutant *esc*. All thoracic and abdominal segments resemble eighth abdominal segment in their hair pattern (A8′), although they often lack the spiracles normally found on A8. The head is not invaginated, forms many of the characteristic head structures but is apparently also partially transformed, since it bears hair patterns resembling those of A8.

symmetrical, leading to the determination of a symmetrical partial pattern of embryonic segments. Other homoeotic mutations transform the patterns made by some of the segments of a normally polarised embryo. For example, deletion *DpP10* (which removes the region of chromosome 3 just to the right of the *bithorax* and *bithoraxoid* loci) results in an embryo of head, thorax and a sequence of segments all resembling the first abdominal segment (Fig. 9*b*). Deletion *DfP9* (which also removes the *bithorax* and *bithoraxoid* loci) gives an embryo of head, prothorax and then a sequence of more-or-less mesothoracic segments (Lewis, 1978), as shown in Fig. 9*c*. Similarly, mutation *esc* (Struhl, 1981*d*) apparently transforms all abdominal and thoracic segments and at least partially transforms some head segments into eighth abdominal segments (Fig. 9*d*). It is likely that these lethal mutations, like the viable mutation *bithorax* (Fig. 8), are changes in interpretation of positional information. This would indicate that the same positional map is set up in most (perhaps all) embryonic segments and that the normal body pattern depends on each segment interpreting the map according to its specific combination of active switch or 'selector genes' (García-Bellido, 1975) such as *Antp*$^+$, *bithorax*$^+$ and *bithoraxoid*$^+$. Normal pattern will also depend on genes such as *esc*$^+$ (Struhl, 1981*d*) necessary for the appropriate combinations of selector genes to be switched on in the different regions of the blastoderm (i.e. in the different segments). However, until chimeric wild type/mutant segments can be analysed (as in Fig. 8), it remains possible that positional maps are different in each segment and mutations such as those in Fig. 9 are merely causing some segments to set up inappropriate maps.

Discussion

Insects are considered to have evolved from ancestors consisting of a large number (probably around 20) of simple, very similar segments, each bearing walking legs (Fig. 1*a*). In extant insects, the number of segments is usually considerably reduced and they differ dramatically in structure. This change must have resulted from modifications to developmental mechanisms.

During the evolution of insects, there seems to have been a change in the manner in which the embryonic segments are formed (Sander, this volume). In general, more recently evolved orders of insects (e.g. Diptera) have changed from *short germ* development (where the embryo is formed from only a small part of the original blastoderm, and segments are added sequentially during growth of a posterior blastema), to *long germ* development (where the embryonic segments are formed early and more-or-less simultaneously, over much of the surface of the blastoderm). Intriguingly, there are many insects with *intermediate germ* development where the more anterior segments are formed early and the posterior ones added gradually during growth, and all three forms of development can be found within a single order (e.g. Coleoptera). Despite the evolution of a different mode of segment development, the *number of segments* initially formed does *not* seem to have changed. The ancestral number (6 head, 3 thoracic and 10 or 11 abdominal) of segment primordia can often be discerned, and the studies of the *Drosophila* head and abdomen indicate that all segments are initially present but specific ones then fail to develop larval or adult structures.

In hemimetabolous insects, the cumbersome, non-functional larval wing grows externally; while in holometabolous development, the wing grows internally, everts, and reorganisation of the relationships between wing epidermis, muscle and nerve occurs in the immobile pupal phase. This has been accompanied (or followed) by other changes between larval and adult stages of holometabolous insects. In extreme cases, such as *Drosophila*, they are different animals, occupying different niches and composed of cells which were derived from different parts of the embryonic segment and responded differently to the conditions of embryonic and postembryonic life (Fig. 3*d*).

Within the insect segment, patterns of structures are formed, with the epidermal cells differentiating as appropriate to their *position*. I have discussed pattern formation in terms of Wolpert's (1969) concept of 'positional information' whereby the cells cooperate to form a simple map, such as a gradient (perhaps with pre-existing boundary cells having a special role). The cells then independently interpret their positional value according to a complex set of rules which will be influenced by their inherited characteristics (e.g. their genotype) and their developmental history. Wolpert (1971) argues that positional maps are simple and that the mechanism of forming them has been highly conserved. The evolutionary divergence of species or of the once-identical segments of an animal will have proceeded by changes in the complex ways that cells interpret the map. An alternative view of pattern-formation, that of 'prepattern', suggests that visible patterns are rather simple reflections of complex underlying landscapes with singularities (e.g. peaks in chemical concentrations) evoking pattern elements (e.g. bristles). Evolution in this case is assumed to have involved changes in the form of the prepatterns.

Most of the major features of insect epidermal pattern formation can be readily understood in terms of positional information. This is strongly supported by the results of grafting between different segments (Fig. 6) and different legs (Fig. 7), and the studies of homoeotic mutants (Figs. 8, 9) which demonstrate that very different visible patterns can form from the same set of spatial cues. However, this does not preclude other mechanisms of pattern formation, e.g. bristles may be spaced by mutual *inhibition*, small bracts may be specifically *induced* by the adjacent bristles, and a repeating pattern such as the segmentation of the blastoderm may well depend on a similarly repeating signal or *prepattern*.

The demonstrations of positional homology between some insect segments and segmental appendages are certainly consistent with Wolpert's contention that evolution of morphology occurs via changes in the interpretation of a universal map of positional information. However, they are very far from showing that the maps in *all* segments are *completely* identical. There are many sorts of differences which would escape detection, even in a fairly rigorous test such as homoeotic transformation (Fig. 8). The map may change by the elimination or insertion of values or an alteration in spacing, while retaining its general form.

Even if positional maps may have changed during evolution of insect segments, it is clear that *interpretations* have changed dramatically. The segments become differently determined at around blastoderm stage, and the results of the homoeotic studies have suggested that this involves turning on particular switch or 'selector' genes (García-Bellido, 1975) in each segment. Thus *bithorax*[+] is turned on in the metathoracic but not in the mesothoracic segment, so if the gene is mutant (and hence ineffective) the metathorax forms wing and second leg. Similarly, if *Antennapaedia*[+] is inactive, the disc where it is normally active (antenna) develops like the mesothoracic leg disc, where it is normally turned off. Most homoeotics transform head, thoracic or abdominal segments to *mesothorax* and have been termed atavic (García-Bellido, 1977), since they seem to transform the insect towards the presumed ancestral form. Some reverse transformations have been interpreted as mutations in control sites, leading the adjacent selector gene to be active in conditions where it would normally be repressed (e.g. in the *Cbx* mutant the adjacent *bithorax*[+] is now active in the mesothorax which thus forms halteres and third legs). The 'bithorax complex' seems to consist of several genes, and Lewis (1978) has suggested that none is active in the mesothorax, *bithorax*$^+$ and *postbithorax*$^+$ are active in the metathorax, these plus *bithoraxoid*$^+$ are active in the first abdominal and so on until all the genes in the complex are switched on in the last abdominal segment. Evolution will have involved the acquisition of various selector genes which are turned on by conditions in the metathoracic and abdominal (García-Bellido, 1977; Lewis, 1978) or head (García-Bellido, 1977) regions of the blastoderm map. These will have effects on other genes and will ultimately shift the interpretation of positional information in these segments away from the 'primitive' mesothoracic state. 'By this evolutionary process homogeneity, such as the sequential repetition of identical metameres, is changed into diversity . . . ' (García-Bellido, 1977).

Of course, after selector genes had evolved, the segments had identities and could change independently. The dramatic effect of inactivating the *bithorax*$^+$ and *postbithorax*$^+$ genes of *Drosophila* is the production of a four-winged fly. Bx^+ and pbx^+ now act as switches between alternative pathways, and there is *no* indication that Diptera evolved suddenly, as 'hopeful monsters' (Goldschmidt, 1952) by the acquisition of these genes. Undoubtedly, when meso- and meta-thorax first became distinct, they bore neither wings nor halteres.

It is likely that the segments have become morphologically very diverse largely by changes in their interpretation of a segmental map, and that this has followed the evolution of control genes active in some or only one of the segments. However, there are considerable difficulties with the 'selector gene' system outlined above, and with the idea of an unmodified mesothoracic state of determination. For example, clones of some deletions of part of the bithorax complex show a transformation of meso- and meta-thoracic leg to prothoracic leg (Morata, 1982), as do some *esc* genotypes (Struhl, 1981*d*), indicating that there are control genes normally off in the prothorax but on in the mesothorax. Similarly, although most alleles of *Antennapaedia* (and other genes such as ss^a) transform antenna to mesothoracic leg, clones of *Antennapaedia* alleles assumed totally to lack activity do not affect the antenna, but transform mesothoracic leg to antenna. This suggests that the gene is normally active in the mesothorax but not in the second head segment (Struhl, 1981*c*). There are probably a large number of position-activated control genes interacting in complex ways to produce the relatively few stable states of determination which characterise the different insect segments and govern their interpretation of the segmental positional map.

My research is supported by the Science and Engineering Research Council and The Royal Society.

I thank Drs Jane Karlsson, Linda Partridge and David Wright for helpful comments and discussions over a long period of time.

References

Adler, P. (1979). Position-specific interaction between cells of the imaginal wing and haltere discs of *Drosophila melanogaster*. *Devl Biol.* **70**, 262–7.

Anderson, D. (1973). *Embryology and phylogeny in annelids and arthropods.* Pergamon Press: Oxford.

Barbier, R. (1966). Etude de l'hypoderme de *Galleria mellonella* L. (Lepidoptère, Pyralidae) à l'aide d'homogreffes larvaires. *C. r. Acad. Sci. Paris* **262**, 2073–6.

Bohn, H. (1965). Analyse der Regenerationsfähigkeit der Insekten-extremität durch Amputatations- und Transplantationsversuche an Larven der afrikanischen Schabe *Leucophaea maderae* Fabr. (Blattaria). II. Mitt. Achensdetermination. *Wilhelm Roux Arch. EntwMech. Org.* **156**, 449–503.

Bohn, H. (1970). Interkalare Regeneration und segmentale Gradienten bei den Extremitäten von *Leucophaea* – Larven (Blattaria). I. Femur und Tibia. *Wilhelm Roux Arch. EntwMech. Org.* **165**, 303–41.

Bohn, H. (1972). The origin of the epidermis in the supernumerary regenerates of triple legs in cockroaches (Blattaria). *J. Embryol. exp. Morph.* **28**, 185–208.

Bryant, P. (1975). Pattern formation in the imaginal wing disc of *Drosophila melanogaster*: fate map, regeneration and duplication. *J. exp. Zool.* **193**, 49–78.

Bullière, D. (1970). Sur la déterminisme de la qualité des régénérates d'appendices chez la Blatte, *Blabera craniifer*. *J. Embryol. exp. Morph.* **23**, 323–35.

Bullière, D. (1971). Utilisation de la régénération intercalaire pour l'étude de la détermination cellulaire au cours de la morphogenèse chez *Blabera craniifer* (Insecte Dictyoptere). *Devl. Biol.* **25**, 672–709.

Dubendorfer, A. (1971). Untersuchungen zum Anlagenplan und Determinationszustand der weiblichen Genital-und Anal primordien von *Musca domestica* L. *Wilhelm Roux Arch. EntwMech. Org.* **168**, 142–68.

Emmert, W. (1972). Entwicklungsleistungen abdominaler Imaginalscheiben von *Calliphora erythrocephala*. Experimentelle Untersuchungen zur Morphology des Abdomens. *Wilhelm Roux Arch. EntwMech. Org.* **169**, 87–133.

Epper, F. (1980). The genital disc of *Drosophila melanogaster*. A developmental and genetic analysis of its sexual dimorphism. Ph.D. Thesis, University of Zurich.

Epper, F. (1981). Morphological analysis and fate map of the intersexual genital disc of the mutant *Double sex-dominant* in *Drosophila melanogaster*. *Devl Biol.* **88**, 104–14.

French, V. (1978). Intercalary regeneration around the circumference of the cockroach leg. *J. Embryol. exp. Morph.* **47**, 53–84.

French, V. (1980). Positional information around the segments of the cockroach leg. *J. Embryol. exp. Morph.* **59**, 281–313.

French, V. (1981). Pattern regulation and regeneration. *Phil. Trans. R. Soc. Lond. Ser. B* **295**, 601–17.

French, V. (1982). Leg regeneration in insects: cell interactions and lineage. *Am. Zool.* **22**, 79–90.

French, V., Bryant, P. & Bryant, S. (1976). Pattern regulation in epimorphic fields. *Science* **193**, 969–81.

French, V. & Domican, J. (1982). The regeneration of supernumerary cockroach antennae. *J. Embryol. exp. Morph.* **67**, 153–65.

García-Bellido, A. (1966). Pattern reconstruction by dissociated imaginal disc cells of *Drosophila melanogaster*. *Devl Biol.* **14**, 278–306.

García-Bellido, A. (1975). Genetic control of wing disc development in *Drosophila*. In *Cell patterning*, Ciba Foundation Symposium vol. 29, pp. 161–82. Associated Scientific Publishers: Amsterdam.

García-Bellido, A. (1977). Homoeotic and atavic mutations in insects. *Am. Zool.* **17**, 613–29.

García-Bellido, A., Ripoll, P. & Morata, G. (1973). Developmental compartmentalization of the wing disc of *Drosophila*. *Nature New Biol.* **245**, 251–3.

Goldschmidt, R. (1952). Homoeotic mutants and evolution. *Acta biotheor.* **10**, 87–104.

Haynie, J. & Bryant, P. (1976). Intercalary regeneration in imaginal wing disk of *Drosophila melanogaster*. *Nature* **259**, 659–62.

Janning, W. (1978). Gyandromorph fate maps in *Drosophila*. In *Genetic mosaics and cell differentiation*, ed. W. Gehring, pp. 1–28. Springer-Verlag: Berlin.

Karlsson, J. (1979). A major difference between transdetermination and homeosis. *Nature* **279**, 426–8.

Kerridge, S. & Sang, J. (1981). Developmental analysis of the homoetic mutation *bithoraxoid* of *Drosophila melanogaster*. *J. Embryol. exp. Morph.* **61**, 69–86.

Kuklova-Peck, J. (1978). Origin and evolution of insect wings and their relation to metamorphosis, as documented by the fossil record. *J. Morph.* **156**, 53–126.

Lawrence, P. (1973). A clonal analysis of segment development in *Oncopeltus* (Hemiptera). *J. Embryol. exp. Morph.* **30**, 681–99.

Lawrence, P. (1981). The cellular basis of segmentation in insects. *Cell* **26**, 3–10.

Lawrence, P., Green, S. & Johnston, P. (1978). Compartmentalisation and growth of the *Drosophila* abdomen. *J. Embryol. exp. Morph.* **43**, 233–48.

Lewis, E. (1978). A gene complex controlling segmentation in *Drosophila*. *Nature* **276**, 565–70.

Lohs-Schardin, M. (1982). *Dicephalic* – a *Drosophila* mutant affecting polarity in follicle organisation and embryonic patterning. *Wilhelm Roux Arch. EntwMech. Org.* **191**, 28–36.

Lohs-Schardin, M., Cremer, C. & Nüsslein-Volhard, C. (1979a). A fate map for the larval epidermis of *Drosophila melanogaster*: localised cuticle defects following irradiation of the blastoderm with an ultraviolet laser microbeam. *Devl Biol.* **73**, 239–55.

Lohs-Schardin, M., Sander, K., Cremer, C., Cremer, T. & Zorn, C. (1979*b*). Localised ultraviolet laser microbeam irradiation of early *Drosophila* embryos: fate maps based on location and frequency of adult defects. *Devl Biol.* **68**, 533–45.

Madhavan, M. & Madhavan, K. (1980). Morphogenesis of the epidermis of adult abdomen of *Drosophila. J. Embryol. exp. Morph.* **60**, 1–31.

Manton, S. (1977). *The Arthropoda: habits, functional morphology and evolution.* Clarendon Press: Oxford.

Morata, G. (1975). Analysis of gene expression during development in the homeotic mutant *Contrabithorax* of *Drosophila melanogaster. J. Embryol. exp. Morph.* **34**, 19–31.

Morata, G. (1982). The mode of action of the bithorax genes of *Drosophila melanogaster. Am. Zool.* **22**, 57–64.

Morata, G. & García-Bellido, A. (1976). Developmental analysis of some mutants of the *bithorax* system of *Drosophila. Wilhelm Roux Arch. EntwMech. Org.* **179**, 125–43.

Nardi, J. (1977). Are cell affinity relationships for insect epidermal tissues evidence for combinatorial control? *Nature* **268**, 665–7.

Nöthiger, R., Dubendorfer, A. & Epper, F. (1977). Gynandromorphs reveal two separate primordia for male and female genitalia in *Drosophila melanogaster. Wilhelm Roux Arch. EntwMech. Org.* **181**, 367–73.

Nüsslein-Volhard, C. (1977). Genetic analysis of pattern formation in the embryo of *Drosophila melanogaster.* Characterisation of the maternal-effect mutant *bicaudal. Wilhelm Roux Arch. EntwMech. Org.* **183**, 249–68.

Palka, J. & Schubiger, M. (1975). Central connections of receptors on rotated and exchanged cerci of crickets. *Proc. natn. Acad. Sci. USA* **72**, 966–9.

Postlethwaite, J. & Schneiderman, H. (1971). Pattern formation and determination in the antenna of the homoeotic mutant *Antennapedia* of *Drosophila melanogaster. Devl Biol.* **25**, 606–40.

Sander, K. (1976). Specification of the basic body pattern in insect embryogenesis. *Adv. Insect Physiol.* **12**, 125–238.

Schubiger, G. & Wood, W. (1977). Determination during early embryogenesis in *Drosophila melanogaster. Am. Zool.* **17**, 565–76.

Schupbach, T., Wieschaus, E. & Nöthiger, R. (1978). The embryonic organisation of the genital disc studied in genetic mosaics of *Drosophila melanogaster. Wilhelm Roux Arch. EntwMech. Org.* **185**, 249–70.

Snodgrass, R. (1935). *Principles of insect morphology.* McGraw-Hill: New York.

Steiner, E. (1976). Establishment of compartments in the developing leg imaginal discs of *Drosophila melanogaster. Wilhelm Roux Arch. EntwMech. Org.* **180**, 9–30.

Struhl, G. (1981*a*). Anterior and posterior compartments in the proboscis of *Drosophila. Devl Biol.* **84**, 372–85.

Struhl, G. (1981*b*). A blastoderm fate map of compartments and segments of the *Drosophila* head. *Devl Biol.* **84**, 386–96.

Struhl, G. (1981*c*). A homoeotic mutation transforming leg to antenna in *Drosophila. Nature* **292**, 635–8.

Struhl, G. (1981*d*). A gene product required for correct initiation of segmental determination in *Drosophila. Nature* **292**, 36–41.

Stumpf, H. (1966). Mechanisms by which cells estimate their location within the body. *Nature* **212**, 430–1.

Stumpf, H. (1968). Further studies on gradient-dependent diversification in the pupal cuticle of *Galleria mellonella. J. exp. Biol.* **49**, 49–60.

Turner, F. & Mähowald, A. (1979). Scanning electron microscopy of *Drosophila melanogaster* embryogenesis. III. Formation of the head and caudal segments. *Devl Biol.* **68**, 96–109.

Underwood, E., Turner, F. & Mähowald, A. (1980). Analysis of cell movements and fate mapping during early embryogenesis in *Drosophila melanogaster. Devl Biol.* **74**, 286–301.

Wieschaus, E. (1978). Cell lineage relationships in the *Drosophila* embryo. In *Genetic mosaics and cell differentiation*, ed. W. Gehring, pp. 97–118, Springer-Verlag: Berlin.

Wieschaus, E. & Gehring, W. (1976). Clonal analysis of primordial disc cells in the early embryo *Drosophila melanogaster. Devl Biol.* **50**, 249–63.

Wieschaus, E. & Nöthiger, R. (1982). The role of the transformer genes in the development of genitalia and analia in *drosophila melanogaster. Devl Biol.* **90**, 320–34.

Wolpert, L. (1969). Positional information and the spatial pattern of cellular differentiation. *J. theor. Biol.* **25**, 1–47.

Wolpert, L. (1971). Positional information and pattern formation. *Curr. Top. devl Biol.* **6**, 183–224.

Wright, D. & Lawrence, P. A. (1981). Regeneration of the segment boundary in *Oncopeltus. Devl Biol.* **85**, 317–27.

Developmental constraints: internal factors in evolution

STUART A. KAUFFMAN

Department of Biochemistry and Biophysics, University of Pennsylvania, School of Medicine, Philadelphia, Pennsylvania 19104, USA

#Ac434

Developmental biologists are renewing their historical interest in problems of evolutionary biology. The immediate cause of renewed attention by developmental biologists is resurgence of interest in the extent to which evolution reflects selection or drift as filtering processes compared to the ways evolution reflects constraints caused by the processes of ontogeny (Bonner, 1982). For example, is the apparent morphological stasis seen among trilobites (Eldredge & Gould, 1972) over many millions of years the result of stabilizing selection? It may be, but that Neo-Darwinian account must assert that stable trilobite morphology in the face of presumptive genetic variability reflects a maximally adapted phenotype in a stable environment. Neither 'maximally adapted', nor 'stable environment' can be reliably assessed, thus the account is weak.

  Does morphological stasis reflect highly 'canalized', rigid developmental constraints? The supposition that developmental constraints may play a role in evolution is certainly not new, nor are attempts to analyze developmental constraints novel. Yet, in an important sense, although the historical traditions of analysis are useful they have not helped strongly to characterize the features or roles of such constraints.

Developmental constraints: an empiricist tradition

At least three intertwined patterns of analysis of the relation between development and evolution have flourished in the past 50 years. These are allometry, heterochrony, and quantitative genetics.

Allometry

Allometry can be seen as a pattern of analysis which derives from comparative morphology, and rests on the observation that one shape can be mathematically transformed into a different but topologically equivalent shape by a sufficient distortion of the axes chosen as the coordinate system in which to draw the shape (Huxley, 1932; Thompson, 1942). This mathematical feature of geometry permits comparison of the shape of an organ at different stages of ontogeny, or between different organisms. In principle, it yields a re-statement of the growth dynamics of the organ in one species in terms of apparent spatially oriented growth laws. Between species, it yields conclusions about the ways in which differential growth would yield the observed differences in ultimate shape. Furthermore, if the femur in a series of species can be shown to be related by the same allometric transformation applied successively, that observation aids in the construction of phylogenies linking the species together in a plausible evolutionary tree.

While useful, analysis of allometric transformations *per se* does not clarify the role of developmental constraints in evolution. The reason is straightforward. Allometric analysis can be utilized to redescribe almost any shape change. It constitutes an advantageous redescription, but poses no constraining explanation. That is, if a specific allometric transformation in head dimensions is observed between related species, that orderliness may help to link them phylogenetically, but offers no account of why the *specific* observed transformation occurred, nor, more importantly, why *arbitrary* alternatives could not have been seen. For example, imagine a spherical ball of putty whose shape is going to deform in some wholly arbitrary way. It is true, but not *predictively* useful, that each stage in the deformation is a small change from the previous stage, hence an allometric 'neighboring' shape, and that a sequence of such neighboring shapes can be related in a 'phylogeny'. This does not imply that 'developmental' constraints exist which limit the possible ways the morphology can change. A second example emphasizes this point: Imagine a loose string whose ends are fixed. Deforming the string randomly results in a complex wave-form in the string which can be decomposed into a weighted sum of sinusoidal functions of different wave-lengths (Fourier decomposition). Construct an algorithm which arbitrarily changes any one or many of the weighting coefficients by small amounts yielding a maximum small change in the deformation shape of the string. A succession of such forms can be cast into a phylogeny of neighboring patterns, which may even branch and leave 'phenotypic' spaces between the morphologies of the distinct 'lineages', yet no constraint exists in the way the next form deviates from any present form. Smooth transformations between neighboring morphologies and 'phenotypic' gaps between branching lineages of forms may illuminate phylogenies, but need not imply developmental constraints. By contrast, if only the fundamental and first Fourier component (or any small number of components) were utilized in generating the shape of the string, a highly constrained family of neighboring forms would be engendered. Similarly, consider the intellectual circumstances in which we might understand the growth mechanisms of a given organ, and thereby understand that alteration in the growth parameters might lead to a *constrained class* of allometric shape transformations. Such a theory would potentially predict the delimited *possible neighboring* morphologies open to the evolving species. Some putative examples will be discussed below.

Heterochrony

Heterochrony is the analysis of changing temporal patterns in ontogeny. The underlying concept is that, if the time of onset, rate, or time of cessation of development of one organ changes with respect to that of other tissues in an organism, the altered ontogeny yields an altered organism. A large number of instances of heterochrony in evolution have been analyzed usefully (de Beer, 1958; Gould, 1977). Among the most useful have been those which show that a suite of traits coordinately co-vary between related species, and can be accounted for by a single heterochronic ontogenetic alteration (Bolt, 1979; Alberch & Alberch, 1981; Alberch, 1982). Further, the hypothesis that genes controlling morphological variation in evolution often do so through control of timing of morphogenetic processes has received considerable attention (Goldschmidt, 1938, 1940; Waddington 1940, 1957).

While obviously useful, the analysis of heterochronies in evolution, like allometry, does not by itself establish the existence of developmental constraints. Consider a balloon twisted into the

shape of a tetrapod, and imagine that, during its inflation, valves
tune the timing and rate of air flow into legs, body, tail, and head.
In this very simple example of form generation, the hypothesis that
alterations occur in the timing and extent to which valves open
leads to a clearly constrained family of neighboring morphologies,
linked by heterochronies. Now weaken the assumptions and
imagine a general tetrapod shape made of putty, in which the
development of morphologies of body, limbs, tail, and head are
reliable but wholly arbitrary, and again are governed by signals for
time of initiation, rate, and cessation of each morphogenetic
programme. As in the case with allometry, a series of small
deformations in the timing will lead to heterochronies in a series of
tetrapods exhibiting coordinated alterations in morphologies which
are illuminating with respect to 'phylogenetic' relations among the
tetrapods. But because the generation of morphology in the putty is
wholly arbitrary, although reliable, the heterochronic alterations
carry no constraining predictions about the character of neighboring morphologies. Finally, further weaken the assumptions such
that the relative timing of morphogenetic programmes is not itself
controlling morphogenesis, but is a byproduct of other ontogenetic
processes, and the analysis of heterochronies remains useful in
constructing phylogenies, but misleading as a general ontogenetic
hypothesis.

These hypothetical examples point to limitations in the implications of heterochronies. First, very many alterations in ontogenies
are likely to be *associated* with alterations in the relative timing of
developmental processes. It does not follow that generation of the
altered ontogenies is generally regulated by controlling relative
timing *per se*. Thus, universal redescription of morphogenesis in
terms of heterochronies leads seductively to an inappropriate
picture of ontogenesis. Secondly, like the putty example, the
concept of hypermorphosis resulting from an elongated period of
development in an organ or tissue is so unconstrained as to be
non-predictive. Without a theory to account for why the observed
coordinated heterochronic effects were seen, and, even more
critically, why wholly *arbitrary* alternatives were not to be expected, the hypermorphic tetrapod limb might differ from the
juvenile form in almost any respect. Coordinated heterochrony *per
se* does not imply developmental constraints. In order to yield
constrained predictions, we need to know how a tetrapod limb is

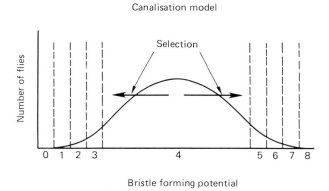

Fig. 1. One theory of canalization. Gaussian distribution shows hypothetical distribution of scutellar bristle forming potential in a *Drosophila* population. Thresholds (vertical lines) separate levels of this potential sufficient to cause 0,1,2 . . . scutellar bristles. Selection is imagined to shift the Gaussian distribution to the left or right. Unequal spacing of thresholds assures a canalization of four bristles in the population.

actually generated, and whether, like the balloon example, there is actually a *limited* family of potential neighboring limb morphologies.

Canalization and quantitative genetics

Geneticists have, for the past number of decades, studied a variety of aspects of development. One tradition in quantitative genetics is based on the concept of serial thresholds (Rendel, 1967). For example, the number of scutellar bristles on *Drosophila melanogaster* is generally four, but is subject to variation and is selectable. The general hypothesis supposes that an underlying Gaussian distribution of genetic factors which contribute to the generation of scutellar bristles exists in the population, and that the number of bristles is determined by a series of thresholds (Fig. 1). Therefore, the number of scutellar bristles in any single fly is determined by its inherited position in the Gaussian distribution of genetic factors, between a given pair of thresholds, while the distribution of bristles in the population is determined by the shape of the Gaussian distribution, and the (generally non-uniform) spacing between thresholds with respect to this Gaussian distribution.

Selection is pictured as shifting the underlying Gaussian distribution upward or downward without altering its shape, while the threshold values remain constant. Where threshold values are far apart (Fig. 1) 'canalization', a resistance to alteration in bristle number with selection, will be seen; conversely, where thresholds are close together, sensitivity of bristle number to selection will be seen.

This tradition of analysis has led to a substantial series of experiments (Dun & Fraser, 1959; Rendel, 1967; Fraser, 1970). Without discussing the merits of the programme further, it seems fair to note that it rests on a universal picture of the control of quantitative traits, with virtually no regard for the actual ontogenetic mechanisms generating bristles, digits, eyes. Arbitrary spacing of thresholds allows the theory to redescribe any distribution of canalized and non-canalized quantitative traits, but affords no theory accounting for the particular quantities which are canalized, nor why arbitrary alternatives are not to be expected. If highly canalized properties are examples of developmental constraints, this theory can offer no insight into those constraints without explaining threshold spacings themselves.

Developmental constraints: a structuralist programme

Recently, Webster & Goodwin (1981) have called attention to the older tradition of the rational morphologists of the eighteenth century, who sought underlying laws of form in biology. Further, these authors have discussed structuralism and its potential application in biology. Their suggestion warrants serious consideration.

Two exemplars typify the structuralist stance: (1) The mathematical theory of the permutation group specifies all possible permutations of a set of N objects onto itself. Each of these $N!$ permutations assigns to each of the N, its serial position in the permutation. The list of all $N!$ possible permutations permits a description of a restricted number of the permutations which are immediate *neighbors* to each permutation, and those which are distant from it. Small alterations in the transformation rules would move between neighboring permutations. (2) The possible patterns of pure resonant vibrations of a freely suspended plate can be described by one or another member from an infinite series of functions

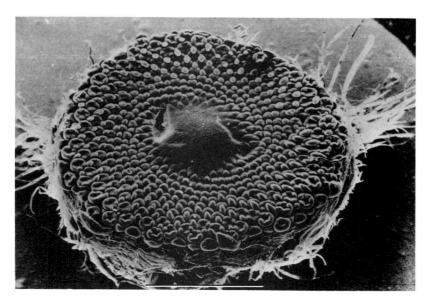

Fig. 2. Scanning electron microscope image of the head of a sunflower, showing 55 : 89 phylotaxis. Bar = 1000 μm. (From R. Erikson.)

(eigenfunctions) describing the standing waves on the plate. Smooth changes in the boundary and initial conditions of the plate (i.e. its shape and initial motion) typically lead to a smooth change in the functions describing its vibrations to a 'neighboring' pattern.

Potential examples of structuralist constraints

Phylotaxis

The phenomenology associated with phylotaxis (Fig. 2) has received rich attention. Pinecones have left- and right-handed spirals of scales. The number of left-handed rows is related to the number of right-handed rows as members of the well-known Fibonacci series: 1,1,2,3,5,8,13,21 . . . , in which the last digit is the sum of the two previous digits. A variety of accounts of this phenomenon have been offered (Church, 1904; Turing, 1959; Adler, 1977; Mitcheson, 1977; R. O. Erikson, personal communication). One proposes that close packing of scale primordia on a conical meristem, guided by diffusion of inhibitors and tissue expansion, suffices to generate the observed Fibonacci ratio of left- and right-handed spirals (Mitcheson, 1977). According to this account, changing the size of primordia with respect to the conical meristem by altering diffusion constants or tissue growth will yield transformation from one to another pair of the Fibonacci series. The issue here is not whether this hypothesis is correct but its flavor. It delimits the possible morphologies open to this tissue to the set of Fibonacci numbers. In contrast to the 'canalizing' tradition in quantitative genetics, which might account for the quixotic skipping from 5–8 to 8–13 pairing seen in pinecones in terms of arbitrarily crowded and spaced thresholds, this hypothesis attempts to afford a natural explanation of a discrete spectrum of phenotypes under a simple transformation rule. It leads to the expectation that, as pinecones evolve, it will be relatively easy to change to neighboring pairs in the Fibonacci series, but difficult, say, to transform to radial rays from top to base.

In a similar vein, Harris & Erikson (1980) have recently provided an account, applicable to microtubules, bacterial flagellae, viral capsids, and other biological structures which are tubular packings of subunits, which describes the possible paths of continuous and discontinuous transitions between close packing arrangements as tube length contracts. Again, discrete alternatives, with restricted transition pathways among them, emerge.

Echinoid morphology

Raup (1968), in a theoretical analysis of echinoid skeletal growth patterns, shows that close packing properties somewhat analogous to two-dimensional soap bubble arrays may account for skeletal plate shapes, their arrangements in columns, and transitions in arrangements, again demarking the world of neighboring morphologies. As D'Arcy Thompson (1942) summarized, on a variety of size scales, general mathematical properties of close packing may constrain organic forms.

Shell patterns

The precise, geometrical pigment patterns on gastropod shells arise from the coordinated activities of secretory cells along the length of the mantle organ as it synthesizes shell (Fig. 3). Recently, J. H.

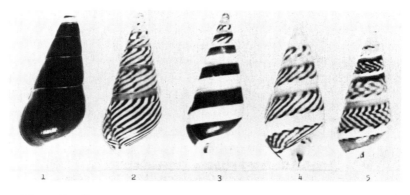

Fig. 3. Banding patterns on *Bankivia fasciata*. (From J. H. Campbell.)

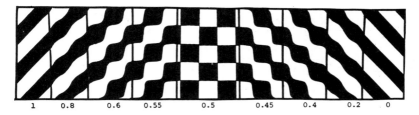

Fig. 4. Standing wave pattern produced by two sine waves of unequal amplitude. $x = a\sin(\theta + T) + b\sin(\theta - T)$, setting a and b to ratios equal to 1,0.8 . . . 0.2,0. Horizontal (θ) axis, and vertical (T) axis are interpreted as length along mantle and length perpendicular to mantle in shell deposition in *Bankivia*. (From J. H. Campbell.)

Campbell (personal communication) has formulated an elegant theory in which coordination is mediated by developmental signals existing as propagating or standing sinusoidal waves, which migrate along the mantle, creating domains by activating pigment-secreting cells as they pass (Fig. 4). The theory appears to account for the overlapping subsets of 10–15 major pattern forms, and the observed transitions between them. Again, the issue is not whether Campbell's account is correct but its flavor. In accounting for the observed patterns in detail with a few simple sinusoidal wave components, it affords a theory which automatically generates a constrained family of neighboring patterns, delimited both from inaccessible alternative neighboring patterns and from very distant patterns whose generation would require drastic modifications of

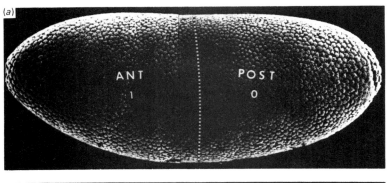

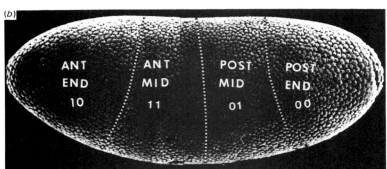

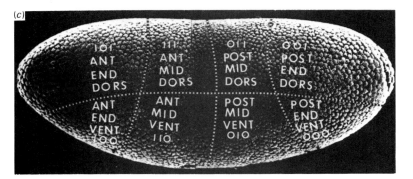

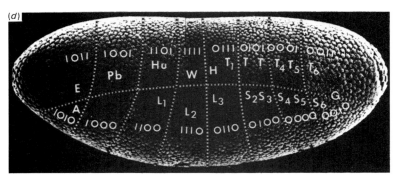

the generative algorithm. Therefore, it leads to restricted expectations about the evolution of neighboring patterns.

Combinatorial code – sequential compartmentalization model of *Drosophila* commitments

Another example of a structuralist theory in development is our own work on *Drosophila* pattern formation (Kauffman, 1973; Kauffman *et al.*, 1978). Somatic commitments to alternate segmental commitments arise early in development. The theory postulates that the egg is first subdivided into anterior and posterior compartmental domains (García-Bellido, 1975), reflecting heritable 'anterior' and 'posterior' somatic commitments (Kalthoff, 1979; Kauffman, 1981) (Fig. 5a). It is further sequentially subdivided into 'end' v. 'middle' commitments, then into 'dorsal' v. 'ventral' commitments, and 'even' v. 'odd' commitments, such that each terminal compartment carries a unique combination of commitments (Fig. 5b–d). Thus, the combination 'anterior', 'middle', 'dorsal', 'even' is an epigenetic code word signifying the dorsal mesothoracic segment region.

This combinatorial theory is explicitly structuralist in character. It defines a set of possible code words corresponding to a finite set of alternative developmental decisions, and a constrained set of transformations among them. Thus, it defines a meaning for 'neighboring developmental programmes' as pairs whose combinatorial code words differ by a single binary decision. For example, antenna and genital, although far apart on the blastoderm fate map, differ by only one, anterior v. posterior, commitment. These properties allow this combinatorial model to account for a fairly significant number of metaplastic transformations seen in

Fig. 5. (*a*)–(*d*) Hypothetical sequential compartmentalization of *Drosophila melanogaster* embryos into 16 cell domains creating a unique combination of binary alternative commitments in each domain which constitute 'epigenetic code words'. In 5d letters show blastoderm fate map positions of anlagen for adult ectodermal structures: E, eye; A, antenna; Pb, proboscis; L1, L2, L3, first to third leg; Hu, humerus; W, wing–mesothorax; H, haltere; T1–T6, abdominal tergites; S2–S6, abdominal sternites; G, genitalia; ANT = anterior; POST = posterior.

homeotic mutants, as well as transdetermination, including homeotic mutants causing transformation between distant areas on the
blastoderm fate map, coordinated homeotic transformations of
domains sharing the same commitment to its binary alternate, and
generation of mirror symmetrical mutants on length scales ranging
from $\frac{1}{2}$ egg to within a segment (Kauffman, 1981). I will suggest
below that this kind of putative combinatorial character to developmental programmes may be almost unavoidable, and reflect a very
general principle of genomic dynamical organization, which acts as
a constraint in evolution.

Correlated with the combinatorial model described above, my
colleagues and I (Kauffman *et al.*, 1978) have proposed a modified
form of the Turing model to account for the positional system
which triggers sequential developmental commitments. We suppose that a system coupling reaction and diffusion occurs in the
egg, and spontaneously establishes chemical gradients whose peak
to trough length depends upon parameters such as diffusion
constants. We assume that these diffusion constants gradually
decrease, such that the chemical system establishes in the egg a
succession of differently shaped gradients of the same chemicals.
We also assume that each pattern has a single threshold level which
triggers alternative compartmental commitments: 'anterior' v.
'posterior', 'end' v. 'middle', 'dorsal' v. 'ventral', 'even' v. 'odd'
(Fig. 6a–d).

While there are both grounds to support this model, and reasons
to think it weak (Kauffman, 1981; Sander, 1981), I describe it
briefly here because of its structuralist approach. The model relates
the possible shapes of gradients to eigenfunctions of the Laplacian
operator on the physical domain in question. It is not an arbitrary
model. It specifies the expected neighboring gradient shapes, and
sequences of shapes, which might be expected if the domain or
other parameters change, and hence leads to constrained evolutionary
expectations about altered spatial and temporal patterns of compartmentation and commitments.

Recently, Murray (1981) has proposed a related reaction–diffusion model to account for stripe and patch coat color markings on a
variety of mammals ranging from the honey badger, white on the
dorsal half and black on the ventral half (Fig. 7a), and the Valais
goat, black in the anterior half and white in the posterior half
(Fig. 7b), to zebras and giraffes. Like the *Drosophila* model, it

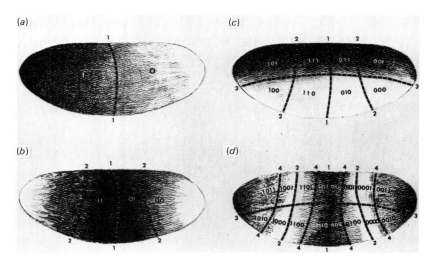

Fig. 6 (a)–(d) Reaction–diffusion model generating a temporal sequence of differently shaped chemical gradient patterns of the same morphogens. A single threshold level triggers the alternative heritable commitments shown in Fig. 5a–d. 0000, 0110, etc. are shorthand re-statements of binary code in Fig. 5a–d.

Fig. 7. (a) Honey badger, (b) Valais goat (From Murray, 1981).

characterizes a constrained set of expected patterns, and expected transitions between them; for example, a tendency for 'spot' patterns to transform distally to stripe patterns on narrowing conical domains such as tails.

These examples illustrate a simple point. The true biological neighborliness of two developmental patterns is not established by raw comparison of the patterns, but implicitly refers to changes required in the developmental mechanism which generates them. The examples emphasize that 'neighbors' may appear quixotically distant until the right generative algorithm is found. Alternatively, as in the observed string and shell pattern examples generated utilizing a few Fourier components, apparently neighboring patterns may not be accessible because of constraints in the generative algorithm. On the other hand, the existence of a generative algorithm *per se* does not imply constraints, as the example of the loose string deformed arbitrarily using all possible Fourier components shows. Once found, the law-like properties of the algorithm *may* define a restricted family of neighboring patterns and expected transitions among them. Properties of close packing, restricted eigenfunction patterns, topological constraints in phase resetting patterns of biological clocks (Winfree, 1980) and of pattern duplication and regeneration (Lewis, 1981), combinatorial properties in developmental programmes, and restricted patterns of bifurcations in non-linear dynamical systems more generally, seem likely sources of ontogenetic constraints. In short, when we genuinely understand the control of shapes, patterns and morphologies, at least some evolutionary transitions between a restricted family of genuinely neighboring forms should fall out as predictions, not mere arbitrary empirical correlations. In this sense, the concept 'developmental constraints' connotes the relative ease of transformation in evolution to a *delimited* set of neighboring organisms. Such constraints are internal factors in evolution channeling the progression of forms.

Ensemble constraints: evolutionarily stable gene regulatory networks

I turn now to a different sense of 'developmental constraint' which may play a critical role in evolution and has received little attention. The detailed mechanisms of gene regulation in eukaryotes are not yet known. Points of regulation include transcription to heterogeneous nuclear RNA, splicing, capping, polyadenylation and transport of mature messenger RNA to the cytoplasm, initiation and termination of translation, and post-translational modifications

(Brown, 1981). Among these, it seems clear that the early expectation of genetic cis-acting sites and trans-acting sites is substantiated. Families of such genetic loci have now been found in yeast, mouse, *Drosophila*, and maize (McClintock, 1957; Sherman & Helms, 1978; Dickenson, 1980*a, b*; Errede *et al.*, 1980). Furthermore, it is now widely appreciated that tandem duplications have arisen in evolution (Rubin *et al.*, 1976; Wilson *et al.*, 1977; Livak *et al.*, 1978), while recent evidence demonstrates that fairly rapid dispersion of some genetic elements occurs through chromosomal mutations (Bush, 1980) including transpositions, translocations, inversions, and recombination (Dooner & Nelson, 1977; Dover, 1979; Young, 1979; Bush, 1980). Dispersal of loci by processes such as these provide the potential to move cis-acting sites to new positions and thereby create novel regulatory connections, opening novel evolutionary possibilities (Krone & Wolf, 1977; Sherman & Helms, 1978; Fritsch *et al.*, 1979; Bush, 1980; Errede *et al.*, 1980). This raises a general question: if duplication and dispersion occur with characterizable probabilities per site, then in the absence of selection, is it possible to build a statistical theory of the expected control structure of a genetic regulatory system after many such transformations?

An initial extremely simple approach to this complex question is illustrated in Fig. 8*a*. Here I have assumed that the genome has only four kinds of genetic elements, cis-acting (Cx), trans-acting (Tx), structural (Sx) and empty (–). All types of elements are indexed. Any cis-acting site is assumed to act in polar fashion on all trans-acting and structural genes in a domain extending on its right to the first blank locus. Each indexed trans-acting gene, Tx, regulates all copies of the corresponding cis-acting gene, Cx, wherever they may exist on the chromosome set. Structural genes, Sx, play no regulatory roles. In Fig. 8*a* I have arrayed 16 sets of triads of cis-acting, trans-acting and structural genes separated by blanks on four 'chromosomes'. A graphical representation of the control interactions among these hypothetical genes is shown in Fig. 9*a*, in which an arrow is directed from each labelled gene to each gene which it affects. Thus, in Fig. 9*a*, C_1 sends an arrow to T_1 and an arrow to S_1, while T_1 sends an arrow to C_1. A similar simple architecture occurs for each of the 16 triads of genes, creating 16 separate genetic feedback loops. By contrast, in Fig. 8*b* each triad carries the indexes (Cx, $Tx + 1$, Sx), while the sixteenth

(a)

CHROMOSOME 1   *C1 T1 S1–C2 T2 S2–C3 T3 S3–C4 T4 S4–*

CHROMOSOME 2   *C5 T5 S5–C6 T6 S6–C7 T7 S7–C8 T8 S8–*

CHROMOSOME 3   *C9 T9 S9–C10 T10 S10–C11 T11 S11–C12 T12 S12–*

CHROMOSOME 4   *C13 T13 S13–C14 T14 S14–C15 T15 S15–C16 T16 S16–*

(b)

CHROMOSOME 1   *C1 T2 S1–C2 T3 S2–C3 T4 S3–C4 T5 S4–*

CHROMOSOME 2   *C5 T6 S5–C6 T7 S6–C7 T8 S7–C8 T9 S8–*

CHROMOSOME 3   *C9 T10 S9–C10 T11 S10–C11 T12 S11–C12 T13 S12–*

CHROMOSOME 4   *C13 T14 S13—C14 T15 S14—C15 T16 S15—C16 T1 S16—*

Fig. 8. (*a*) Hypothetical set of 4 haploid chromosomes with 16 kinds of cis-acting (C_1, C_2, . . .), trans-acting (T_1, T_2, . . .), structural (S_1, S_2, . . .) and 'blank' genes arranged in sets of 4 loci C(x), T(x), S(x),–. See text.
(*b*) Similar to (*a*), except the triads are C(x), T(x + 1), S(x),–. See text.

is (C_{16}, T_1, S_{16}). This permutation yields a control architecture containing one long feedback loop (Fig. 9*b*).

To begin to study the effects of duplication and dispersion of loci on such simple networks, I ignored questions of recombination, inversion, deletion, translocation, and point mutations completely, and modelled dispersion by using transposition alone. I used a simple programme which decided at random for the haploid chromosome set whether a duplication or transposition occurred at each iteration, over how large a linear range of loci, between which loci duplication occurred, and into which position transposition occurred.

Even with these enormous simplifications, the kinetics of this system is complex and scantily explored. Since, to simplify the model, loci cannot be destroyed, the rate of formation of a locus depends upon the number of copies already in existence, and

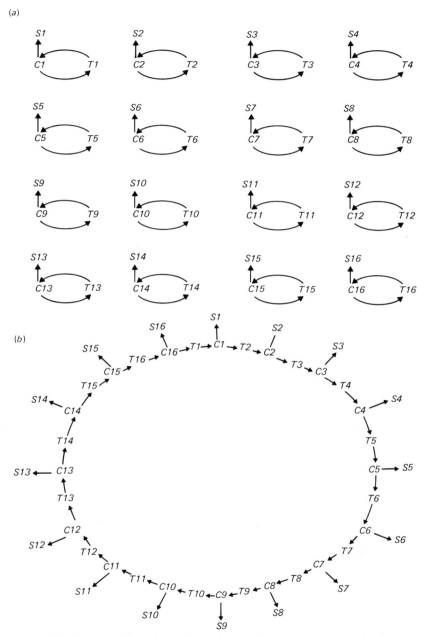

Fig. 9. (a) Representation of regulatory interactions according to rules in text, among genes in chromosome set of Fig. 8a. (b) Regulatory interactions of chromosome set in Fig. 8b.

approximately stochastic exponential growth of each locus is expected. This is modified by the spatial range of duplication, which affords a positive correlation of duplication of neighboring loci, and is further modified by the frequency ratio of duplication to transposition. The further assumption of first-order destruction of loci would decrease the exponential growth rates. However, the kinetics are not further discussed here, since the major purpose of this simple model is to examine the regulatory *architecture* after many instances of duplication and transposition have occurred. I show the results for the two distinct initial networks in Fig. 10*a*, *b* for conditions in which transposition occurs much more frequently than duplication, 90 : 10, and 2000 iterations have occurred. The effect of transpositions are to randomize the regulatory connections in the system. Consequently, while the placement of individual genes differ, the overall architectures of the two resulting networks look fairly similar after adequate transpositions have occurred.

This result raises the possibility of studying more formally the structural properties of such randomized regulatory networks to seek their statistically stable features. Conceptually, this study would construct the ensemble of all possible regulatory networks with a specified number of copies of each type of cis-acting, trans-acting and structural gene, then form averages, over the ensemble, of those network properties which are of interest.

Random directed graphs

Even the simple model in Fig. 8*a*, *b* is complicated. The effects of duplication and transposition are to create new copies of old genes, and by their dispersal to new regulatory domains, generate both more, and novel, regulatory couplings among the loci. A minimal initial approach to the study of the ensemble properties of such systems is to study the features of networks in which N genes are coupled completely at random by M regulatory connections. This kind of structure is termed a randomly directed graph (Erdos &

Fig. 10. (*a*) Regulatory interactions from chromosome set in Fig. 8*a* after 2000 transpositions and duplications have occurred, in ratio 90 : 10, each event including 1 to 5 adjacent loci.

(*b*) Similar to (*a*), after random transposition and duplication in chromosome set from Fig. 8*b*.

(a)

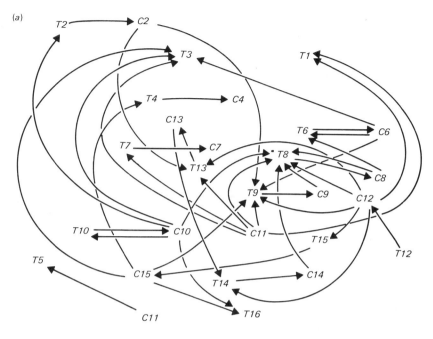

(b)

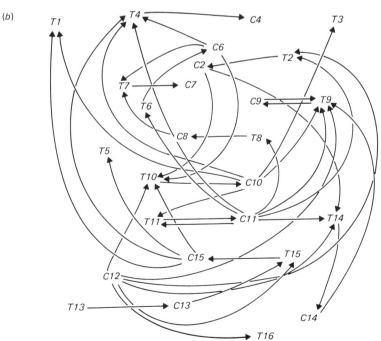

(a)

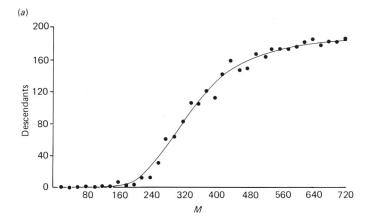

(b)

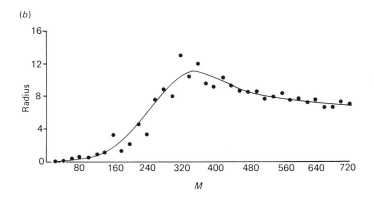

(c)

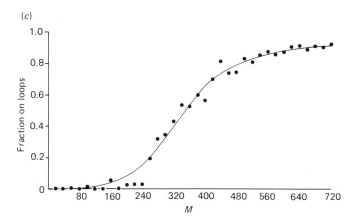

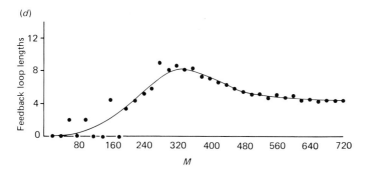

Fig. 11. (*a*) Descendant distribution, showing the average number of genes each gene directly or indirectly influences as a function of the number of genes (200) and regulatory interactions (*M*).

(*b*) Radius distribution, showing the mean number of steps for influence to propagate to all descendants of a gene, as a function of *M*.

(*c*) Average number of genes lying on feedback loops, as a function of *M*.

(*d*) Average length of shortest feedback loops that genes lie on, as a function of *M*.

Renyi, 1959, 1960; Berge, 1962) in which nodes represent genes and arrows represent regulatory interactions.

To analyze such randomly directed graphs, I employed a computer algorithm which generated at random *M* ordered pairs chosen among *N* 'genes', assigned an arrow running from the first to the second member of each pair, then analyzed the following features of the resulting directed graph: (1) The number of genes directly or indirectly influenced by each single gene, termed the *descendants* from each gene; (2) the *radius* from each gene, defined as the minimum number of steps for influence to propagate to all its descendants; (3) the fraction of genes lying on genetic feedback loops among the total *N* genes; (4) the length of the smallest feedback loop for any gene which lies on a feedback loop.

As would be expected, these properties depend upon both the number of genes, *N*, and the number of regulatory arrows, *M*, which connect them. Figure 11*a*–*d* shows the results for networks with 200 genes and regulatory connections, *M*, ranging from 0 up to 720. Figure 11*a* shows the mean number of descendant genes in a network of 200 genes. As *M* increases past *N*, a crystallization of large descendant structures occurs, and some genes begin directly or indirectly to influence a large number of other genes. The curve is sigmoidal, and by $M = 3N$, on average any gene can directly or indirectly influence 0.87 of the genes. Figure 11*b* shows the mean radius as a function of M. It is a non-monotonic function of the number of regulatory connections, low when connections are few and each gene can influence all of its descendants in a few steps, maximum when the number of connections is between $1.5N$ and $2N$, and each gene can influence a large number of other genes through a still relatively sparse network, then declining as additional regulatory connections provide shorter routes from each gene to influence all its descendants. As would be intuitively expected, the fraction of genes lying on feedback loops parallels the mean number of genes descendant from each gene. As M increases, the chance of forming closed loops goes up (Fig. 11*c*). Similarly, the lengths of the smallest feedback loops on which genes lie parallels the radius distribution, reaching a maximum for M between $1.5N$ and $2N$, then declining as M increases.

This brief analysis of the structural properties of random genetic networks suggests how robust such statistically typical features are expected to be. These first, simplest models of the expected regulatory architecture among the genetic elements in a eukaryote under the actions of duplication, transposition, translocation, recombination, deletion, and point mutations, of course, are no more than the schemata of a theory. Their purpose, at this stage, is to indicate that such a theory should be constructable, and indeed would characterize the statistically robust structural properties of large, evolving genetic regulatory networks in the absence of selection. Simultaneously, such a theory would demonstrate the form of genetic network toward which those in organisms would fall in the absence of selection, and perhaps allow an assessment of how strong selection would have to be to cause a difference of given degree from the average. In short, the mean statistical features of such networks are constraints in the evolution of the genomic regulatory system which selection must continuously overcome.

Genetic regulatory networks are of importance because their organized dynamical behavior coordinates the activities of genes and their products in the regulation of cell differentiation. If it is possible to envision a theory which characterizes the expected *architecture* of gene regulatory systems under the actions of duplication, transposition, recombination, deletion, and point mutations, in genomic evolution, then it becomes of interest to assess the expected dynamical behavior of such regulatory systems. My own previous work (Kauffman 1969, 1971, 1974), as well as that of a number of other authors (Walker & Ashby, 1966; Newman & Rice, 1971; Glass & Kauffman, 1972, 1973; Aleksander, 1973; Babcock, 1976; Cavender, 1977; Thomas, 1979; Sherlock, 1979a, b; Walker & Gelfand, 1979, and personal communication; A. E. Gelfand, personal communication; F. Fogelman-Soulie, E. Goles-Chacc & G. Weisbuch, personal communication) has already begun to shed some light on this problem. Again, analysis has largely been carried out only for the simplest kinds of dynamical models of regulatory systems, in which each gene is assumed to be a simple binary, on–off device, regulated by input arrows from other binary genes in a random genetic network, in which M arrows ($M = 2N$) connect N genes. The behavior of each binary gene is governed by randomly assigning to it one of the possible binary Boolean 'switching' rules (Kauffman, 1969, 1971, 1974). For example, a given gene might be active at the next moment if any of its cis- and trans-acting regulatory genes are currently active or if any is inactive. Although formulated in terms of transcriptional regulation, the approach can include regulation of transcript processing and transport, by cis-acting sequences on transcripts, and trans-acting intranuclear signals (Davidson & Britten, 1979).

The quite surprising result of these studies is that such randomized genetic networks, rather than exhibiting uselessly disordered behavior, spontaneously exhibit extraordinarily constrained, ordered dynamical properties, reminiscent in many respects to those found in contemporary cells. First, among the N genes, approximately 60–70% settle into fixed 'on' or fixed 'off' states. The fundamental consequence of this is simply pictured. This 60–70% constitutes a large 'subgraph' of the entire genetic network, whose elements are fixed active or inactive. This large 'forcing structure' subgraph blocks the propagation of varying inactive and active 'signals' from genetic loci which are not part of the large dynamically fixed forcing structure. Those remaining genes form smaller interconnected clusters, each of which is *functionally* isolated from influencing the other clusters by the large forcing structure. Each functionally isolated subsystem has a small number of alternative dynamical modes of behavior into which it settles: alternate steady states or oscillatory patterns of gene activities among the genes in that cluster. Since the clusters are functionally isolated, the alternative dynamical properties of the entire genetic system are constrained to the *possible combinations* of the *J* alternative modes of behavior of the first cluster, the *K* modes of the second functionally isolated cluster, the *L* modes of the third . . .. That is, such networks inherently exhibit *combinatorial* behaviors, comprising possible combinations of the alternate modes of the functionally isolated subsystems.

This allows initial construction of a picture of the coordinated gene activities corresponding to a cell type. The genes fixed in the active state would correspond to and satisfy the need in genetic reulatory systems for housekeeping genes ubiquitously active in all cell types (Kleene & Humphreys, 1977; Chikaraishi *et al.*, 1978; R. Goldberg, personal communication). Each different cell type would comprise a unique combination of the alternative modes of behavior of the functionally isolated subsystems. This picture is clearly reminiscent of the combinatorial code model for *Drosophila* commitments described above, and similar combinatorial properties in early mouse embryogenesis (Gardner, 1978). Although space does not permit an adequate discussion here, it seems likely that such combinatorial properties may both crystallize spontaneously, and also be critical for the further selective evolution of cell types. Selective evolution requires the capacity to accumulate partial successes sequentially. Were the genome organized such that a small change in connections could alter coordinated dynamical patterns of gene activities throughout the network, preservation of past favorable combinations of activities would be difficult. Accumulation of partial successes requires either genuinely isolated subsystems, hard to maintain in a scrambling genome, or functionally isolated subsystems which are otherwise coupled, as arise inevitably in these model genomes. Selective modification of the combinations of gene activities in one functionally isolated subsystem would not alter the dynamics of the remaining system, hence allows piecewise evolution of favorable new cell types. This suggests that combinatorial properties, resulting from functionally isolated subsystems of genes, may be common in differentiation. If so the structuralist 'combinatorial code' theme with restricted neighboring developmental programmes may be widely relevant and reflect a basic developmental constraint in metazoan evolution.

The analysis of the typical dynamics in this class of model genetic regulatory systems can be utilized to calculate a number of expected properties which are then predictions, many of which are now open to test: (1) How many genes are expected to be ubiquitously expressed in common on all cell types in an organism? (2) How different are the gene expression patterns of any two cell types expected to be? (3) If a single gene's activity state is temporally reversed by inducing stimuli, how many 'descendant' genes will typically alter their activities? How many steps will be required for this influence to propagate? (4) After such transient reversals of activity of any single gene by internal noise or external signals, how often do cell types stably return to their former pattern of gene activities? How often are they induced to undergo differentiation transitions to other cell types? (5) How many subsequent cell types can any single cell type differentiate into by transiently signaling and regulating the behavior of any single gene? (6) If a regulatory gene is deleted from the network, or a new (e.g. viral) gene inserted, typically how many genes will alter their dynamical behavior? How many cell types will be altered? Will new cell types be created? How similar are such novel cell types to cell types existing in the dynamical repertoire of the genomic network prior to the mutation? (7) In the pathways of differentiation transitions among cell types, are there typically 'isolated' cell types which cannot be 'reached' from any remaining cell type by differentiation signals? Such cell types could constitute atavisms, like teeth in birds (Kollar & Fisher, 1980), or preadaptations available in evolution for utilization by opening new pathways to reach them.

Without redescribing here in more detail the behaviors of these model systems, which have been discussed elsewhere in depth (Walker & Ashby, 1966; Kauffman, 1969, 1971, 1974; Sherlock, 1979a, b; Walker & Gelfand, 1979, and personal communication; A. E. Gelfand, personal communication), I would again stress two points. First, construction of such ensemble theories is possible and should become increasingly realistic as we learn more about genetic regulatory mechanisms and the modes of genomic evolution. Secondly, insofar as duplication and dispersion of genetic loci throughout the genome tend toward a particular statistically robust class of regulatory architectures, their typical structural and typical dynamical features constitute developmental constraints in evolution which may be very powerful. Very powerful selection might be

required to deviate far from 'typical'. It is, therefore, not unin-
teresting that many of their typical dynamical properties already
appear to be fairly similar to those which actually appear to occur in
contemporary metazoans and already provide many of the features
required for adaptive selection of new cell types within those
constraints. Perhaps the constraints have in fact constrained.

Orthogenesis: asymmetric ontogenetic changes induced by mutations

The last topic I want to discuss is orthogenesis, a vision of
directionality in evolution cast out as a vitalistic expression of
purpose by nature. Yet the issue is not vitalism. Briefly stated, the
question is whether the alterations in ontogeny caused by the next
'random' mutation can be biased in particular directions, leading to
an average tendency to change ontogenies in oriented ways.

   An example chosen from the analysis of randomized genetic
networks suggests that oriented ontogenetic responses to random
mutations may be typical. In the network models, each cell type
consists of one recurrent pattern of activities of all the genes.
Differentiation is modeled as transitions between cell types triggered by signals which transiently alter the activity (on or off) of
any single gene. The networks show that, typically, any cell type is
stable to most such perturbations, but occasionally is triggered to
differentiate into a restricted number of other neighboring cell
types. If there are L cell types, an $L \times L$ matrix lists all L^2 possible
differentiation transitions between L cell types, and is typically a
sparse matrix with many o elements, and non-zero (i, j) elements
which show the probability of differentiating from the Ith to Jth
cell type.

   As noted above, a robust property of these model gene networks
is that 60–70% of the loci fall to fixed active or inactive states.
Therefore, deletion of a gene which is *inactive* in all cell types will
not alter the cell types of the system. However, such a deleted gene
may be transiently active in differentiation transitions between cell
types. Its deletion may change the $L \times L$ matrix of differentiation
probabilities. To examine this, I randomly removed any single gene
belonging to this class of fixed inactive genes, and examined the
$L \times L$ matrix. Interestingly, any such 'mutant' affected only a few
of the (non-zero) differentiation steps. Thus, this class of mutants

preserves cell types, yet causes selective alterations in some of the developmental pathways among them. The capacity for piecewise alteration of developmental pathways with preservation of favorable cell types must be critical in metazoan evolution, and be related to homoeotic mutants and re-expression of atavisms like birds' teeth. More surprisingly, however, most of the mutants in this class affected much the same small number of differentiation steps in the $L \times L$ matrix, and, typically, a given (i, j) element was more likely to increase than to decrease, or vice versa, in response to the next random mutation. Thus, the response to the next random mutant in this special class of mutants, is on average, an *oriented* change in the pathways of differentiation, increasing flow down some pathways, inhibiting others. Under the drive of random mutants, the responding system tends, non-isotropically, in a given direction. Furthering the parallel with homoeotic mutants, it is interesting that several of these in *Drosophila* transform eye to wing, none causes the reverse; several transform antenna to leg, none causes the reverse (Ouweneel, 1976). Mutants generating metaplasias between two tissues may commonly arise with asymmetric frequencies in the two opposite directions.

This result should not be surprising. Only ignorance in general would have persuaded us that the response of any integrated complex dynamical system to random alterations in its parameters or structure would be fully isotropic. The tendency toward asymmetry may be common, and impose an internal directional force of unknown magnitude in evolution. Thus, orthogenetic tendencies may be common, if perhaps hard to recognize.

Finally, it is at least interesting to wonder whether transitions from 'orthogenetic' asymmetries to their absence may also occur, and play a role in macroevolution. For example, it might be the case that, in the model gene systems, the cell differentiation matrix responds non-isotropically to random mutants in the class specified, until some 'stationary' region is found, where alterations become isotropic. Transitions from non-isotropic orthogenetic to isotropic responses might lead to marked differences in evolutionary rates, and hence play a role in radiation and stasis.

This work is partially supported by grants from A.C.S., numbers CD-30 and CD-149, NIH number GM22341 and NSF number PCM8110601–01.

References

Adler, I. (1977). The consequence of contact pressure in phylotaxis. *J. theor. Biol.* **65**, 29–77.

Alberch, P. (1982). Developmental constraints in evolutionary processes. In *Evolution and development*, ed. J. T. Bonner, pp. 313–32. Springer-Verlag: Berlin.

Alberch, P. & Alberch, J. (1981). Heterochronic mechanisms of morphological diversification and evolutionary change in the neotropical salamander, *Bolitoglossa occidentalis* (Amphibia; Plethodontidae). *J. Morph.* **167**, 249–64.

Aleksander, I. (1973). Random logic nets: stability and adaptation. *Int. J. Man/Machine Studies* **5**, 115–31.

Babcock, A. K. (1976). Logical probability models and representation theorems on the stable dynamics of the genetic net. PhD thesis, State University of New York, Buffalo.

Berge, C. (1962). *The theory of graphs and its applications*. Methusena: London.

Bolt, J. R. (1979). *Amphibamus grandiceps* as a juvenile dissorophid: Evidence and implications. In *Mazon Creek fossils*, ed. M. Nitecki, pp. 529–63. Academic Press: New York.

Bonner, J. T. (ed.) (1982). *Evolution and development*. Report of the Dahlem Workshop on Evolution and Development, Berlin 1981. Springer-Verlag: Berlin.

Brown, D. B. (1981). Gene expression in eukaryotes. *Science* **211**, 667–74.

Bush, G. L. (1980). In *Essays on evolution and speciation*. Cambridge University Press: Cambridge.

Cavender, J. A. (1977). Kauffman's square root law: possible correlations. *J. theor. Biol.* **65**, 791–3.

Chikaraishi, D. M., Deeb, S. S. & Sueoka, N. (1978). Sequence complexity of nuclear RNAs in adult rat tissues. *Cell* **13**, 111–20.

Church, A. H. (1904). *On the relation of phylotaxis to mechanical laws*. Williams and Nargate: London.

Davidson, E. H. & Britten, R. J. (1979). Regulation of gene expression: Possible role of repetitive sequences. *Science* **204**, 1052–9.

de Beer, G. (1958). *Embryos and ancestors*, 3rd edn. Oxford University Press: Oxford.

Dickenson, W. J. (1980a). Tissue specificity of enzyme expression regulated by diffusible factors: evidence in *Drosophila* hybrids. *Science* **207**, 995–7.

Dickenson, W. J. (1980b). Complex cis-acting regulatory genes demonstrated in *Drosophila* hybrids. *Dev. Genet.* **1**, 229–40.

Dooner, H. K. & Nelson, O. E. (1977). Controlling element-induced alterations in UDP glucose:flavonoid glucosyltransferase, the enzyme specified by the *bronze* locus in maize. *Proc. Natn. Acad. Sci.* **74**, 5623–7.

Dover, G. A. (1979). The evolution of DNA sequences common to closely-related insect genomes. In *Insect cytogenetics*, eds. R. L. Blackman, G. M. Hewett and M. Ashburner, pp. 174–93. Royal Entomological Society London Symposium no. 9. Blackwell: Oxford.

Dun, R. B. & Fraser, A. S. (1959). Selection for an invariant character, vibrissa number in the house mouse. *Aust. J. biol. Sci.* **12,** 506–20.

Eldredge, N. & Gould, S. J. (1972). Punctuated equilibria: an alternative to phyletic gradualism. In *Models in paleobiology*, ed. T. J. M. Schopf, pp. 82–115. Freeman, Cooper and Co.: San Francisco.

Erdos, P. & Renyi, A. (1959). In *On the random graphs 1*, vol 6, Inst. Math., Univ. De Breceniensis: Debrecar, Hungary.

Erdos, P. & Renyi, A. (1960). In *On the evolution of random graphs*, Publ. No 5, Math. Inst. Hung. Acad. Sci.

Errede, B., Cardillo, T. S., Sherman, F., Dubas, E., Deshuchamps, J. and Waine, J-M. (1980). Mating signals control expression of mutations resulting from insertion of a transposable repetitive element adjacent to diverse yeast genes. *Cell* **22,** 427–36.

Fraser, A. S. (1970). An epigenetic system. In *Towards a theoretical biology*, vol. 3, ed. C. H. Waddington, pp. 57–62. Edinburgh University Press: Edinburgh.

Fritsch, E. F., Lawn, R. M. & Maniatis, T. (1979). Characterization of deletions which affect the expression of fetal globin genes in man. *Nature* **279,** 598–603.

García-Bellido, A. (1975). Genetic control of wing disc development in *Drosophila*. In *Cell patterning*, Ciba Foundation Symposium No. 29, pp. 161–182. Associated Scientific Publishers: Amsterdam.

Gardner, R. L. (1978). The relationship between cell lineage and differentiation in the early mouse embryo. In *Results and problems in cell differentiation*, vol. 9, ed. W. J. Gehring, pp. 205–41. Springer: Berlin.

Glass, L. & Kauffman, S. A. (1972). Cooperative components, spatial localization and oscillatory cellular dynamics. *J. theor. Biol.* **34,** 219–37.

Glass, L. & Kauffman, S. A. (1973). The logical analysis of continuous non-linear biochemical control networks. *J. theor. Biol.* **39,** 103–29.

Goldschmidt, R. (1938). *Physiological genetics*. McGraw-Hill: New York.

Goldschmidt, R. (1940). *The material basis of evolution*. Yale University Press: New Haven, Conn.

Gould, S. J. (1977). *Ontogeny and phylogeny*. Belknap Press: Cambridge, Mass.

Harris, W. F. & Erikson, R. O. (1980). Tubular arrays of spheres: geometry, continuous and discontinuous contraction, and the role of moving dislocations. *J. theor. Biol.* **83,** 215–46.

Huxley, J. S. (1932). *Problems of relative growth*. Methuen: London.

Kalthoff, K. (1979). Analysis of a morphogenetic determinant in an insect embryo (*Smittia* spec., Chironomidae, Diptera). In *Determinants of spatial organization*, eds. St Subtelny and I. Konigsberg, pp. 97–126. Academic Press: New York.

Kauffman, S. A. (1969). Metabolic stability and epigenesis in randomly constructed genetic nets. *J. theor. Biol.* **22,** 437–67.

Kauffman, S. A. (1971). Gene regulation networks: a theory for their global structure and behavior. In *Curr. Top. devl. Biol.* **6,** 145–82.

Kauffman, S. A. (1973). Control circuits for determination and transdetermination. *Science* **181,** 310–18.

Kauffman, S. A. (1974). The large-scale structure and dynamics of gene control circuits: an ensemble approach. *J. theor. Biol.* **44,** 167–90.

Kauffman, S. A. (1981). Pattern formation in the *Drosophila* embryo. *Phil. Trans. R. Soc. Ser. B.* **295**, 567–94.

Kauffman, S. A., Shymko, R. M. & Trabert, K. (1978). Control of sequential compartment formation in *Drosophila*. *Science* **199**, 259–70.

Kleene, K. C. & Humphreys, T. (1977). Similarity of hRNA sequences in blastula and pluteus stage sea urchin embryos. *Cell* **12**, 143–55.

Kollar, E. J. & Fisher, C. (1980). Tooth induction in chick epithelium: expression of quiescent genes for enamel synthesis. *Science* **207**, 993–5.

Krone, W. & Wolf, V. (1977). Chromosome variation and gene action. *Hereditas* **86**, 31–6.

Lewis, J. (1981). Simpler rules of epimorphic regeneration: the polar coordinate model without polar coordinates. *J. theor. Biol.* **88**, 371–92.

Livak, K. J., Freund, R., Schweber, E. R., Wensink, P. C. & Meselson, M. (1978). Sequence organization and transcription at two heat shock loci in *Drosophila*. *Proc. natn. Acad. Sci.* **75**, 5613–17.

McClintock, B. (1957). Controlling elements and the gene. *Cold Spr. Harb. Symp. quant. Biol.* **21**, 197–216.

Mitcheson, G. J. (1977). Phylotaxis and the Fibonacci series. *Science* **196**, 270–5.

Murray, J. (1981). On pattern formation mechanisms for lepidopteran wing patterns and mammalian coat markings. *Phil. Trans. R. Soc. Ser. B.* **295**, 473–96.

Newman, S. A. & Rice, S. A. (1971). Model for constraint and control in biochemical networks. *Proc. natn. Acad. Sci.* **68**, 92–6.

Ouweneel, W. J. (1976). Developmental genetics of homeosis. *Adv. Genet.* **18**, 179–241.

Raup, D. M. (1968). Theoretical morphology of echinoid growth. *J. Paleont.* **42**, 50–63.

Rendel, J. M. (1967). *Canalization and gene control.* Logos Press: London; Academic Press: New York.

Rubin, G. M., Finnegan, D. J. & Hogness, D. S. (1976). The chromosomal arrangement of coding sequences in a family of repeated genes. *Prog. Nucleic Acid Res. Biol.* **19**, 221–6.

Sander, K. (1981). Pattern generation and pattern conservation in insect ontogenesis – problems, data and models. In *Forschritte der Zoologie*, vol. 26, pp. 101–19. Gustav Fischer Verlag: Stuttgart.

Sherlock, R. A. (1979a). Analysis of the behavior of Kauffman binary networks. I. State space description and the distribution of limit cycle lengths. *Bull. math. Biol.* **41**, 687–705.

Sherlock, R. A. (1979b). Analysis of the behavior of Kauffman binary networks. II. The state cycle fraction for networks of different connectivity. *Bull. math. Biol.* **41**, 707–24.

Sherman, F. & Helms, C. (1978). A chromosomal translocation causing over-production of iso-2-cytochrome *c* in yeast. *Genetics* **88**, 689–707.

Thomas, R. (1979). Kinetic logic: A Boolean approach to the analysis of complex regulatory systems. *Lecture Notes in Biomathematics* no. 29, ed. R. Thomas. Springer-Verlag: New York.

Thompson, D'A. T. (1942). *On growth and form*, 2nd edn. Cambridge University Press: Cambridge.

Turing, A. M. (1959). In *Alan M. Turing, mathematician and scientist*, S. Turing, pp. 67–74. Heffer: Cambridge.

Waddington, C. H. (1940). *Organizers and genes*. Cambridge University Press: Cambridge.

Waddington, C. H. (1957). *The strategy of the genes*. Allen and Unwin: London.

Walker, C. C. & Ashby, W. R. (1966). On temporal characteristics of behavior in certain complex systems. *Kybernetik* **3**, 100–8.

Walker, C. C. & Gelfand, A. E. (1979). A system theoretic approach to the management of complex organizations: management by exception, priority, and input span in a class of fixed-structure models. *Behav. Sci.* **24**, 112–20.

Webster, G. & Goodwin, B. C. (1982). The origin of species: A structuralist approach. *J. Social biol. Struct.* **5**, 15–47.

Wilson, A. C., Carlson, S. S. & White, T. J. (1977). Biochemical evolution. *A. Rev. Biochem.* **46**, 573–639.

Winfree, A. T. (1980). *The geometry of biological time*. Springer-Verlag: Berlin, New York.

Young, M. W. (1979). Middle repetitive DNA: A fluid component of the *Drosophila* genome. *Proc. natn. Acad. Sci.* **76**, 6274–8.

Comparative anatomy of cuticular patterns in the genus *Drosophila*

ANTONIO GARCÍA-BELLIDO

Centro de Biologia Molecular, Consejo Superior de Investigaciones Cientificas,
Universidad Autonoma de Madrid, Canto Blanco, Madrid – 34, Spain

#6435

The first question which the study of variation may be expected to answer relates to the origin of that discontinuity of which species is the objective expression. Such discontinuity is not in the environment, may it not then, be in the living thing itself? (Bateson, 1894, p. 17).

Summary

We have studied the microscopic organization of several cuticular patterns in head, thorax and abdomen in both sexes of imagos from 82 species of the genus *Drosophila* (subg. *Pholadoris, Dorsilopha, Sophophora* and *Drosophila*). There are quantitative and qualitative pattern differences between groups and species. The comparative analysis of qualitatively different patterns suggests that (1) there are discontinuous variations between groups and species; (2) the pattern variations cannot easily be correlated with the established phylogeny of the group, based on other characteristics; (3) sexual differences in certain patterns are exclusive of certain groups. We discuss possible developmental genetic mechanisms underlying these variations and the possible role of 'natural selection' in their modulation and fixation.

Introduction

For comparative anatomists of the last century it was obvious that ontogeny and evolution were intimately related. Since then we have learned that this relation is based on the transmission of the same hereditary material through somatic cell divisions and through generations and species. However, we still have very vague ideas of the underlying mechanisms that cause, between species, the conservation of certain characters and the variation of others.

Two general theories try to explain both stability and variation during evolution and development. For the former, adaptation is considered to be the moulding force for the selection and fixation of genotypes which generates new forms. In development, global properties, such as positional information, developmental constraints, etc. would constitute the modulating element to which genotypes respond to generate spatial and temporal differentiation. In both theories the genome is contemplated as a plastic and continuous source of variation.

In multicellular organisms, genetic variations are expressed at an epigenetic level, as the result of genetic and developmental interactions. Therefore the discontinuous phenotypes caused by mutations may simply reflect the existence of discontinuous sets of phenotypic responses. Therefore, the 'wildtype' alleles of these genes cannot be said to control the 'normal' phenotype. However, their allelic variations are precisely the material bases for these developmental differences. Moreover, in organisms where genetic analysis is more advanced, such as *Drosophila*, genes are known whose function is specifically to determine developmental alternatives, possibly controlling the activities of other genes. Thus developmental geneticists now envisage a more particulate picture of the genome: i.e. it consists of sets of genes with related functions, which, expressing themselves in individual cells, determine the organization of those cells into systems. It is in this sense that genes are considered to determine development and their allelic variations evolutionary diversity. To dismiss their specific role in both processes would be formally equivalent to ignoring the role of atoms in determining the specificity of molecules.

I will try to discuss how this concept of the genome can be applied to the analysis of morphological variations in species of *Drosophila*. This analysis is facilitated by the existence of a very consistent phylogenetic tree, based on cytogenetic, biochemical and molecular data. Thus, this comparative study is expected to give us some hints as to which are the constraints and the rules that underly morphological variation in the genus. It should also help developmental geneticists working on *Drosophila melanogaster* to gain a certain perspective in distinguishing the pecularities from the generalities of its development.

Genetic factors in phenotypic variation in *Drosophila melanogaster*

In an evaluation of phenotypic variations, we have to make the formal distinction between (*a*) variations in quantitative parameters, such as length, number of cells, etc., related to size and shape and (*b*) qualitative characters, such as appearance of different cell types or their spatial arrangement in patterns. The distinction between these two types of characters is necessarily subjective and, in principle, at the genetic level, both would be explained by variations in amount and time of gene action. However, in any particular species and because of variable growth conditions, quantitative characters are exposed to wide variations, but patterns remain extremely constant. Moreover, sporadic quantitative variations are as a rule not heritable, or are multigenic in origin, whereas pattern variations are usually based on allelic forms of discrete genes.

I will now summarize some modern notions in developmental genetics of *D. melanogaster* to serve as background information for the interpretation of phenotypic variations within the genus *Drosophila* (see García-Bellido & Ripoll, 1978, for references).

Genetic, cytological (chromomeres) and molecular (mRNA) data in *D. melanogaster* suggest that the number of genes (units of function) in this species is around 5000. Extrapolation to the cytological data of *D. hydei* and *Chironomus* sp. suggests a similar number of genes in these species. This number is only about double that calculated for bacteria. Thus, increase in morphological complexity may not be accomplished by correlative increase in the number of genes. Pattern variations most possibly originate by allelic changes in pre-existing genes or changes in their interactions.

Most of the mutations corresponding to insufficient function are recessive. The most extreme alleles of about 90% of the loci are lethal in homozygous condition, but they can be analyzed phenotypically in mitotic recombination clones. The analysis of lethals in clones of homozygous cells indicates that only 10% of the zygotic lethals are lethal in proliferating tergite epidermal cells; a higher fraction is lethal in proliferating cells of the imaginal discs and about 80% are lethal in the female germ line cells. These differences probably reflect the cellular genetic requirements of these systems, but indicate that most genes are needed in most cells. The

classical notion of stage and tissue specificity of genes, based on the
analysis of the developmental crises of lethals, needs revision,
because of the great importance of maternal gene products during
zygotic development. Thus the inference of sets of genes for
embryonic, larval and imaginal development is possibly an oversim-
plification.

Viable alleles of lethal loci or mutants in viable loci may have
morphological phenotypes. However, these loci form only a small
fraction. Most of the morphological mutants analyzed in mitotic
recombination clones show a cell autonomous phenotype. Only a
small proportion of the lethals studied in clones in somatic cells
have morphological phenotypes. We can classify mutants which
affect patterns into four types: (1) those which specifically suppress
pattern elements, like mutants in the *achaete–scute* gene complex,
which remove chaetae in certain positions; (2) those which add
elements, like *hairy, polychaetus, extra-macrochaetes*, etc., which
cause the appearance of extra chaetae in certain positions; (3) those
which substitute one pattern for another – included here are
homoeotic mutants like those of the *bithorax* gene complex, and
mutants which transform secondary sexual characters; (4) disrup-
tive mutants which disorganize patterns in many ways, changing
the orientation of the elements, removing entire regions of the
cuticle, causing extra growths, etc.

The last one usually has multiple pleiotropic effects and is the
most frequent type. Possibly the alterations are caused by non-
specific perturbations of growth processes. The three former types
seem to be specific in their effects. Different pattern mutations
affect only one subset of the pattern, leaving the rest unchanged.
Moreover, double mutant combinations are additive in their pattern effects. From the consideration of these mutants, patterns
appear, as in a kaleidoscope, as super-positions of features or of
types of elements, each independently affected by mutants in
certain loci. However, only in a few of the cases studied can we
state, with any degree of certainty, that the corresponding genes are
specifically related to the appearance or organization of these
elements.

These few cases correspond to some homoeotic and sex differentiation genes and to *achaete–scute*. Several arguments indicate
that these genes are gene complexes containing several related
functions. Mutations of both insufficiency and excess of function in

these complexes have systemic effects, i.e. they change the overall pattern or particular subsets of its elements, depending on the nature of the mutation. This behavior suggests that these genes control the activity of other genes which implement their systemic signals in developmental terms. Their activity is in turn controlled by other regulatory genes which define their spatial specificity of action (see Lewis, 1978; Baker & Ridge, 1980; Capdevila & García-Bellido, 1981; García-Bellido, 1981). We do not know how general is this complex hierarchical dependence between genes, nor whether we can extrapolate these conclusions to other genes involved in morphogenesis. It is important, nevertheless, to recognize that patterns can be modulated by the activity of a few, functionally related genes.

Patterns appear as discrete entities or fixed inventories of segments and compartments, i.e. they have distinct cell lineage origins. Since they appear in a metameric or reiterative way in the fly, it follows that patterns in different segments and compartments are developmentally homologous. But they are also genetically homologous because they can be changed into one another by mutation in specific homoeotic genes. Double mutant combinations of homoeotic genes cause additive pattern transformations suggesting that patterns are under the control of specific combinations of homoeotic genes. These inferences have evolutionary implications because we can envisage mutations causing pattern variations in specific compartments without affecting others. The same inferences can be applied to any developmental pathway which is controlled by the activity of few genes, like in the sexual dimorphic patterns which depend on sex differentiation genes.

Experiments in directed selection have established the existence of genetic modifiers, both major and minor in their effects, which can change patterns (chaeta patterns, for example, see Sondhi, 1963). Major modifiers for wing vein mutants (Thompson, 1973) have been identified which correct extra-vein or vein-interruption mutants in a generic way. These findings support the notion that small subsets of genes may have major effects on patterns.

Phylogenetic variations in the genus *Drosophila*

Descriptive anatomy of internal organs and of cuticular features in eggs, larvae, pupae and adults of species of the genus *Drosophila* have been reported by many authors (Sturtevant, 1942; Wheeler, 1949; Patterson & Stone, 1952; Okada, 1956; Throckmorton, 1962; Bock & Wheeler, 1972). We will consider in this survey only pattern variations in the cuticle of adults of both sexes. The wealth of types and arrangements of cuticular elements (chaetae, sensillae and trichomes) and the fact that they are single-cell derivatives permits. accurate and objective descriptions. Moreover chaetae are innervated elements (mechano- and chemo-receptors) which send axons to the central nervous system, and so are presumably important elements in sensorial perception. From each species, we have studied several specimens of both sexes, dissected into head, dorsal thorax, legs, and abdomen, mounted in Euparal and scored under the light microscope. The specimens were obtained from stock cultures maintained by L. Throckmorton in his laboratory. They correspond to 82 species from the major groups of the genus. In addition we have studied certain patterns mentioned in the text, under the dissecting microscope, in pinned specimens of species of other genera of Throckmorton's collection. The phylogenetic tree used as reference in Figs. 11 and 12 follows that of Throckmorton (1962), which takes into consideration the most comprehensive anatomical, cytogenetic and biochemical data available for all the groups. The inspection of that phylogenetic tree will permit the reader directly to compare our data with those of other authors for other anatomical features. We will follow the anatomical terminology of Ferris (1950) in the following description.

In this survey we will only consider those patterns which differ qualitatively between species. The sample studied is very small when we compare it with the 1500 or so species described just in the genus *Drosophila* (Wheeler, 1982). Although most of the major groups are represented, the pattern variations found in this study surely represent a minor part of the existing ones. Quantitative variations such as organ size, number of cells, number of cuticular elements, etc., and small variations in the distribution of the elements, are left for a more specialized report.

A few general considerations should precede the description of particular patterns. The species studied differ largely in size although they are more or less uniform within groups. Thus the *victoria* group contains the smallest flies and the *repleta* group the largest, those of the *melanogaster* group being intermediate. However, shapes are constant and individual organs proportional to body size, throughout the genus. There is a five-fold difference in the area of the wing between the smallest species studied, *D. latifasciaeformis*, and *D. gibberosa* (*annulimana* gp), the largest. This difference can be accounted for by both the number of cells (3 times) and the adult size of these cells (1.6 times). Flies with larger cells usually have larger cuticular elements. Flies with a higher number of cells also have more reiterative elements in a pattern (i.e. chaetae in tergites or in the notum). The patterned distribution of elements, however, remains very constant throughout the genus.

Head

Only two features in the chaeta pattern of the head are distinctly different in different species of the genus. In all the studied species of the *saltans* group there is a cluster or about ten chaetae in the prefrons (Pfr, Table 1) between the third joint of both antennae (Fig. 1*b*) which is absent in other species (Table 1). This pattern is within the anterior compartment of the head.

Another major pattern difference occurs in the premandibular plate (see PMN in Table 1) of the occipital region at both sides of the neck (Fig. 1*a*). The pattern consists of a group of scattered chaetae which appear only in the subgenus *Scaptodrosophila* (= *Pholadoris*) and in all the species of the subgenus *Sophophora* (Table 1).

Thorax

In the pre-episternum of the prothorax, close to the joint of the coxa, all the species studied of the subgenus *Scaptodrosophila* have a single long 'propleural' chaeta (Ppl Fig. 1*c*) which is absent in all other groups (Table 1).

The pattern of macrochaetae in the notum varies largely between different genera of acalipterate diptera (Fig. 2). In the genus *Drosophila* it is very constant. But in the subgenus *Scaptodrosophila* all the studied species have two long chaetae, of the macrochaete type, in the notum at both sides of the middle line, after the dorsocentral chaetae and close to the scutelar suture (prescutelar chaeta, Psct) (Fig. 1*d*, Table 1). Interestingly the same chaetae can be found in species of other genera related to *Drosophila* (Fig. 2).

	N	Pfr	PMN	Ppl	Psct	Com	MW×3	FeID	Ta II III	FeI	S×A	Ta I	St1	An(tr)	St6	T6(tr)	AG=T7	♂>♀ pigm
		Head					*Thorax*			*sex diff.*			*Abdomen*			*in mates*		
SCAPTODROSOPHILA																		
gp *victoria*	2	−	+	+	+	−	−	−	−	+	−	+	+	+	+	+	−	−
coracina	4	−	+	+	+	−	−	−	−	2	−	+	2	+	+	+	−	−
latifasciaeformis	1	−	+	+	+	−	−	−	−	−	−	−	−	+	+	+	−	+
DORSILOPHA																		
buscki	1	−	−	−	−	−	−	−	−	−	−	+	−	+	−	+	−	−
SOPHOPHORA																		
gp *obscura, obscura*	3	−	+	−	−	−	−	−	−	−	+	+	−	−	−	−	−	+
affinis	1	−	+	−	−	−	−	−	−	−	+	•	−	−	−	−	−	+
melanogaster, mel.	2	−	+	−	+	−	−	−	−	−	+	•	−	−	−	−	−	+
ananassae	2	−	+	−	−	−	−	−	−	−	+	•	−	−	−	−	−	+
montium	2	−	+	−	−	+	−	−	−	−	+	•	−	−	+	1	+	−
suzukii	1	−	+	−	−	+	−	−	−	−	+	•	−	−	−	+	+	−
willistoni	8	+	+	−	−	+	−	−	−	−	−	•	−	+	+	+	+	−
saltans, saltans	6	+	+	−	−	+	−	+	−	−	−	•	−	+	−	+	+	+
sturtevanti	1	+	+	−	−	+	−	−	−	−	−	•	−	+	−	−	+	−
DROSOPHILA																		
gp *virilis*	9	−	−	−	−	−	−	−	−	−	−	+	−	+	+	+	−	−
robusta	2	−	−	−	−	−	−	−	−	−	−	+	−	+	+	+	−	−
melanica	5	−	−	−	−	−	−	−	−	−	−	−	−	+	+	+	−	−
annulimana	1	−	−	−	−	−	−	−	+	−	−	+	−	+	+	+	−	−
mesophragmatica	2	−	−	−	−	+	−	−	+	−	−	+	−	+	+	+	−	−
repleta, mulleri	6	−	−	−	−	+	−	−	+	−	−	+	−	+	+	+	−	−
mercatorum	1	−	−	−	−	+	−	−	+	−	−	+	−	+	+	+	−	−
melanopalpa	4	−	−	−	−	+	−	−	−	−	−	+	−	+	+	+	−	−
hydei	3	−	−	−	−	−	−	−	−	−	−	×	−	+	+	+	−	−
acanthoptera	1	−	−	−	−	−	−	−	−	−	−	+	−	+	−	−	−	−
immigrans	6	−	−	−	−	−	+	−	+	−	+	+	+	+	+	+	+	+
funebris	2	−	−	−	−	−	−	−	−	−	−	+	−	−	+	+	−	+
quinaria	1	−	−	−	−	−	−	−	+	−	−	•	−	−	−	−	−	−
guarani	1	−	−	−	−	−	−	−	+	−	−	•	−	+	+	−	−	−
pallidipennis	1	−	−	−	−	−	−	−	+	−	−	+	+	+	+	+	+	−
tripunctata	1	−	−	−	−	−	−	−	−	−	−	•	−	+	+	−	−	−
tumiditarsus	1	+	−	−	−	−	−	−	−	+	+	+	−	+	+	+	+	−

Table 1. *Distribution of patterns in different subgenera (capitals),*
groups (gp) and subgroup of species

N: number of species studied; Pfr: Prefrontal chaetae; PMN: premandibular chaetae; Ppl: propleural chaeta; Psct: prescutelar chaeta; Com:
costa media with bracts; MW×3: wing margin "hexa-type" of pattern; Fe
I D: femur foreleg, 'Dornen'; Ta II III: tarsi in second and third legs; Fe
I: femur, foreleg; SxA: sexcomb area; Ta I: tarsi foreleg; St1: first
sternite with chaetae; An(tr): analia with trichomes; St6: sixth sternite;
T6(tr): sixth tergite with trichomes; AG = T7: genital arch similar to
seventh tergite; Pigm: pigmentation. Patterns (+: fully presented, •
slightly so) described in text. 2: only two of the studied species show the
pattern.

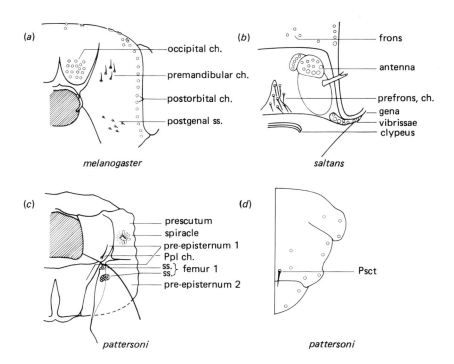

Fig. 1. Chaeta patterns in the head (*a, b*) and in the thorax (*c, d*) that show
variations in the genus *Drosophila* (see Table 1). The variable elements are solid
(black). ss: sensillae, ch: chaetae. Ppl: propleural, Psct: prescutelar.

236 *Antonio García-Bellido*

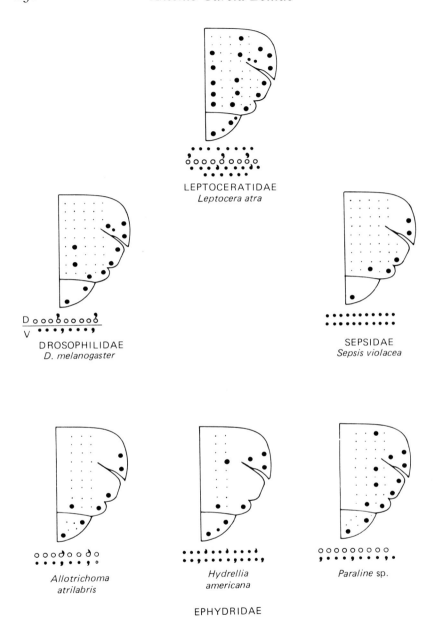

Fig. 2. Chaeta patterns in the notum and wing margin of species of Drosophilidae and other families of acalipterate diptera. Heavy dots: macrochaetae, small dots: microchaetae. Patterns of chaetae in the wing margin as in Fig. 4. D = dorsal chaetae; V = ventral chaetae.

In the wing, I will consider three major variations. In *D. melanogaster* the marginal chaetae of the proximal costa have a 'bract' near the base of their sockets. Bracts are trichome modifications, presumed to serve the adjacent mechano-receptor chaetae. This pattern is general in the genus, with the exception of species of the groups and subgroup of the subgenus *Sophophora* shown in Table 1 (Com), which have, in addition, a similar pattern of bracteated chaetae in the medial costa (Fig. 3) in both dorsal and ventral compartments.

The marginal chaetae of the anterior wing margin constitute what is designated as the 'triple row' in *D. melanogaster* (see Fig. 4). This pattern consists of three rows of chaetae: two in the dorsal compartment, the medial (M) and dorsal (D) rows of chaetae, and a single row of chaetae in the ventral compartment. However, the last of these rows contains two types of chaetae: the majority (V) are straight, appearing at intervals similar to those of the medial row, and another type, called here ventral-prima (V'V'') are slender and bent, appearing at intervals similar to those of the dorsal chaetae. In other species with the same formula, these V' chaetae appear separated from the ventral row, either between the medial and ventral rows, or outside the ventral row (Fig. 4). Dorsal and ventral-prima chaetae are, possibly, sensory elements different from medial and ventral chaetae (Palka *et al.*, 1979). This 'quadruple row' formula is the most common and known as the tetra-type. The *repleta* group has a different marginal chaeta pattern (Table 1) that we will call hexa-type. It has in addition to the two dorsal and two ventral rows, two extra rows of chaetae, similar to the ventral type, externally to the dorsal and ventral-prima rows. We have found only one species, *D. briegeri*, which has an intermediate formula (Fig. 4). It is interesting to note that the tetra- and hexa-types appear in other genera of Diptera (Fig. 2). The wing margin pattern of chaetae appears then as a mirror-image duplication of rows of chaetae at both sides of the dorso-ventral compartment border. This pattern has suffered small modifications in its elements and so the medial and ventral chaetae are usually different in size and shape. However, in *D. gibberosa* (*annulimana* gp) and in *Hydrellia americana* (Ephydridae) both rows have the type of chaetae found in the medial row.

Most of the species studied here have uniformly unpigmented, transparent wings. In some species of separated groups, spots of

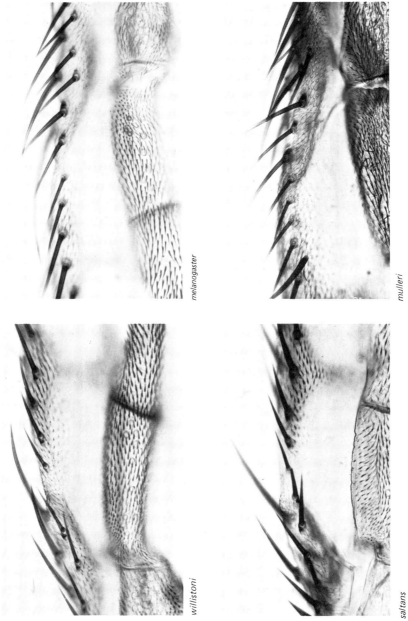

Fig. 3. Pattern of chaetae of the proximal and medial costa in different species. Note the accompaning 'bracts' in the chaetae of the proximal costa in *willistoni* and *saltans* and their absence in the medial costa of *melanogaster* and *mulleri*.

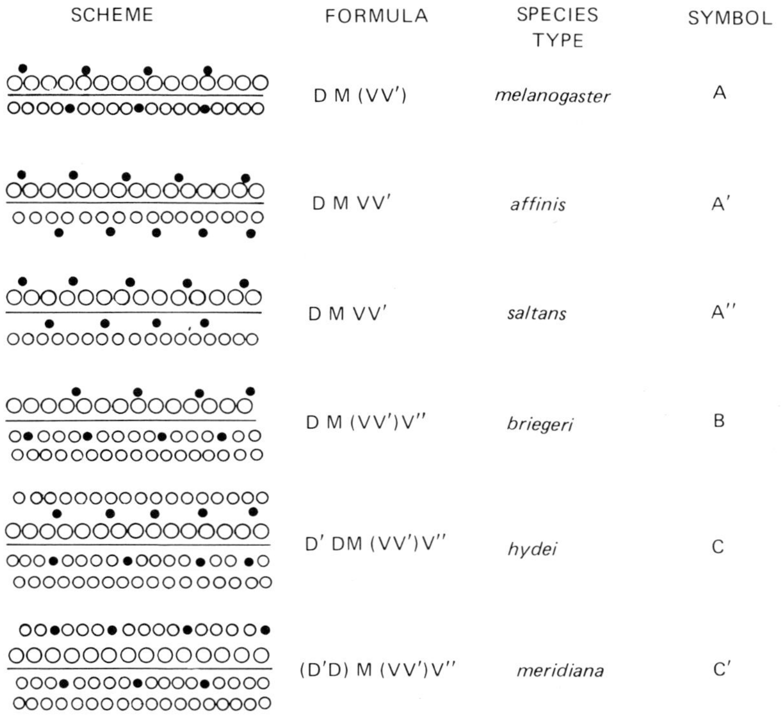

SCHEME	FORMULA	SPECIES TYPE	SYMBOL
	D M (VV')	*melanogaster*	A
	D M VV'	*affinis*	A'
	D M VV'	*saltans*	A''
	D M (VV')V''	*briegeri*	B
	D' DM (VV')V''	*hydei*	C
	(D'D) M (VV')V''	*meridiana*	C'

Fig. 4. Schemes and formulae of the chaeta pattern of the anterior wing margin of different species of the genus *Drosophila*. D and D': dorsal, M: medial and V, V' V'': ventral chaetae.

pigment appear in certain vein junctions (*D. subbadia, D. quinaria*) or intervein regions (*D. suzukii* ♂♂). The richest and most typical variations in wing pigmentation patterns are found in Hawaiian species (Hardy, 1966; Carson *et al.*, 1970).

All three pairs of legs have very complex chaeta patterns, which are very constant throughout the genus. Two major differences should be mentioned. In all the species of the *saltans* subgroup and in the *immigrans* group there is a row of very thick and bracteated chaetae in the femur ('Dornen') running along the anterior compartment (Fig. 5). These chaetae replace thin or average size chaetae in most other species (Fe I D, Table 1). The second variation relates to the pattern of chaetae running through the tarsi of the second and third pairs of legs at both sides of the anterior–posterior compartment boundary. The shape and organization of

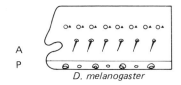

Fig. 5. Chaeta patterns in the distal femur. Full drawn chaetae are the variable elements. A = anterior compartment; P = posterior compartment.

these chaetae is very different from that of the general *D. melanogaster* type in all the species of the *immigrans* group and in the single species studied of the *quinaria, quarani* and *pallidipennis* groups which show variations to that pattern. *Drosophila gibberosa* again shows other differences in these patterns.

Abdomen

Possibly all the species of the genus have adult representatives of both dorsal and ventral anlagen of all the abdominal segments. These adult structures can be more or less conspicuous. There are, however, clear differences in the presence or absence of chaetae in the ventral derivatives (sternites) of different groups. The *D. melanogaster* type is represented in Fig. 6. There are neither sternital plates nor chaetae in the first segment. Some species of the subgenus *Scaptodrosophila* have in both males and females a first sternite with chaetae (Fig. 6; St1, Table 1).

Another clear pattern variation found in the genus consists of the appearance or absence of trichomes in the anal plates of both sexes. In *D. melanogaster* these plates are void of trichomes. However, as seen in Table 1 (An(tr)), this is rather the exception in the genus.

There are notorious differences within the genus with respect to the pattern of trichomes and pigmentation in the abdominal

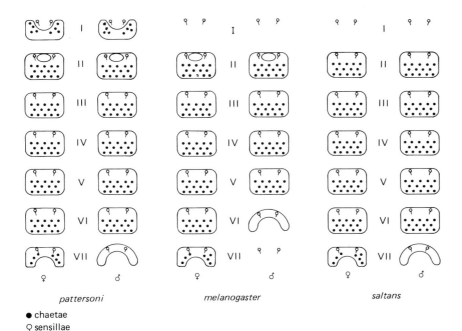

pattersoni   *melanogaster*   *saltans*

● chaetae
♀ sensillae

Fig. 6. Schematic representation of the sternital plates of males and females in three major types of species.

tergites. Difficulties in the classification of these variations, and, in many cases, their gradual and quantitative nature, suggest their exclusion from this survey. Figure 7, however, shows some of these pigmentation patterns and possible relationships between them.

Sexual differences

It is interesting that variations in secondary sexual characters are the most obvious and most numerous of all the pattern variations found in the genus. Sexual dimorphism has been detected in the wing, the legs, the abdomen and the terminalia (genitalia and analia). Most of these differences have been reported in taxonomic studies, because of their wealth of variation. Among the present sample of species we have found sexual dimorphism in the wing in *D. suzukii*. Male wings have a pigmented spot at the end of vein II. This character also appears in other species of the *melanogaster* group (see Bock & Wheeler, 1972).

242   *Antonio García-Bellido*

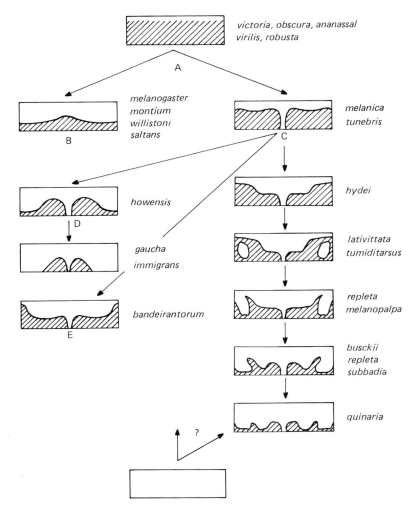

Fig. 7. Different pigmentation patterns in tergites (IV) in different species and hypothetical derivations (arrows).

Variations in sexually dimorphic patterns may appear in several regions of the thoracic legs. It is important to point out that the pattern variations only occur in the male sex of the different species; female patterns are very similar throughout the genus. The most affected is the first leg. We have found major differences in the femoral pattern of species of the *victoria* and *coracina* groups

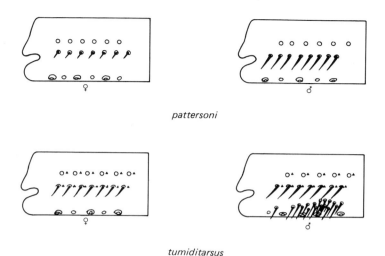

pattersoni

tumiditarsus

Fig. 8. Sexually dimorphic patterns in the femur of different species.

(subgenus *Scaptodrosophila*). These differences are in the appearance of heavy long chaetae in the anterior compartment of the male leg. A similar modified pattern appears in *D. tumiditarsus* (Fig. 8, Table 1).

The most common sexual differences occur in the foreleg basitarsus and tarsal segments. In most species, included *D. melanogaster* males have several scattered, thin, yellowish, non-bracteated chaetae bent outward and perpendicular to the longitudinal axis of the leg, in the posterior compartment. The number and length of these chaetae may vary with the species and the group. They are very numerous, long and bent, in species of the *immigrans* group (Fig. 10). Other differences appear in the sexcomb area (anterior compartment) typical of *D. melanogaster*. Sexcomb teeth may be present and there may be reduction of the transverse row of chaetae in the basitarsus and first tarsal segments. The variation in the number of teeth, their shape, and their orientation relative to the leg axes and the tarsal joints affected is extraordinary (Fig. 9, see Bock & Wheeler, 1972). Sexcomb teeth only appear in the *obscura* and *melanogaster* subgroups (Table 1). Other pattern variations appear in species of the *immigrans* group. In these there is a substitution of the chaetae of the transverse rows of the basitarsus by brushes of long, thin chaetae (Fig. 10). These

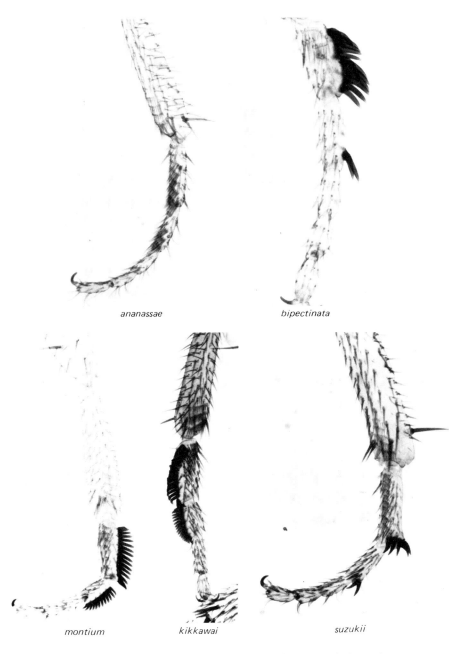

Fig. 9. Sexually dimorphic patterns in the tarsi of species of the *melanogaster* group. Male forelegs.

rubida

pararubida

albomicans

formosana

sulfurigaster

immigrans

Fig. 10. Sexually dimorphic patterns in the tarsi of species of the *immigrans* group. Male forelegs.

brushes differ in size, density and length of chaetae in the different species. Other sexual modifications appear in *D. tumiditarsus*, consisting of brushes at the end of the basitarsus and first tarsal joint, but with a different location and configuration from those of the *immigrans* species group.

Species of the *immigrans* group also show sexual dimorphism differences in the second pair of legs. Bent and slender chaetae, similar to those in the first leg, appear in the tarsal segments of males.

Sexual differences in the dorsal and ventral segmental derivatives of the abdomen appear throughout the genus. *Drosophilia melano-gaster* males do not have a sternital plate in the seventh segment and that of the sixth is reduced and void of chaetae (see Fig. 6). This is the general pattern in *Drosophila* species but there are variations in certain groups. In all the studied species of the subgenus *Scaptodrosophila* and all of the *willistoni* and *saltans* groups of *Sophophora*, males have a reduced seventh sternite and a normal sixth with chaetae (Fig. 6) (see Wheeler, 1960, for variations in Drosophilidae and Throckmorton, 1962, for variations in the genus *Drosophila*).

Drosophila melanogaster has no tergital plate corresponding to the VII segment (Santamaria & García-Bellido, 1972). This is a general feature of all the species of the genus. In *D. melanogaster*, as in most other groups, the sixth tergite is longer than other tergites and has, laterally, the spiracles of both VI and VII segments. Moreover, in *D. melanogaster* this VI tergite is mostly void of trichomes; but in most other groups trichomes like those in other tergites cover its entire surface (see T6(tr), Table 1). The relative length of this tergite also varies. It is reduced to that of other segments in *D. montium* and in species of the *willistoni* and *saltans* groups and in *D. tumiditarsus*. In all these species, however, the usually thin genital arch of the male, is now thickened and carries chaetae and pigment in a pattern similar to that of the sixth tergite (Table 1).

Sexual dimorphism in pigmentation in the abdomen appears in several groups of species. In all cases the end of the male abdomen is more pigmented than that of the females (Pigm, Table 1).

A similar sexual dimorphism occurs in the analia of all the studied species. In males the anus is covered by two plates, located symmetrically on either side of the fly. In females the two plates are asymmetric and located dorsally and ventrally to the anus. We now know from clonal analysis (Dübendorfer & Nöthiger, 1982) that these plates derive from two laterally symmetric anlage, which remain independent in males, but are split in two halves which subsequently meet in the middle line in females. In fact in species of the *willistoni* group the medial fusion of the dorsal half anlagen in females is incomplete.

The most conspicuous variations in sexual characters occur in the male genitalia. The female external genitalia, genital arch and vaginal plates, are extremely constant in shape and patterns of sensillae and chaetae throughout the genus. By contrast the male genitalia are extremely variable between species of the same group and between groups. They may vary in all parts, in shape, type and number of elements, and in the arrangement in patterns. The configuration of the male external genitalia is used for taxonomic purposes and drawings can be found in specialized papers (see Bock & Wheeler, 1972, for the species related to *D. melanogaster*).

Considerations about the genetic bases of morphological variation

Of the numerous pattern variations found in species of the genus *Drosophila* we have paid special attention to qualitative variations. We have chosen those variations which appear as discontinuous changes present or absent as whole patterns, in certain groups of the genus. Small variations in those patterns, in the number or distribution of the elements, can be seen in species of the same group. It can be argued that incipient patterns exist in species not yet studied or in already extinct ones. On the other hand, the possibility exists that these patterns arose suddenly, as integrated units. The distribution of these patterns along the phylogenetic tree suggests that patterns appear and are maintained as such units. Especially interesting are the cases where the same patterns appear in branches unrelated to each other. This is the case, for example, of the appearance of the prescutelar chaetae in species of some genera related to *Drosophila* and in certain species of the subgenus *Scaptodrosophila* (Fig. 11) and the hexa-type of marginal chaetae in species of the *repleta* group and again in other genera of Diptera (Fig. 2 and 11). A similar situation is found in many cuticular patterns within the genus *Drosophila* (see Table 1, Figs. 11 & 12)

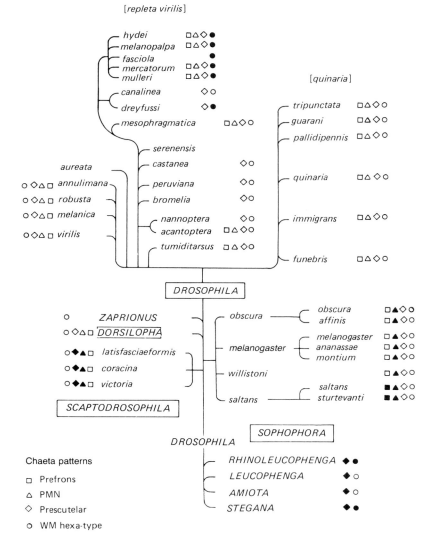

Fig. 11. Phyletic distribution of some chaeta patterns in different groups of species of the genus *Drosophila* and related genera. Symbols: solid, present; open, absent; no symbols, not studied. Sections in parenthesis; subgenera boxed in capitals, genera in capitals; PMN = premandibular chaetae; WM = wing margin pattern.

and in the morphology of several internal organs studied by Throckmorton (1962). There is no obvious way to order all these variations in a consistent phylogenetic tree; any phylogenetic

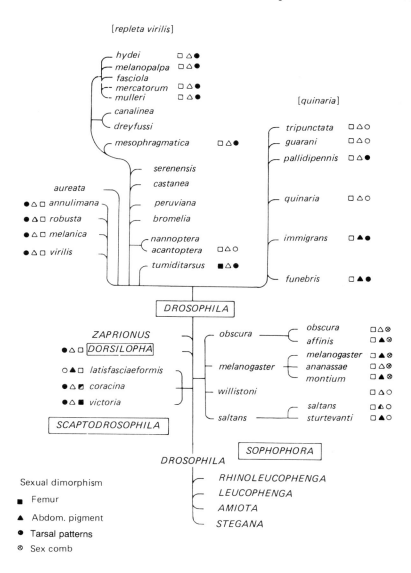

Fig. 12. Phyletic distribution of some sexually dimorphic patterns in different groups of species in the genus *Drosophila*. Symbols as in Fig. 11.

relationship for one particular feature, inferred to derive by lineage, would be inconsistent with that made to accommodate many others.

If these variations cannot be explained by lineage, should they then be explained by 'convergent evolution'? In genetic terms, lineage reflects the activity of the same subset of genes and convergence can be based on the independent appearance of different new combinations of genes leading to the same or very similar phenotypes. In the absence of experimental genetic evidence, to favor one or the other model depends on the notion we may have *a priori* of the plasticity of the genome to invent new genotypes. Mutational data in *D. melanogaster* do not suggest a high versatility of the genome to generate new patterns. Two arguments from the present comparative studies support the idea that this plasticity is limited. First, these patterns exposed to variation are very stable within groups or phylogenetically related groups. Secondly, it is a general trend within the genus *Drosophila* that more primitive forms (subgenera *Scaptodrosophila* and *Sophophora*) have patterns which disappear in more evolved groups (Table 1). It is not impossible, and studies in other genera will show it, that most new patterns found in evolved groups of *Drosophila* are ancestral patterns. In fact, homoeotic mutation causes the appearance of atavic patterns (see García-Bellido, 1977). This stability, or phenotypic inertia, supports then the notion that similar patterns are caused by similar genotypes.

The 'balance hypothesis' advanced by Throckmorton (1962) to account for the appearance of similar phenotypes in not directly related groups of species is very plausible. He envisages the recurrence of similar types as resulting from the existence of discrete genetic combinations (genotypes) that can be expressed or hidden depending on whether they are in 'homozygous' or 'heterozygous' conditions. In developmental genetic terms, this hypothesis could mean that patterns result from the activity of discrete combinations of genes which, without losing their identity as genes needed for other physiological functions, would be morphologically expressed or not depending on the particular combinations of alleles. Obviously small variations in the pattern are caused by the different genetic contexts of the species. What I consider specially important in this theory is that it brings morphological variation closer to a developmental genetic understanding. It suggests that patterns result from the activity of a few (major) genes working in an integrated way to give finite and discontinuous phenotypic solutions.

In the preceding discussion we have contemplated morphological variations as a consequence of genetic variation. It immediately begs the question of why these phenotypes and the corresponding genotypes change between species. The classical answer is that species differ in their physiology and habitat, and in any particular species the observed phenotype is the best adapted. This interpretation runs into difficulties when we try to apply it to morphological variations. Patterns like the ones we have considered here are complex entities made out of several elements arranged in particular spatial configurations. Those in which chaetae are the fundamental element have a neurological and presumably a behavioral representation. Thus, if they occur in a given species it is possibly because of their adaptive value. Therefore, if it is difficult to explain how they appear, it is even more difficult to explain how they can disappear in more evolved species. We should wonder why these patterns, maintained within groups of species and between related groups, are absent in others, and can reappear in unrelated species.

A more obvious paradox is that found in variations in sexually dimorphic patterns. We have seen that, within the genus *Drosophila*, these variations occur in patterns in the wing, legs and different abdominal segments. Moreover, some species may differ in the appearance of several of these organs whereas in others sexual dimorphism is restricted almost to the genital organs (see Table 1, Fig. 12). Again sexual dimorphism in a certain pattern is very conspicuous in some species and very slight in others of the same group (sexcomb teeth in the *melanogaster* group, or sex brushes in the *immigrans* group). Even more extreme variations exist in the male genitalia of the different species of the genus. It is certainly relevant that the sexual dimorphic patterns vary only in the male sex; patterns in females are very constant throughout the studied species. Sexual dimorphism like that found in the genus *Drosophila* relates mainly to sexual recognition in the courtship stages and to mating. The question which already worried Darwin still remains. What is the adaptive value of sexually dimorphic characters, which can be extremely luxurious in certain species and almost absent in other closely related ones, if all are supposedly fit to perform the same function?.

The habitats and ecological parameters of the Drosophilidae and of the species of the genus *Drosophila* are very similar (Throckmorton, 1975). However, we are ignorant of subtle differences which may have permitted natural selection, working through many generations, to mold the observed patterns and pattern variations. I want, however, to consider the possibility that morphological variations are not, or are only a little, modulated by selection. This is in contrast with physiological variations, such as enzyme polymorphisms used as paradigms of evolutionary processes in population genetics. Obviously, most of the severe morphological changes caused by single mutations would be discarded by selection. It is conceivable that slight allelic substitutions of genes with morphological effects remain selectively neutral in the populations. Morphological changes or new patterns would then appear as the result of the combinatorial effect of these new alleles. Whereas most combinations would have no phenotypic effect or be deleterious, certain ones would give rise to new patterns. Patterns may then appear suddenly, not by gradual selection, as integrated entities, and can be maintained through species, disappear or reappear in far related species as the 'same' entities. In this hypothesis, the main constraints and the main generative forces of morphological variation reside in the genome itself, rather than in the moulding forces of selection. The apparently sudden appearances of new morphs in the fossil record, which have lead to the hypothesis of 'punctuated equilibrium', as opposed to the traditional 'phyletic gradualism' (Eldredge & Gould, 1972) could be explained along these lines. It is not necessary to invoke macromutation to explain these changes if patterns and morphs result from discontinuous, combinatorial, effects of genes and if selection is not very stringent.

Again, as shown by the fossil record, sudden changes are followed by periods of stasis. We have seen in this report that new patterns remain stable in related species. If we reduce the importance of selection as a formative force, we are left with the difficulty of explaining the fixation of new morphs or patterns and their stability in species. One of the most intriguing features of morphological variation when comparing related species is the extreme constancy in minute details of patterns within species, and the variations of these patterns between species. This intraspecific stability is actually what suggests its adaptive origin. Thus, for example, although most species of the *melanogaster* group have sexcomb teeth (see Fig. 9), they differ in the number of teeth, their shape and spatial distribution. However, the peculiarities of the particular species are constant. We have examined specimens of *D. melanogaster* and of *D. simulans* cultures (Throckmorton's collection) from seven and four localities, respectively, in each of Europe, America, Asia and Australia. They are indistinguishable within the species but differ in the same details between species.

Those differences can be formally explained by 'foundational effects' leading to genetically homogeneous populations. However, given the high degree of allelic polymorphism present in populations the original homogeneity would soon disappear because of the segregation of the alleles which determine patterns. We have no estimate of the number of genes involved in the described pattern variations, but this could be small. Moreover, we cannot exclude the possibility that these pattern variations are caused by allelic variations in one or a few controlling genes. Thus, genetic variants which affect the activity of the *achaete–scute* complex mimic chaeta patterns found in Drosophilidae (García-Bellido, 1981), and these which affect the *bithorax*-complex uncover primitive abdominal patterns (Lewis, 1978).

Possibly other mechanisms play a role in the stability of morphological phenotypes. Unfortunately, this aspect of population genetics has been somehow neglected in favor of studies of enzyme polymorphisms more amenable to quantitative analysis. Mechanisms such as changes in dominance and epistasis, selection for pleiotropic phenes and genetic canalization may play important roles in the fixation and stabilization of morphological variants. To know the role of these intrinsic mechanisms in morphological stability is important because this stability is an attribute of adaptation.

It has been my intention, in this brief survey, to bring morphological characters, as we contemplate them in developmental genetic terms, to the consideration of evolutionary geneticists. After all, morphological variations are caused by genes and morphological changes are the paradigm of evolution (Bonner, 1982).

I am most indebted to Dr L. Throckmorton for allowing me to take samples of the species of *Drosophila* cultures and studying pinned specimens of other genera of his collection and by his comments on the manuscript. Comments and discussions on this work by P. Alberch, W. R. Baker, P. A. Lawrence and the colleagues in the laboratory are greatly appreciated. I thank E. Reoyo for her skilful help preparing the specimens for microscopic study. This work was carried out by grants of the Spanish FIS and CAICYT.

References

Baker, B. S. & Ridge, K. A. (1980). Sex and the single cell. I. On the action of major loci affecting sex determination in *Drosophila melanogaster. Genetics* **94**, 383–423.

Bateson, W. (1894). *Materials for the study of variation.* Macmillan: London.

Bock, I. R. & Wheeler, M. R. (1972). The *Drosophila melanogaster* species group. *Stud. genet. VII. Univ. Texas Publ.* No. **7213**, 1–102.

Bonner, J. J. (ed.) (1982). Evolution and development. Dahlem workshop Reports. *Life Science* No. 22, Springer Verlag: Berlin.

Capdevila, M. P. & García-Bellido, A. (1981). Genes involved in the activation of the bithorax complex of *Drosophila. Wilhelm Roux Arch. EntwMech. Org.* **190**, 339–50.

Carson, H. L., Hardy, D. E., Spieth, H. T. & Stone, N. S. (1970). The evolutionary biology of the Hawaiian Drosophilidae. In '*Essays in evolution and genetics in honour of Th. Dobezhansky*, eds M. K. Hecht and W. C. Steere, pp. 151–80. Appleton-Century-Crofts: New York.

Dübendorfer, K. & Nöthiger, R. (1982). A clonal analysis of cell lineage and growth in the male and female genital disc of *Drosophila melanogaster. Wilhelm Roux Arch. EntwMech. Org.* **191**, 42–55.

Eldredge, N. & Gould, S. I. (1972). Punctuated equilibrium: an alternative to phyletic gradualism. In *Models in paleobiology,* ed. T. J. M. Schopf, pp. 82–115. Freeman, Cooper and Co.: San Francisco.

Ferris, G. F. (1950). External morphology of the adult. In *Biology of* Drosophila, ed. M. Demeree, pp. 368–416. Wiley: New York.

García-Bellido, A. (1977). Homoeotic and atavic mutations in insects. *Am. Zool.* **17**, 613–29.

García-Bellido, A. (1981). From the gene to the pattern: chaeta differentiation. In *Cellular controls in differentiation,* eds. C. W. Lloyd and D. A. Rees, pp. 281–301. Academic Press: New York.

García-Bellido, A. & Ripoll, P. (1978). Cell lineage and differentiation in *Drosophila.* In *Genetic mosaics and cell differentiation,* ed. W. J. Gehring, pp. 119–56.

Hardy, D. E. (1966). Description and notes on Hawaiian Drosophilidae (Diptera). *Univ. Texas Publ.* No. **6615**, 111–262.

Lewis, E. B. (1978). A gene complex controlling segmentation in *Drosophila. Nature* **276**, 565–70.

Okada, T. (1956). *Systematic study of Drosophilidae and allied families of Japan.* Gihodo Co.: Tokyo.

Palka, J., Lawrence, P. A. & Hart, H. S. (1979). Neural projection patterns from homoeotic tissue of *Drosophila* studied in *bithorax* mutants and mosaics. *Devl. Biol.* **69**, 549–75.

Patterson, J. T. & Stone, W. S. (1952). *Evolution in the genus* Drosophila. Macmillan: New York.

Santamaria, P. & García-Bellido, A. (1972). Localization and growth pattern of the tergite anlage of *Drosophila. J. Embryol. exp. Morph.* **28**, 397–417.

Sondhi, K. C. (1963). The biological foundations of animal patterns. *Rev. Biol.* **38**, 289–327.

Sturtevant, A. H. (1942). The classification of the genus *Drosophila*. *Univ. Texas Publ.* No. **4213**, 5–51.

Thompson, J. N. Jr (1973). General and specific effects of modifiers of mutant expression. *Genet. Res. Camb.* **22**, 211–15.

Throckmorton, L. H. (1962). The problem of phylogeny in the genus *Drosophila*. *Stud. Genet. II. Univ. Texas Publ.* No. **6205**, 415–87.

Throckmorton, L. H. (1975). The phylogeny, ecology and geography of *Drosophila*. In *Handbook of genetics*, vol. 3, ed. R. C. King pp. 421–69. Plenum Press: New York.

Wheeler, M. R. (1949). Taxonomic studies on the Drosophilidae *Univ. Texas Publ.* No. 4920, 157–206.

Wheeler, M. R. (1960). Sternite modification in males of the Drosophilidae (Diptera). *Ann. ent. Soc. Am.* **53**, 133.

Wheeler, M. R. (1982). The Drosophilidae: a taxonomic overview. In *The genetics and biology of* Drosophila, vol. 3*a* , eds. M. Ashburner, H. L. Carson and J. N. Thompson Jr, pp. 1–99. Academic Press: New York.

Morphological and genomic variation in plants: *Zea mays* and its relatives

VIRGINIA WALBOT

Department of Biological Sciences, Stanford University, Stanford, CA 94305, USA

#6436

Contrasting patterns of plant and animal development

In higher animals, development of the major organ and tissue systems of the body is completed during embryogenesis. Early in fetal life, the major organs of the body are differentiated, and the remainder of embryo and juvenile development is concerned primarily with growth. Further development is restricted to replacement of some tissues from stem cell populations, such as in erythropoeisis, and the maturation of pre-existing organs after birth, e.g. gonadal development at puberty. With a fixed body plan and size, animals respond to a changing environment by behavioral and physiological adaptations. In the short term, animals can avoid an environmental threat by flight and can shelter themselves from extreme seasonal change by hibernation or migration.

The life strategy of higher plants is quite different from that of animals. During embryonic development in plants, the major growth axis of the plant is established by the differentiation of meristems at the root and shoot apices. In the embryo, these meristems produce examples of adult organs, the primary root and the seedling leaves. However, most of the plant body is produced after germination. The meristems act as organizing centers specifying the number and placement of organs during the adult life of the plant. The reproductive tissue is produced by conversion of a vegetative apical meristem to a flowering apex, and this transition occurs during adult life.

Plants have many physiological adaptations, as animals do, to overcome specific unfavorable environmental circumstances. Although animals respond to a variable environment almost entirely through physiological and behavioral adaptation, plants utilize both development and physiology. As plants are fixed in a specific place, they are unable to move away from threatening environmental conditions. As the life strategy of plants is the continuous production of new organs, the long-term adaptation to a variable environment involves the production of organs specifically suited to new environmental conditions. For example, many plants are capable of making both 'sun' and 'shade' leaves (Allsopp, 1967). A plant grown in the shade will produce large, dark-green leaves which can efficiently capture sunlight. Stem elongation will be enhanced to spread the leaves over a wider area. If this shade-grown plant is now switched to a sunny environment, the shade leaves are not well adapted to the increased light and heat load. New leaves are produced that are small and closely spaced on the stem, and the structure of the leaf will be modified to prevent water loss, the primary constraint on leaf design in bright light situations. The change from sun to shade leaves involves changes in leaf shape, morphology, anatomy and cellular structure (cf. Dengler, 1980). Features such as the thickness of the leaf cuticular wax, the number and disposition of the chloroplasts in the photosynthetic cells, the number and spacing of the stomatal pores on the leaf epidermis, the number of cell layers in the leaf, and the overall shape and disposition of the leaves on the stem will change.

The ability of plants to respond to the environment by developing new kinds of organs is a fundamental feature of plant adaptation (Bradshaw, 1965). Examination of plants of a single species or race along an environmental cline of temperature, altitude or stress (such as salt-spray) demonstrates that graded adaptation of plant organ structure occurs. The developmental adaptations possible can be so great as to mask the basic similarity of forms. Thus, classification of plants at the species level can be a much more difficult task than is encountered with animal groups. Indeed the developmental adaptations can be great enough to obscure close taxonomic relationships; collections can vary in many structural characteristics yet remain competent to produce fertile offspring.

One area of adaptation that held great fascination for Charles Darwin was the variation in plant reproductive strategy. Many plants are capable of reproducing by both vegetative and sexual means. In addition, many species produce several kinds of flowers.

As noted by Darwin in his book *The different forms of flowers on plants of the same species* (1889), some plants can produce incomplete flowers containing only megagametophytes or only microgametophytes as well as complete flowers in which self-pollination is possible. In many species, flowers produced early in the reproductive season are adapted to favor outcrossing: the flowers may be single sexed or have structural or temporal barriers to self-pollination within the flower or on that plant in general. These flowers are often showy to attract insect pollinators or open to allow wind pollination. Later in the season, cleistogamous flowers may be produced. These are typically smaller, inconspicuous flowers which self-pollinate. The presumed advantage of this dual mode of reproduction is that outcrossing is promoted, but should that fail, self-fertilized progeny will still be produced.

Genetic and structural components of form

The information required to program the development of plants and required to allow the great variation in form is contained in the plant genome and the structural or epigenetic inheritance of form. Perhaps even more so than in animals, plants have a structural component to the inheritance of form, because the cells are forever locked together by their cell walls. In horticulture, plants are selected for propagation if their traits are 'true to type'; that is, the traits can be propagated vegetatively, as in cuttings of roses and fruit trees. 'Sports', which may be structural variants, chimeras or mutants, often arise giving a new type, such as a pink grapefruit or a novel apple variety. Such sports are typically propagated by grafts to maintain the 'true to type' nature of the material, because a meristem 'set' to produce a particular form usually continues to do so. Also, many sports and horticultural novelties cannot be maintained through sexual crosses, as the epigenetic factors contributing to meristem function and required to maintain the novel phenotype are lost during the process of normal development from a zygote (Shamel & Pomeroy, 1936; Hartman & Kester, 1975).

Formation of a zygote provides a specific structural starting point for plant development. Embryonic development is presumed to be quite regular, buffered from environmental stress by the vegetative plant. But the outcome of adult plant development depends on a continued interplay between the varying environment and the capacity of the plant to produce organs best adapted to each situation. Major changes in form, such as switches in phyllotaxy from a spiral to opposite leaves, may occur rarely, but once established, the pattern is self-reinforcing. Both patterns of leaf placement are supported by the same genome although one pattern is inherited and the other arises during development. This can be thought of either as flexibility in expression of the genome allowing plasticity in pattern or as an environmentally modulated assembly of gene products into a structure independent of the direct activity of the genome.

It is also possible for a plant to produce several morphological types as progeny. As the germ line arises at the time of conversion of a vegetative shoot apex to a floral primordium, a branch of unusual morphology resulting from genetic change may produce gametes carrying the mutation. In higher animals, the germ line is set aside early and is relatively inactive in terms of producing structures. However, the shoot apex of plants produces the vegetative body of the plant as well as the flowers. Consequently, mutations arising in the apical cells are tested for their ability to organize both vegetative and reproductive structures. Severely deleterious mutations in apical cells may be prevented from entering the germ line, because such cells may be unable to proliferate and differentiate during vegetative development. Furthermore, both the egg and sperm are genetically active cells so the competence of each haploid cell is also tested prior to zygote formation. The requirement of genetic activity and structural differentiation by precursors of plant germ cells may explain why there is so little embryo lethality in plants and why haploid plants can survive.

Another aspect of the developmental plasticity of plants is totipotency. Many plant organs, tissues and in some species even cells retain the ability to regenerate an entire plant. Totipotency does not imply that all of the cells of the plant are necessarily genetically equivalent, but rather that various states of differentiation do not prevent plant cells from reorganizing meristems or an embryo. The totipotency of plant cells can be viewed as an extension of the life strategy of plants, in which parts of the plant, as well as the plant as a whole, are able to respond to the environment by production of new organs.

Additional variation in plant morphology can be achieved by placing explants in tissue culture. For example, potatoes are vegetatively propagated by planting the 'eyes' which are shoot meristems. Under these conditions there is high crop uniformity. However, when individual cells of the potato leaf are put into culture by digesting away the cell walls to liberate the protoplasts within, the resulting cultures give rise to plants with many novel characteristics not found in the original plant (Wenzel *et al.*, 1979; Shepard *et al.*, 1980). Such variation, called somaclonal variability, is found in many higher plants (reviewed by Skirvin, 1978; Larkin & Scowcroft, 1981) and is a source of new characters such as disease resistance, salt tolerance and other agronomically important traits for plant breeders.

Some of the plants regenerated from tissue culture are widely aberrant compared to the starting material. Many of these grossly aberrant plants are aneuploids or have sustained major changes in their karyotype. What is surprising to those more familiar with animal development is that any plant is produced after major karyotypic change. The totipotency of plant cells can be expressed in monosomics, trisomics, aneuploids of undefined character, haploids, tetraploids and in interspecific crosses generated by protoplast fusion. These results suggest that there are few genetic changes resulting in embryo lethality or a non-functional meristem. The plant genome can organize a structure we recognize as a plant from a genome that has been rearranged, added to or subtracted from. This situation is in sharp contrast to that found in the mammals in which even minor unbalanced karyotypic change typically results in embryo lethality and only a few karyotypic abnormalities allow development to proceed.

In plants, both in the laboratory and in nature, interspecific crosses often produce viable progeny even when the parents had different base chromosome numbers. Such plants may have an uneven chromosome number which would result in an aberrant meiosis, but this difficulty is sometimes overcome by a complete doubling of the chromosome number, for example, by creating a hexaploid from a triploid.

In summary, many plants are capable of great variation in form in response to a changing environment or by spontaneous epigenetic or genetic change. Novel forms arise and can be propagated vegetatively because of the epigenetic component of structure specification. Some of the novel forms can be demonstrated to have a genetic basis, but others disappear during sexual reproduction.

Further variation can be induced by challenging progressively
smaller and presumably less structurally constrained plant explants
to develop in culture. We will now examine how structure changes
over evolutionary time by considering the domestication of maize
from a wild grass. Of particular interest is the discussion of the
genetic basis of morphological change.

Morphological and genetic studies on maize and its relatives

Because maize is a domesticated plant of great commercial importance in the New World and the higher plant for which we have the
most sophisticated genetic knowledge, there have been numerous
investigations of its origin. Several theories have been proposed
(reviewed by Beadle, 1972, 1980; Galinat, 1982) and there is still
considerable controversy about the precise origin and manner of
domestication of this crop (Weatherwax, 1954, 1955; Mangelsdorf,
1974). The most widely accepted view is that maize is a domesticated form of Mexican teosinte *Zea mexicana* Kuntze, an extant
annual grass (Beadle, 1980; Galinat, 1982). However, there are
other possible candidates as progenitors of maize. Based on similarity of chromosome number, appropriate distribution, and fertility
with maize, most New World Maydeae have been proposed as
possibilities. The properties of the New World Maydeae, the
closest relatives of maize, are summarized in Table 1.

The major difference between domesticated maize and its wild
relatives is in the structure of the reproductive parts, particularly
the ear. The modern ear of maize is a large structure with the seeds
firmly attached. In fact the reproduction of maize is now dependent
on human intervention as the seeds are so tightly held on the cob
that natural propagation is not possible. The domestication process
probably began about 7500 years ago when Indian populations in
Mexico or Central America harvested the grain of wild teosinte.
The seeds of teosinte are borne individually surrounded by a strong
cupule, and there are only a few seeds in each reproductive branch
(Fig. 1). The seeds of teosinte 'shatter' – they are propelled off the
plant as a means of seed dispersal. Some tendency toward seed
retention is assumed to have allowed Indians to collect teosinte seed
and initiate casual propagation of the selected types. During
domestication strong selection is proposed to account for acquisi-

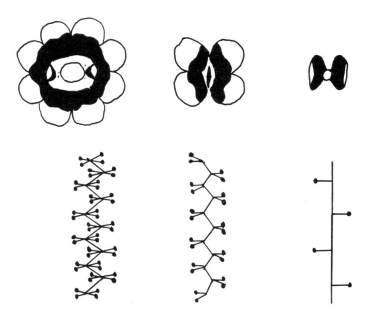

Fig. 1. Diagrams of teosinte and evolving maize ears. On the right the distichous arrangement of teosinte single spikelets is shown; this arrangement results in a two-ranked ear as seen in cross-section above. The shading represents the extent of cupule formation. In the middle a primitive four-ranked maize ear is shown resulting from the production of paired spikelets at each node and the condensation of the ear rachis. The cupules are much reduced. Further reduction in the spacing of the kernels results from doubling the number of ear rows at each node, as shown on the left. This produces an eight-rowed ear. (Adapted from Galinat, 1982.)

tion of other characters making the plant more useful to people. Many characters affecting seed size and cob size were changed, resulting in larger seeds borne in greater numbers on a greatly enlarged cob. The wide spacing between seeds in teosinte cobs has been greatly reduced in maize, and this condensation of seed spacing allows a higher packing density of seeds on the ear.

Galinat (1970, 1982) has examined the morphological evolution of maize from teosinte in detail. The major features of the transition from the teosinte to maize cob and seed structure involve a doubling of parts. In both maize and teosinte, spikelets are produced in pairs along the cob. In teosinte, however, the pediculate spikelet is repressed after organ initiation so only a single

Table 1. *The New World Maydeae*

Plant	Base chromosome number	Knobs	Distribution	Major floral characteristics	Ability to cross with maize
Zea mexicana					
var. Chalco	10	Internal & terminal/large number	Valley of Mexico/ cultivated areas only	Large robust, maize-like plant/development parallels maize	Easily hybridizes/ F_1s are found in field at 0.1%
var. Huehuetenango	10	Terminal only	Northern province of Huehuetenango in N. Guatemala/abandoned fields	Similar to Balsas	Easily hybridizes/ crosses freely in nature
var. Balsas	10	Internal & terminal	Rio Balsas drainage of Guerrero & Michoacan/ grows wild	Least maize-like of Mexican teosintes/ small seeds	Crosses with maize less than Huehuetenango or Chalco/ reproductively isolated
Zea luxurians	10	Terminal only	S. Guatemala/small, wild, isolated populations	Highly tillering/ primitive/seeds are trapezoidal in shape	Hybridizes poorly/ irregular meiosis/ 32% pollen infertility
Zea perennis	20	Terminal	Central Mexico	Perennial	Not fertile
Zea diploperennis	10	Terminal only	Jalisco, Mexico/ highly local populations	Perennial, robust, maize-like	Crosses with maize in the laboratory

Tripsacum spp.	18 36	Terminal	Widespread distribution in N. hemisphere	Primitive/perennial	Crosses poorly/ no F_1s found in nature F_1 females partially fertile, males completely sterile
Zea mays var. Ladyfinger popcorn	10	Internal	Old popcorn variety	Indistinguishable vegetatively from Midwestern Dent	Crosses easily
var. Midwestern Dent	10	Mainly internal/ strain variability	Midwest of USA		

Adapted from Hake (1980) and references therein.

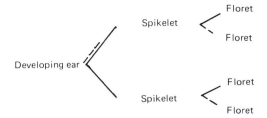

Fig. 2. Diagrammatic representation of female reproductive structure development. Pairs of spikelets are initiated in both teosinte and maize but the pediculate spikelet is repressed in teosinte (dashed line). Each spikelet initially contains two florets bearing both staminate and pistillate structures. In both teosinte and maize only the upper floret typically develops and this only contains mature pistillate structures. An exceptional maize variety, Country Gentleman, shows development of all florets resulting in a crowded ear with irregular rows.

competent spikelet is produced (Fig. 2). Within each spikelet, there are two florets; however, only the upper floret matures. As a consequence, there is a single seed produced by the paired spikelets. This seed is tightly encased in a hard outer tissue, the cupule, which surrounds the spikelet pair. The cupule protects the seed from environmental damage; cupule tissue can even survive digestion by birds, ensuring seed survival (Galinat, 1970). In maize, both of the spikelet primordia develop completely, resulting in twice as many potential seeds (Fig. 2). As in teosinte, only the upper floret of each maize spikelet develops, so two seeds are produced by the spikelet pair. Production of two kernels within a single cupule destroys the protective function of this tissue as the cupule is split open (Galinat, 1982). Thus, development of two spikelets is counterproductive for the survival of a wild plant such as teosinte. In modern maize the cupule is reduced to a narrow rim of tissue at the base of the kernel and no longer serves any protective function.

Further increases in seed number on the cob are accomplished in maize by increasing the density of seeds. In teosinte the spikelets are distichous (Fig. 1) resulting in two kernel rows per ear. In maize, development of both sister spikelets results in a four-rowed cob as the developing kernels of sister spikelets are adjacent. Further increases in seed number result from doubling the number of ear rows from four to eight or more. At each increase, the girth of the cob increases to provide space for the kernels.

Although the major differences between maize and the various teosinte species are in reproductive morphology, maize and teosinte differ by important vegetative characters as well. Most teosinte species and races are highly tillered producing numerous branches from the basal nodes. Second ranks of branching are also possible from the primary lateral branches. The large number of branches gives teosinte a 'grassy' appearance (Fig. 3d). Modern maize has only a single main stem with few, if any, tillers. The leaves of most teosinte races are several-fold narrower than maize leaves and much shorter as well. The total number of leaves produced by a maize plant is rarely more than 20 but many teosinte plants produce 50–100 leaves in a single season (personal observation). The morphology of teosinte is quite variable within a single species. For example, in the Mexican teosintes there are races, such as Chalco, which is a maize mimic and races with a highly tillered habit. Today teosinte is still found throughout Mexico and Central America but typically in small populations, and we may be sampling only a small part of the original diversity of the species. In fact, the recently discovered diploid perennial *Z. diploperennis* was first reported in 1979 and exists only as a very small population (Iltis *et al.*, 1979).

Because maize, teosintes and tripsacums can form fertile hybrids in certain combinations, several investigators have examined the genetic basis for the observed morphological differences between the species (Beadle, 1972, and references therein). Hybrid F_1 progeny of maize and teosinte are produced and self-pollinated to allow the segregation of parental traits. The frequency of recovery of 'pure teosinte and maize morphologies indicates how many genes or blocks of closely linked genes are involved in specifying the major aspects of morphology. These studies with Mexican teosinte and maize indicated that teosinte or maize morphology was recovered in about 1/256 cases. This frequency indicates that there are four separate factors in the genome $[(1/4)^4 = 1/256]$ specifying the morphological distinctions. That so few genes or gene blocks are involved in the morphological leap from teosinte to maize is initially quite surprising, but perhaps less so when we consider plant development is so variable to begin with. By changing a few parameters, such as doubling competent spikelet number, numerous additional changes might naturally occur, such as increase in cob size to hold the extra kernels. The adaptiveness of plant development to changes in one structure is probably a sufficient explanation for the apparently 'coordinated' changes in evolving cobs. Intermediate forms have not been preserved in nature, probably because of their deleterious consequences on seed survival (Galinat, 1982). In addition, some interspecific sterility factors help maintain the integrity of the genomes and plant types in natural populations so intermediate forms are not common. But, the relative ease of transforming teosinte to maize, and vice versa, suggests that the morphological differences are readily reversed.

Over the past 100 years, maize breeding programs have accelerated the specialization of the plant for agriculture by selecting for specific nutritive characters, one large rather than several smaller ears per plant, uniformity of plant growth habit, adaptation for modern farming practices such as mechanical plowing and harvesting, etc. During the initial stages of domestication, it is assumed that a portion of each crop was set aside for next year's planting but that no special provisions were made for seed corn. This situation persisted into the early twentieth century, when most farmers were still raising both a crop and next year's seeds simultaneously. The discovery of hybrid vigor at the turn of the century, however, eventually led to the seed production industry. Hybrid vigor demonstrated that combining highly inbred lines yielded progeny with characteristics superior to either parent. Seed companies produced inbred lines with favorable agronomic traits and then combined these traits in progeny sold as seed for crop production. The yield of hybrid corn was higher than the traditional varieties and displaced them in the 1930s and 1940s (Brown & Goodman, 1977).

The various inbred lines maintained by seed companies and by maize geneticists have very uniform traits and little intra-line morphological variability. This is the result expected from an inbreeding program, because 'rogue' or off-type plants are culled from the population to maintain stand uniformity. Among the various inbred lines there is great quantitative variation for traits such as plant height, days to flowering, leaf width, color, growth rate, etc. However, the basic morphological plan of each inbred type is reasonably fixed. One of the consequences of the production of maize inbred lines may be that the potential diversity of morphological types previously possible in individual plants in nature has been lost. We can suggest that inbreeding has removed some of the capacity of the plant to respond to environmental conditions by the production of an altered morphology.

Single gene changes in maize morphology

Morphological mutants of maize are commonly recovered by geneticists and provided most of the material for the construction of the original linkage map of the genome (Emerson *et al.*, 1935). The impact of a single gene change can be quite striking. For example, *tassel seed* mutations, either dominant or recessive, convert the pollen-producing tassel to an ear (Fig. 3*a*). Spikelets of maize are initiated with complete flowers. Normally repression of pistil development occurs in the tassel, and repression of anther development occurs on the ear. The tassel seed mutations allow the partial-to-complete choice of pistillate development on the tassel. The reciprocal mutation *anther ear* converts ear spikelets to staminate development.

The dominant gene *Corngrass* (Fig. 3*b*) confers a teosinte-like morphology with narrow leaves and extreme tillering; small ears with only a few seed rows are produced. The dominant *Teopod* genes also cause high tillering, narrow leaves, a reduction in ear size, and a tassel consisting of only a single branch. Other genes such as *teosinte branched* can have a similar impact on many characters simultaneously. Dominant and recessive dwarfing genes also have pleiotropic effects (Fig. 3*c*). In some dwarves, the ear has staminate flowers: an andromonecious dwarf. The dwarfing is not proportional; stem and leaf sheath expansion are greatly reduced, but the leaf blade is often of nearly normal size. In some dwarves a single treatment with the plant hormone gibberellic acid at an early seedling stage results in normal morphological development (Coe & Neuffer, 1977). Such results suggest that underproduction of the hormone is responsible for the numerous morphological changes. Furthermore, the non-proportionality of the dwarf phenotype suggests that there is either an asymmetric distribution of the hormone within the plant or that different plant parts require different concentrations of hormone to support normal growth. In either case, a genetic change in hormone concentration or reception activity can be expected to create different effects on different plant parts. Further examples of single gene effects are illustrated in *The mutants of maize* (Neuffer *et al.*, 1968).

Fig. 3. Morphological mutants of maize compared to teosinte. (*a*) Tassel seed; (*b*) Corngrass next to normal plant (bagged). (*c*) dwarf plant is 46 cm tall; (*d*) Teosinte, *Zea diploperennis*. Plant is 30 cm tall except for flowering branch 51 cm high.

The conclusion from examination of these various mutants is that single gene mutations can have a major impact on morphology. These morphological changes are not lethal, although in some cases the plant produces only one type of gamete. If single gene changes can confer so altered a morphology without significant deleterious effects, then the explanation that the morphological differences between maize and Mexican teosinte stem from only four major genetic changes gains great support. However, the rate and pattern of speciation is difficult to assess from the morphological record, and we have turned to another type of analysis.

Use of molecular change to study maize evolution

The angiosperms are the dominant green plant group on earth today; a status achieved during the past 150 million years. However, we know little about the pattern of this rapid and expansive radiation into all major habitats, because the fossil record for plants is quite poor. Furthermore, plant morphology can be highly variable, and the fossil record may not be as accurate for plants as for animals in judging morphological specializations and their origin in underlying genetic change. A further complication is that modern day plant species form interspecific crosses, even intergeneric hybrids, so that the classical definition of a species does not really apply to many groups of higher plants. *False ✗*

An alternative to morphological studies for assessing the distance between taxa is to compare the genomes by DNA hybridization. Such studies can be at the level of comparison of individual genes or comparisons of the unique copy DNA presumed to contain most of the genes or of the whole genome in which repetitive as well as unique copy sequence relationships are assessed. It has been proposed that the gradual change in sequence composition will be a 'molecular clock' in which the time frame for speciation of individual animal taxa can be calculated from the amount of sequence divergence with reference to dated fossilized evidence of speciation events (reviewed by Wilson *et al.*, 1977). Most studies of molecular evolution in animals have assumed that a 'zero time point' exists; that is, a specific time point at which speciation was completed and no further hybridization occurred between the two new species. Furthermore, it is often assumed that the taxa were not polymorphic at the time of speciation for the gene(s) studied at a molecular

✗ The premises are false and the conclusions don't follow.

Table 2. *DNA hybridization studies of maize and its relatives.*
Measurement of the percent sequence divergence

Test DNA	Standard DNA			
	Inbred maize W64A	Z. mexicana var. Balsas	Z. luxurians	Z. diploperennis
Z. mays				
inbred	0.4	—	3.7	—
W64A				
Mo17	0.2	—	—	—
B37	0.5	—	—	—
Z. mexicana				
var. Chalco	0.9	0.2	3.2	—.
var. Balsas	2.7	—	3.6	—
var.				
Huehuetenango	1.4	1.9	2.9	—
Z. luxurians	4.2	—	—	—
Z. dipolperennis	4.1	4.6	3.4	—
Tripsacum laxum	8.8	8.0	3.8	6.1

Whole genomic data or isolated single copy DNA were used for these comparisons. Further details in Hake & Walbot (1980) and Hake (1980).

level. These situations probably do not apply to many groups of higher plants, because polymorphism is high and interspecific hybridization does occur. An aspect of our studies is to determine whether molecular studies will be useful in deriving phylogenies and in assessing the time of speciation events.

We have examined the extent of sequence divergence among maize and its relatives by determining the percentage of sequence mismatch. Divergence is determined by a depression in the thermal stability of DNA hybrids formed from two samples as compared to self-reassociation stability (for technical details, see Hake & Walbot, 1980). From comparison of whole genomes and single copy sequences, it appears that there is limited diversity among modern inbred maize varieties, about 0.5% mismatching (Table 2). This divergence may be caused by scatter inherent in the method or represent real differences among inbred lines (Hake, 1980). A primitive popcorn variety, Ladyfinger Pop, demonstrates about 3%

divergence from inbred lines of maize (Hake, 1980), and this result may be a better indication of the actual diversity present in maize genomes. The modern inbred lines of maize have an extremely narrow genetic base having been derived from only a few varieties chosen at the turn of the century (Brown & Goodman, 1977). The wild Mexican teosinte races show considerable intraspecific diversity (Table 2) comparing Chalco, Balsas and Huehuetenango using Balsas DNA as the standard. These races represent but a few of the geographical locations in which Mexican teosinte is found today, and the full range of genomic diversity has not been assessed.

Using an inbred maize line as the standard, we have compared its hybridization behavior to those of all of the maize relatives; selected reciprocal tests were made using Mexican, Guatemalan and perennial teosinte as well (Table 2). Each 1 °C in melting point depression is approximately equal to 1% sequence mismatch (Hake & Walbot, 1980, and references therein). The divergence between maize and Mexican teosinte is about 1–3%, about 4% from Guatemalan and perennial teosintes, and about 8% from *Tripsacum*. The extent of sequence divergence among maize and its relatives is much higher than that reasonably expected from the assumed age of the group and the apparent rate of sequence change in nuclear genomes of animals; in animals 1% mismatch accumulates in ~ 5 million years. The monocots arose relatively late in angiosperm evolution, perhaps 90–100 million years ago, with the primitive Poales, such as bamboo, appearing perhaps 70 million years ago (Raven & Axelrod, 1974). Evidence for modern grass genera is very scanty in the fossil record, but it is likely that modern species are only a few million years old at most. Thus, the great divergence, up to 8% (?40 million years), among the New World Maydeae is quite unexpected. The intraspecific divergence is even more surprising as it suggests that species characters do not depend on having a nearly identical genome, but that such characters may depend on a few key factors retained throughout the group.

If the genome as a whole is diverging rapidly, what is the rate of change of individual genes? We have extended our studies to examine the divergence of the nuclear ribosomal RNA genes using restriction endonuclease digestion mapping. Thus far our results demonstrate that ribosomal RNA genes are evolving by both base change and insertion–deletion events (E. A. Zimmer, C. Rivin, J. Swanson & V. Walbot, unpublished results). The pattern of

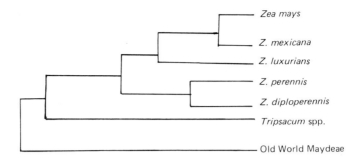

Fig. 4. Phylogenetic relationships in the New World Maydeae based on DNA hybridization homologies and restriction endonuclease mapping of the ribosomal RNA genes. Modern maize is most closely related to the Mexican teosintes. Not included in the diagram is the diversity within maize and the Mexican teosintes.

ribosomal DNA divergence is consistent with the whole-genome studies; the emerging phylogenetic picture is given in Fig. 4. The rate of change for ribosomal DNA is less than for the genome as a whole or for single copy DNA. Although we have not yet finished our analysis, it is clear that different classes of sequences can evolve at different rates.

Many studies suggest that maize and the Mexican teosintes are very closely related. Our studies show that within each species there is genomic diversity measured by DNA hybridization and ribosomal DNA structure. The diversity within maize may reflect multiple origins of maize from various teosintes. Maize may not have evolved from a single founder individual, but rather from acquisition of a few novel morphological traits introduced into an already diversified teosinte genetic background. The genomic diversity could also reflect the high polymorphism, found in plants, for enzyme traits (Allard & Kahler, 1971). An important element of plant adaptation is variation in form and physiology, and maintenance of this ability to adapt may require enzymatic and other gene polymorphisms. We may measure this genetic diversification as DNA mismatch in our experiments. And, as already mentioned, species characteristics may involve only a few genes or gene blocks held in common with the remainder of the genome divergent.

Other kinds of genomic variation in maize

The visible chromosomal morphology of maize varies with respect to the presence and number of knobs (Table 1). The knobs are heterochromatic regions of the chromosomes, and they are found at more than 20 specific locations in *Z. mays* (Carlson, 1977). Knobs may be essentially undetectable or very prominent features of individual chromosomes. Maize knobs are typically internal, while knobs in teosinte are usually terminal. Maize also contains a variable number, from 0 to over 10, of supernumerary B chromosomes which carry no essential functions but do influence the behavior of A chromosomes (Carlson, 1977).

The controlling elements of maize constitute another type of genomic variation. These transposable elements can influence gene expression in a manner coordinate with the developmental program of the plant (McClintock, 1956; Fincham & Sastry, 1974). Transposable elements in bacteria control such important functions as mating type (Simon *et al.*, 1980), and movable genetic elements are similarly implicated in yeast mating type expression (Hicks *et al.*, 1979). Many cases of variable gene expression are known in maize that are attributed to presently described or unknown controlling elements (Coe & Neuffer, 1977). These elements are part of the genome and represent a programmed form of variation.

Conclusions

Plants, including maize and its relatives, demonstrate great variation in form. The adaptation of plants to a changing environment rests in part on the ability to produce organs suited to particular environmental conditions. Although modern inbred lines of maize are quite uniform in morphology, single gene changes can result in dramatic alterations in morphology. Mexican teosinte and maize, which differ greatly in morphology, are still interfertile in most cases, and the introduction of four genes or gene blocks from maize can convert teosinte to a maize-like vegetative and reproductive morphology.

The morphological and genetic studies supporting the origin of maize from domestication of Mexican teosinte are supported by DNA hybridization tests. These molecular studies suggest a phylogenetic tree in which these two species are most closely related.

However, the extent of divergence based on hybridization is much greater than would be expected based on the likely time of origin of maize about 7000 to 8000 years ago. There is also substantial intraspecific genome diversity. We propose that the variation seen within maize varieties and teosinte races is an important aspect of plant biology. The apparent diversity of genomic content may be a fundamental source of the information required to maintain plant morphological variation.

Many thanks are due M. G. Neuffer and D. A. Hoisington for providing the plant photographs. The data on maize–teosinte DNA hybridization was provided by Sarah Hake, and the unpublished data on ribosomal DNA organization in nuclear DNA by Elizabeth Zimmer, Jean Swanson, and Carol Rivin. W. C. Galinat provided valuable information on primitive maize and teosinte morphology. Research support was obtained from grant DEB 81-15322 of the National Science Foundation and CRGO grant 59-2066-22734 from the US Department of Agriculture.

References

Allard, R. & Kahler, A. (1971). Allozyme polymorphisms in plant populations. In *Stadler Genetic Symposium*, vol. 3, ed. G. P. Redei, pp. 3–24. University of Missouri Press: Columbia.

Allsopp, A. (1967). Heteroblastic development in vascular plants. In *Advances in morphogenesis*, vol. 6, eds. M. Abercrombie and J. Brachet, pp. 127–71. Academic Press: New York and London.

Beadle, G. W. (1972). The mystery of maize. *Field Mus. nat. Hist. Bull.* **43**, 2–11.

Beadle, G. W. (1980). The ancestry of corn. *Scient. Am.* **242**, 112–19.

Bradshaw, A. D. (1965). Evolutionary significance of phenotypic plasticity in plants. In *Advances in genetics*, vol. 13, eds. E. W. Caspari and J. M. Thoday, pp. 115–55. Academic Press: New York.

Brown, W. L. & Goodman, M. M. (1977). Races of corn. In *Corn and corn improvement*, ed. G. F. Sprague, pp. 49–88. Am. Soc. of Agronomy, Inc.: Madison, Wisconsin.

Carlson, W. R. (1977). The cytogenetics of corn. In *Corn and corn improvement*. ed. G. F. Sprague, pp. 225–304. Am. Soc. of Agronomy, Inc.: Madison, Wisconsin.

Coe, E. H., Jr & Neuffer, M. G. (1977). The genetics of corn. In *Corn and corn improvement*, ed. G. F. Sprague, pp. 111–223. Am. Soc. of Agronomy, Inc.: Madison, Wisconsin.

Darwin, C. (1889). *The different forms of flowers on plants of the same species*. Appleton Press: New York.

Dengler, N. G. (1980). Comparative histological basis of sun and shade leaf dimorphism in *Helianthus annus. Can. J. Bot.* **58**, 717–30.

Emerson, R. A., Beadle, G. W. & Fraser, A. C. (1935). *A summary of linkage studies in maize*. Cornell Univ. agric. exp. Sta. Mem. No. 180.

Fincham, J. R. S. & Sastry, G. R. K. (1974). Controlling elements in maize. *A. Rev. Genet.* **8**, 15–50.

Galinat, W. C. (1970). The cupule and its role in the origin and evolution of maize. *Mass. agric. exp. Sta. Bull.* **585**, 1–18.

Galinat, W. C. (1982). The origin of corn from teosinte by domestication. In *Prehistory and ecology of cropping systems in the Americas*, Proceedings XIII Internat. Bot. Cong. Symp. 1981. Sydney, Australia (in press).

Hake, S. C. (1980). The genome of *Zea mays*. Ph.D. Thesis, Washington University, St Louis, Missouri.

Hake, S. & Walbot, V. (1980). The genome of *Zea mays*, its organization and homology to related grasses. *Chromosoma* (Berl.) **79**, 251–70.

Hartman, H. T. & Kester, D. E. (1975). *Plant propagation: principles and practices*, 3rd edn. Prentice-Hall: Englewood Cliffs, New Jersey.

Hicks, J., Strathern, J. & Klar, A. J. S. (1979). Transposable mating type genes in *Saccharomyces cerevisiae*. *Nature* **82**, 478–83.

Iltis, H. F., Doebley, F. J., Guzman, R. & Pazy, B. (1979). *Zea diploperennis* (Gramineae): a new teosinte from Mexico. *Science* **203**, 186–8.

Larkin, P. J. & Scowcroft, W. R. (1981). Somaclonal variation – a novel source of variability from cell cultures for plant improvement. *Theor. appl. Genet.* **60**, 197–214.

McClintock, B. (1956). Controlling elements and the gene. *Cold Spr. Harb. Symp. quant. Biol.* **21**, 197–216.

Mangelsdorf, P. C. (1974). *Corn, its origin, evolution and improvement*. Belknap Press, Harvard Univ. Press: Cambridge, Mass.

Neuffer, M. G., Jones, L. S. & Zuber, M. S. (1968). *The mutants of maize*. Crop Sci. Soc. of America: Madison, Wisconsin.

Raven, P. H. & Axelrod, D. J. (1974). Angiosperm biogeography and past continental movements. *Ann. Missouri bot. Garden* **61**, 539–673.

Shamel, A. D. & Pomeroy, C. S. (1936). Bud mutations in horticultural crops. *J. Hered.* **27**, 487–94.

Shepard, J. F., Bidney, D. & Shahin, E. (1980). Potato protoplasts in crop improvement. *Science* **208**, 17–24.

Simon, M., Zieg, J., Silverman, M., Mandel, G. & Doolittle, R. (1980). Phase variation: evolution of a controlling element. *Science* **209**, 1370–4.

Skirvin, R. M. (1978). Natural and induced variation in tissue culture. *Euphytica* **27**, 241–66.

Weatherwax, P. (1954). *Indian corn in Old America*. Macmillan: New York.

Weatherwax, P. (1955). Early history of corn and theories as to its origin. In *Corn and corn improvement*, ed. G. F. Sprague, pp. 1–16. Academic Press: New York.

Wenzel, G., Schieder, O., Przewozny, T. Sopory, S. K. & Melchers, G. (1979). Comparison of single cell culture derived *Solanum tuberosum* L. plants and a model for their application in breeding programs. *Theor. appl. Genet.* **155**, 49–55.

Wilson, A. C., Carlson, S. S. & White, T. J. (1977). Biochemical evolution. *A. Rev. Biochem.* **46**, 573–639.

What are the developmental underpinnings of evolutionary changes in protozoan morphology?

JOSEPH FRANKEL

Department of Zoology, University of Iowa, Iowa City, Iowa 52242, USA

4437

Introduction

Two major questions in contemporary evolutionary biology are (1) to what degree do organizational constraints limit the potential for evolutionary change, and (2) to what extent can pattern and form undergo sudden reorganization? These questions have been explored very little in unicellular organisms, which generally lack a good fossil record and commonly fail to exhibit sufficient organizational complexity. However, ciliates and some groups of flagellates, although not well represented in the fossil record, do express a high degree of intracellular complexity. Ciliates are particularly favorable for analysis. The extensive comparative anatomical study of ciliates has revealed a basic unity of structural organization (Fauré-Fremiet, 1950; Corliss, 1979, pp. 62–63) suggestive of common ancestry (Hanson, 1977, Chapter 10). In certain specific ciliates, morphological analysis together with study of genetics, development, and molecular constitution have supplied sufficient information to allow meaningful consideration of the major problems raised above.

The special characteristic that is of the greatest importance in considering ciliate evolution is the elaboration of complex cell-surface patterns from units organized around ciliary basal bodies (Lwoff, 1950; Grain, 1969; Lynn, 1981). These 'ciliary units' tend to be organized into longitudinal assemblies, or ciliary rows, that grow by interdigitation of new units and become transversely subdivided just prior to cell division. This tandem, basically segmental mode of division (Chatton & Villaneuve, 1937; Frankel & Nelsen, 1981) creates an opportunity for direct inheritance of supramolecular patterns (Sonneborn, 1970), an opportunity that is enhanced by the fact that in the typical ciliate sexual process, conjugation, nuclei are exchanged between cells that retain their structural individuality.

Although unique in many ways, ciliates resemble other organisms in possessing a germ line (in this case nuclear rather than cellular) and in undergoing meiosis and fertilization (Sonneborn, 1947). Their breeding patterns can be placed in a relatively conventional inbreeding–outbreeding continuum (Sonneborn, 1957). They also have certain specialized cell-surface structures that are formed anew in each cell generation, by mechanisms that involve induction (Tartar, 1956), gradients (Uhlig, 1960), and fields (Nanney, 1966a); these mechanisms are analogous and possibly homologous to corresponding processes in multicellular organisms (Frankel, 1974, 1982). Hence, we might expect that in ciliates, as in other organisms, evolutionary transformations should be brought about, in part at least, by changes in the nature and expression of genes that regulate developmental processes.

Evaluation of these possibilities is complicated by the fact that the 'chosen few' ciliates (Nanney, 1980) that have been investigated sufficiently tend to exhibit a remarkable combination of morphological conservatism and molecular divergence (Nanney, 1982a, 1982b). I will consider here the nature of the constraints responsible for this morphological conservatism, and then ask how evolutionary change in morphology can nonetheless occur. This will lead to a consideration of the mechanisms of change, where I will attempt to evaluate both the relative roles of the two levels of inheritance (genetical and supramolecular) encountered in ciliates, and the degree to which transformations are sudden or gradual.

Evolutionary mechanisms in ciliates: four propositions

1. Evolution of pattern is at least partially uncoupled from evolution of the molecules that make up the pattern

Molecules and morphology

In reviewing the remarkably small differences in proteins of humans and chimpanzees, King & Wilson (1975) suggested that evolution of macromolecules and of organismal form need not be closely coupled. In mammals, particularly primates, anatomical evolution is rapid relative to molecular change; in sharp contrast, certain other organisms maintain their form for long periods while undergoing molecular evolution at the usual rates (Wilson *et al.*, 1977). The '*Tetrahymena pyriformis*' sibling-species swarm (Nanney & McCoy, 1976) is an excellent example of a group of structurally complex organisms in which great molecular differences have arisen (see Nanney, 1982*a*, 1982*b*) despite very limited diversification in size (Elliott, 1973; Nanney *et al.*, 1978), proportions (Gates & Berger, 1976), and cell surface pattern (Nanney & Chow, 1974; Nanney *et al.*, 1980; Williams & Bakowska, 1982). *Tetrahymena* thus exhibits a dissociation that is the obverse of the one encountered in primates.

There is one important way in which knowledge of the juxtaposition of evolution of molecules and morphologies has been carried further in *Tetrahymena* than in other organisms. In this ciliate, a favorite of molecularly oriented cell biologists, the evolutionary conservatism of pattern can be contrasted to the lesser conservatism of the molecules that make up the pattern. Within the '*Tetrahymena pyriformis*' sibling-species swarm, ciliary proteins (Seyfert & Willis, 1981), major cytoskeletal proteins (Vaudaux *et al.*, 1977), proteins of the oral cytoskeleton (N. E. Williams, personal communication), and external surface membrane proteins labelled by lactoperoxidase-catalyzed iodination (Williams *et al.*, 1980) all differ substantially among species. Especially noteworthy is a detailed analysis of two of these sibling species, *Tetrahymena thermophila* and *Tetrahymena pyriformis* (*sensu stricto*), which showed that the oral anatomy of the two species manifests an amazing similarity down to fine details (Williams & Bakowska, 1982), while 80% of the proteins extracted from oral cytoskeletal preparations differ in molecular weight; although the tubulins coincide, members of a complex of major proteins that are specifically enriched in oral fractions differ (N. E. Williams, personal communication).

Most striking of all is a double comparison recently carried out by Buhse & Williams (1982), involving *Tetrahymena pyriformis* (*sensu stricto*) and *T. vorax* (strain V_2S). *Tetrahymena vorax* is not a member of the '*Tetrahymena pyriformis*' sibling-species swarm, and is prominently dimorphic: when feeding on bacteria it greatly resembles the members of the '*Tetrahymena pyriformis*' sibling-species swarm in most respects (Kidder *et al.*, 1940; Corliss, 1973), and has a small 'microstome' oral apparatus that is very similar in detailed organization to that of *T. pyriformis* (Smith, 1982*a*; Williams & Bakowska, 1982). But when suitable prey – such as *T. pyriformis* – become available, *T. vorax* transforms into a larger 'macrostome' form, with a gigantic oral apparatus (Williams, 1961; Buhse, 1966*a*, 1966*b*). Buhse & Williams (1982) found that both cytoskeletal proteins and iodinated surface-membrane proteins of *T. vorax* microstomes were indistinguishable from those of macrostomes in one-dimensional polyacrylamide gels. By contrast, the cytoskeletal proteins of the morphologically very similar *T. vorax* microstome form and *T. pyriformis* differed greatly (Buhse & Williams, 1982), as is consistent with earlier results (Vaudaux *et al.*, 1977; Williams *et al.*, 1980). Hence 'To a first approximation . . . *T. pyriformis* cells and *T. vorax* microstomes maintain the same morphology with different proteins, whereas *T. vorax* microstomes and *T. vorax* macrostomes maintain different morphologies with the same proteins . . .' (Buhse & Williams, 1982). Thus, the evolutionary discordance between proteins and higher-order phenotypes holds true even when the proteins being analyzed contribute materially to the phenotypes under consideration. Our understanding of the depth of this discordance is, however, very incomplete. As suggested by N. E. Williams (personal communication), it is possible that critical polypeptide domains necessary for proper assembly and function of oral proteins are conserved, while other less essential parts of proteins may be free to vary.

How is pattern constrained?

We now need to confront the question of what it is that keeps a cell-surface pattern so uniform within the '*Tetrahymena pyriformis*' sibling-species group. As repeatedly pointed out by Nanney (e.g. in Nanney *et al.*, 1978), some constraint(s) must be operating within this group of organisms to maintain their relative constancy of scale and pattern despite their great molecular divergence. There are, however, at least two ways of thinking about this constraint. One is to suppose that the constraint is imposed by natural selection. Once a form and pattern suited to a particular environment is attained, then, if the relevant environment remains the same, the only form of natural selection at work would be stabilizing selection, keeping the phenotype constant. It is this type of constraint that is strongly implied in Nanney's treatment of the problem (Nanney *et al.*, 1978). The alternative is to imagine that there is some additional constraint inherent in the developmental process itself: once a particular developmental pathway is selected, very large changes in underlying parameters may be necessary to 'dislodge' the course of development into a new pathway. This notion, implicit in Waddington's concept of developmental 'canalization' (Waddington, 1957), has recently become quite popular among paleontologists (Gould, 1977; Alberch, 1980).

The way to probe the nature of the constraints is, first, to describe in detail the pattern in question and how it develops and, secondly, to observe to what degree and in what manner this pattern and its development can vary. If it is very easy to bring about variation in a feature that is highly constant in nature, then the basis of the constancy is probably selective. If, on the other hand, a particular feature is specially resistant to environmentally or mutationally induced change, then a developmental constraint underlying that feature is at least suggested, particularly if such a putative constraint can be related to an aspect of the early development of the structure. This analytical strategy has been applied to the pattern of basal bodies in the *Tetrahymena* oral apparatus, to which we now turn.

The anatomy of *Tetrahymena* is shown schematically in Fig. 1. The cell typically possesses 18 to 21 longitudinal ciliary rows, and has an oral apparatus (OA) near its anterior end. The OA includes four assemblies of closely spaced ciliary rows, three on the cell's left called *membranelles* (Fig. 1a, M1, M2, and M3) and one on the right known as the *undulating membrane* (UM). These are located in a depression known as the *buccal cavity*, with the mouth-opening or *cytostome* near its posterior end. The ciliary assemblies probably help sweep food particles, normally bacteria, toward the cytostome.

The process of development of the oral ciliature, as analyzed in *T. thermophila*, is remarkably complex (McCoy, 1974; Nelsen, 1981; Bakowska *et al.*, 1982*b*). In cells that are preparing to divide, an oral primordium (OP) first appears near the equator of the cell, to the cell's left of the right-most of the two ciliary rows that end anteriorly at the old OA (Fig. 1*b*). The development of the oral ciliature is illustrated diagrammatically in Fig. 2. The four ciliary assemblies are constructed in a stereotyped manner. For the three

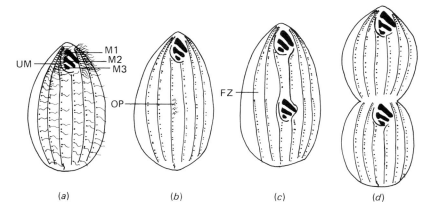

Fig. 1. Diagrams of the oral (ventral) surface of *Tetrahymena thermophila*, (*a*) before the onset of oral development, (*b*) during early oral development, (*c*) just after formation of the fission zone (FZ), and (*d*) during cell division. Oral structures and six ventral ciliary rows are shown in each diagram. Oral structures include three membranelles (M1, M2, M3) and an undulating membrane (UM). These oral structures develop from an oral primordium (OP) which appears in the equatorial region of the cell, to the cell's left of the right postoral ciliary row (diagram *b*). Basal bodies of ciliary rows (dots) and the longitudinal microtubule bands located to the cell's right of each of these rows (lines) are indicated, while the cilia themselves are shown only in diagram (*a*). Note that the mode of cell division permits the perpetuation of the number and arrangement of the pre-existing ciliary rows.

membranelles, this involves production of oriented basal body couplets within the oral field, side-by-side alignment of these couplets into 'promembranelles' each made up of two rows of basal bodies, and subsequent addition of a third (and locally a fourth) row of basal bodies. The UM is constructed by an alignment of single basal bodies into a somewhat irregular longitudinal row, followed by the formation of a second row to its right. After all of the basal bodies of the membranelles and UM have been aligned into regular rows (Fig. 2, stages 5*a* and 5*b*), oral ribs grow out to the left of the UM, and some of the membranellar basal bodies become dislocated posteriorly from their original positions. This dislocation, which immediately precedes the hollowing out of the buccal cavity, generates the characteristic 'sculptured' pattern that involves the right ends of M1 and M2, and all of M3, totally obscuring the latter's original three-rowed regularity.

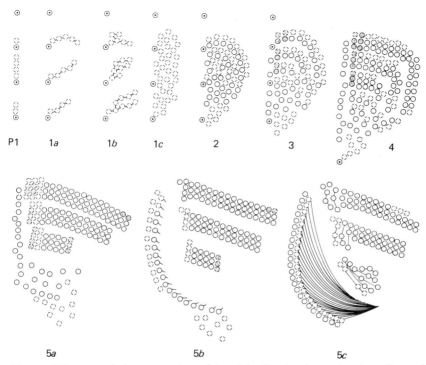

P1 1a 1b 1c 2 3 4

5a   5b   5c

Fig. 2. Diagram of the patterning of basal bodies in the oral primordium of *Tetrahymena thermophila*. Basal bodies are indicated by circles. Those bounded by solid lines $\boxed{(\bigcirc,\odot,\circledast)}$ indicate ciliated basal bodies: basal body of ciliary row, $\boxed{\odot}$; permanent basal body of oral field, $\boxed{\bigcirc}$; basal body of oral field that is destined for resorption, $\boxed{\circledast}$. Circles bounded by dashed lines $\boxed{(\bigcirc,\circledast)}$ indicate unciliated basal bodies, with the stippling having the same meaning as before. The numerals below the diagrams indicate the stages of oral development (Bakowska *et al.*, 1982*b*). Aspects of this process that are of importance here are: (*a*) the formation of an oral (stomatogenic) field (stages P1 to 1*c*), (*b*) the formation of oriented couplets of basal bodies, one of them ciliated and the other not, within the field (stages 2 & 3), (*c*) the assembly of promembranelles by the side-by-side alignment of basal body couplets (stages 3–4), (*d*) generation of a third basal body row in each membranelle by addition of a basal body anterior to each of the aligned couplets (a very short fourth row is also added at the cell's right end of each membranelle) (stages 4–5*a*), (*e*) alignment of the undulating membrane as a single row of basal bodies followed by subsequent addition of a second row (stages 4–5), (*f*) sculpturing of the right ends of the membranelles (stage 5*c*). Sculpturing takes place coincident with the formation of the oral ribs, shown as curved lines to the cell's left of the undulating membrane. The short line-segments between the basal bodies in the sculptured regions of the membranelles do not indicate actual structures, but rather the probable paths of movement of the membranellar basal bodies during the sculpturing process. (Slightly modified from Fig. 27 of Bakowska *et al.*, 1982*b*, with publisher's permission.)

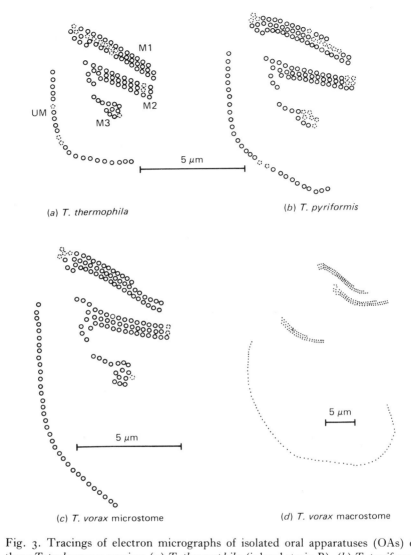

Fig. 3. Tracings of electron micrographs of isolated oral apparatuses (OAs) of three *Tetrahymena* species: (*a*) *T. thermophila* (inbred strain B), (*b*) *T. pyriformis (sensu stricto)* (strain GL-C), (*c*) and (*d*) *T. vorax* (strain V_2S), microstome and macrostome forms, respectively. Only the outlines of basal bodies that ordinarily are ciliated are shown, and other features are omitted. Diagrams (*a*), (*b*), and (*c*) are traced from scanning electron micrographs, and basal bodies are shown schematically in end view as circles. Solid circles indicate basal bodies visible in the specific preparations that were traced, dotted circles those obscured in the traced preparations but with positions inferred from other similar preparations. Diagram (*d*) is traced from a transmission electron micrograph, and basal bodies are

While the new oral primordium is completing its development, the old anterior OA changes as well. The buccal cavity regresses, the old oral ribs disappear, and the membranelles become flush with the surface at stage 5*a*; then at stage 5*c* the oral ribs and buccal cavity re-form. It is particularly interesting that while there is no indication of loss or re-formation of membranellar basal bodies during this process, the sculptured pattern of the membranelles largely disappears when the oral ribs are lost, and becomes re-established when the oral ribs reappear (Bakowska *et al.*, 1982*b*). The displacements of basal bodies that generate the 'sculptured' pattern are thus reversible, and perhaps related to the formation of the buccal cavity.

Whatever the mechanism that generates the sculptured pattern, its end result is remarkably constant within a population (Bakowska *et al.*, 1982*a*). This process endows each membranelle with a shape that is as irregular, stereotyped, and diagnostic as that of any bone of a mammal.

The bacteria-feeding OAs of all *Tetrahymena* species examined to date are remarkably similar (Fig. 3*a*–*c*; the variations in curvature of the UMs are artefacts of preparation). The ciliary patterns of OAs of *T. thermophila* (Fig. 3*a*) and of *T. pyriformis* (Fig. 3*b*) are virtually identical, while those of the microstome OA of *T. vorax* (Fig. 3*c*) are only slightly different from those of the monomorphic tetrahymenas. The pattern of sculpturing of membranelles is similar among these forms. The major differences are in the size of the OAs, which average about 140 basal bodies in *T. thermophila* (Bakowska *et al.*, 1982*a*), 160 in *T. pyriformis* (J. Bakowska, N. E. Williams & J. Frankel, unpublished observations), and 170 in *T. vorax* microstomes (Smith, 1982*a*) (these counts include unciliated basal bodies not shown in Fig. 3). These size differences are achieved by varying the number of basal bodies within the rows that make up the membranelles and UM, while the number of rows remains rigidly constant. The differences in OA

indicated by heavy dots. The locations of the three membranelles (M1, M2, M3) and of the undulating membrane (UM) are labelled in diagram (*a*). Diagrams (*a*), (*b*), and (*c*) are all to the same scale, while diagram (*d*) is to a much smaller scale (compare the scale markers). Diagram (*a*) was traced from Fig. 3 of Bakowska *et al.* (1982*a*), (*b*) and (*c*) from Figs. 6 and 7, respectively, of Williams & Bakowska (1982), and diagram (*d*) from Fig. 10 of Smith (1982*b*).

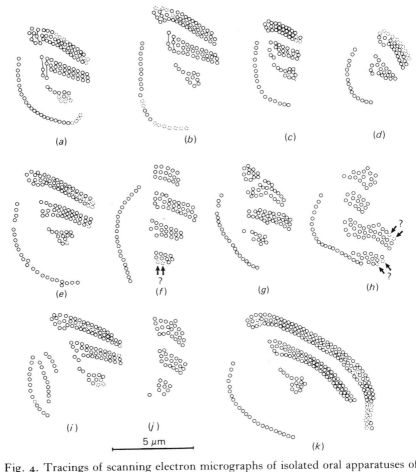

Fig. 4. Tracings of scanning electron micrographs of isolated oral apparatuses of *Tetrahymena thermophila*, inbred strain B, all magnified to the same scale (marker at bottom). Illustrative conventions are as in Fig. 3, with line-segments in diagrams (*a*) and (*b*) indicating probable paths of basal body movement during sculpturing. Question marks indicate uncertainties in inferring positions of obscured basal bodies. (*a*) to (*d*) are diagrams of OAs of wild-type cells, (*e*) to (*k*) of mutants. (*a*) and (*b*): OAs of cells grown in nutrient medium at 28 °C and 38 °C respectively; (*c*) and (*d*): OAs of cells starved for 4 days in a non-nutrient medium at 38 °C. OAs of mutant cells shown in (*e*) to (*j*) are all from cells grown in the same nutrient medium at 28 °C. (*e*) and (*g*): the 'membranellar pattern' (*mp*) segregant of Kaczanowski (1975), now designated *mpA*; (*f*) and (*h*): a 'membranellar pattern' mutant (*mpB*), not allelic to *mpA*; (*i*): the 'misaligned UM' mutant, *mum*; (*j*): secondary OA of the 'janus' mutant, *jan*; (*k*): OA of one of the 'pseudomacrostome' mutants, *psmA1*, isolated following 6 h of culture at 40 °C. Diagrams (*c*) and (*d*) are tracings of preparations shown in Figs. 9 and 13,

size are roughly parallel to differences in cell size among these species (Corliss, 1973; J. Frankel, unpublished observations).

The macrostome form of *T. vorax* (Fig. 3*d*) has undergone obvious alterations in the size and organization of the OA. The huge space between the membranelles and the UM is there in part to accommodate the very large cytostome required to take in ciliate prey (Buhse, 1966*b*; Smith, 1982*b*). The UM and membranelles retain the same basic configuration as in the microstome OA but have increased numbers of basal bodies; the average macrostome OA has about 360 basal bodies, more than twice as many as the microstome (Smith, 1982*b*). Further, the sculpturing of the membranelles is quite different: particularly noteworthy is the retention by M3 of the regular three-row pattern, which in microstome OAs is obliterated during the formation of the buccal cavity. This absence of sculpturing of M3 in the macrostome OA might well be related to alteration of its position relative to the UM and/or the cytostome.

This comparative survey tells us three things relevant to possible constraints in the design of the *Tetrahymena* oral apparatus: first, both the three-rowed architecture of the membranelles and the organization of the UM are invariant; secondly, the *length* of these structures, measured in number of units, is free to change; and thirdly, the pattern of sculpturing of membranelles is amazingly constant in all *Tetrahymena* OAs designed for feeding on bacteria, but changes greatly as soon as the 'design' is modified in association with drastic changes in mode of feeding.

The capacity of *T. thermophila* to produce aberrant OAs was tested by varying culture conditions and inducing appropriate mutations. Some typical results are shown in Fig. 4, which represents tracings of scanning electron micrographs of isolated OAs prepared by J. Bakowska and E. M. Nelsen. Growth in nutrient medium near the upper physiological temperature limit, i.e. 38 to 40 °C, brought about abnormalities in the sculpturing of membranelles, most readily interpreted as an 'undersculpturing'

respectively, of Bakowska *et al.*, (1982*a*). The remaining diagrams are tracings of unpublished micrographs of oral cytoskeletal preparations isolated by J. Bakowska and E. M. Nelsen following the procedures of Vaudaux & Williams (1979), slightly modified (Bakowska *et al.*, 1982*a*). All tracings are from single preparations, except for diagram (*k*), which is a composite with membranelles drawn from one preparation and the undulating membrane from another.

resulting from reduced displacement of basal bodies from their
original positions within rows (compare Fig. 4*a* and *b*). Severe
starvation (at 38 °C) led to formation of OAs with a reduced
number of basal bodies in M1 and M2 and in the UM, without
affecting the three-rowed configuration of the membranelles or the
structure of the UM, and without materially altering the pattern of
sculpturing of the membranelles (Fig. 4*c*, *d*) (Bakowska *et al.*,
1982*a*). Genic alterations at about a dozen known loci bring
about abnormalities in the number and organization of membranelles and/or the UM; examples are shown in Fig. 4*e–k*.
However, despite some disruption in certain mutants, membranelles tend to retain a three-rowed configuration, and the basic
internal structure of the UM is preserved even within UM fragments (T. Lansing, personal communication). But sculpturing
of membranelles is abnormal in all mutants, although in *psmA1*
(Fig. 4*k*) the abnormality is probably caused by the high temperature used to elicit maximum expression; in this mutant, sculpturing
is likely to be normal at lower temperatures.

   This survey permits us to evaluate the relative contribution of
developmental and selective constraints to the evolutionary stability
of the bactivorous *Tetrahymena* OA. The evaluation is necessarily
highly tentative, in part because our mutant selection procedure,
based on changes in cell size and shape, may yield a biased sample
of oral-pattern abnormalities. Subject to this qualification, it seems
that the internal organization of membranelles and the UM might
be developmentally constrained. Since each membranelle originates
developmentally as a two-rowed promembranelle generated by the
side-by-side alignment of basal body couplets (Fig. 2), a one-rowed
membranelle is probably an impossibility. It is perhaps surprising
that mutants with two-rowed membranelles have not been found.
The absence of such mutants could be interpreted as indicating that
the additional basal body proliferation that generates the third row
(Williams & Frankel, 1973) is deeply ingrained in the oral developmental sequence. Such a conclusion is, however, inherently pro-
visional: it is always possible that the next mutant might suppress
this additional basal body proliferation and give rise to membranelles with two rows of basal bodies, such as are actually
observed in other ciliates (Peck *et al.*, 1975; Mulisch *et al.*, 1981).

   Further rounds of proliferation to generate four or more rows of
basal bodies have not yet been found in *Tetrahymena* (except at the

extreme right end of a membranelle), but must certainly be counted as an evolutionary possibility, since members of the related genus *Glaucoma* can produce membranelles with up to 15 rows of basal bodies (de Puytorac *et al.*, 1973) even though they began in development with the usual two-rowed promembranelles (Frankel, 1960; Peck, 1974; de Puytorac *et al.*, 1973).

In the other dimension, the number of basal bodies along the length of each membranelle and of the UM is certainly not subject to any inherent developmental constraint. The number of basal body couplets that are initially produced and mobilized into promembranelles can be adjusted to fit with the size of the cell and may be remarkably increased in certain mutants (e.g. *psmA1*, Fig. 4*k*).

The analysis of environmentally induced and genic variants also revealed a normally latent capacity to make fewer (Fig. 4*d*) or more (Fig. 4*f–h*) than the usual three membranelles, strongly suggesting that the evolutionary stability of the three-membranellar state is probably ascribable to natural selection rather than to any inescapable developmental constraint. Multiple membranelles are characteristic of ciliates of the class Polyhymenophora (Corliss, 1979), which includes well-known forms such as *Stentor* and *Euplotes*.

Most remarkably, the pattern of sculpturing of membranelles, which is so constant across *Tetrahymena* species, is highly sensitive to both environmental and mutational disturbance. This pattern, which appears very late during the ontogeny of the OA (Fig. 2), is thus conservative without being highly canalized. Although this suggests maintenance by rather strong stabilizing selection, we must be careful in drawing conclusions concerning *what* is selected. The visible pattern of sculpturing might be an epiphenomenon (see G. C. Williams, 1966; Gould & Lewontin, 1979), and selection might really be for something else, such as a particular degree of local deformation of the cell surface during shaping of the buccal cavity: basal bodies located in the region of this deformation would then be carried along by the 'streaming' surface, and stop when the stream stops. Such an interpretation would account for the relatively uncanalized nature of the sculpturing process, since any slight difference in the degree of surface deformation would affect the sculptured pattern. This same interpretation would also be consistent with the evolutionary constancy of the pattern, since there is presumably an optimal buccal configuration for efficient feeding on bacteria. When members of dimorphic *Tetrahymena* species remodel their OAs to take in food particles of about three orders of magnitude greater volume, the shape of the buccal cavity will come to differ drastically and so would the pattern of membranellar sculpturing.

Thus, it is likely that both selective and developmental constraints have contributed to the conservatism of oral patterning in microstome tetrahymenas and that considerable although not unlimited latitude has existed for evolutionary change in oral patterns, perhaps associated with a changed mode of feeding as in *T. vorax*. It is especially remarkable that pattern change, when it does occur, involves little or no change in the proteins that make up the pattern (Buhse & Williams, 1982). It is fairly evident that what natural selection 'sees' in this case is a biological function, and thus of necessity the structure required to carry it out, and *not* the individual protein molecules that make up the structure. Of course, if we assume a genic foundation for structural differences (see pp. 297–300) then the structural level of organization must somehow be related to its molecular underpinnings. For the dimorphic *Tetrahymena* species, we might presume that proteins outside the structure regulate the form and pattern of the structure, perhaps rather directly as with the 'scaffolding' protein of bacteriophage P22 (King, 1980), possibly more indirectly by mechanisms at which we can hardly guess.

2. A cytotactically propagated pattern transition might occasionally be the first step in evolutionary change

Inferences concerning how ciliary patterns might change in ciliate evolution depend upon an understanding of the nature and inheritance of variations in patterns that have evolutionary potential. We will delay consideration of variations based on conventional genic changes to the next two sections, and will consider here the more unconventional cases of intraclonal diversities generated in the absence of transmissible genic differences. Many of the known intraclonal diversities are based on epigenetic (Nanney, 1958; Sonneborn, 1977) or genetic (Orias, 1981) changes in the somatic macronucleus and will disappear when the macronucleus is replaced during conjugation. Such changes could be of evolutionary significance only in asexual species (for a possible example of such changes in *T. vorax*, see Williams (1961) and Shaw & Williams (1963)).

Of possibly more general significance are non-genically inherited differences in structural configurations that are propagated longitudinally during clonal growth. These include the difference between the singlet and doublet biotype (Fauré-Fremiet, 1945, 1948; Tartar, 1961; Sonneborn, 1963; Nanney, 1966b; Nanney et al., 1975); differences in number of ciliary rows (Nanney, 1966b; Frankel, 1973, 1980); and the difference between normally oriented and inverted (180°-rotated) ciliary rows (Beisson & Sonneborn, 1965; Ng & Frankel, 1977). All of these differences are clonally propagated, can persist through conjugation, and have been demonstrated to be independent of differences in nuclear genes (Sonneborn, 1963; Beisson & Sonneborn, 1965; Nanney, 1966c; Frankel, 1980). This type of direct perpetuation of pattern probably reflects a continuity at a supramolecular level; Sonneborn (1964) termed this 'cytotaxis', defined as '. . . ordering and arranging of new cell structure under the influence of pre-existing structure'. The longitudinal extension and segmental subdivision characteristic of ciliate growth are especially favorable for cytotactic propagation, and both locally mediated patterns and large-scale developmental fields can be thus perpetuated (Aufderheide et al., 1980).

At first glance, it might be supposed that cytotaxis has substantial evolutionary potential. There are, however, two limiting considerations. One is that cytotactic variants such as the doublet biotype and ciliary-row inversions are easily generated in the laboratory but are not found among ciliates in nature. Substantial variation in number of ciliary rows does, however, occur in nature among forms that are otherwise anatomically similar (Borror, 1980) and that in some cases are known to belong to the same biological species (Genermont et al., 1976; Machelon, 1978). A second limitation is that, although cytotactically propagated differences are sufficiently stable to allow demonstration of a distinctive mode of cellular inheritance, they are transient in evolutionary terms (Nanney, 1977): experimental analysis shows that, after the differences have been generated, the clonal phenotype tends to drift toward a 'stability center' (Nanney, 1966b) that is genically controlled (Heckmann & Frankel, 1968), and at most allows a limited range of stable variation (Frankel, 1973, 1980; Nelsen & Frankel, 1979).

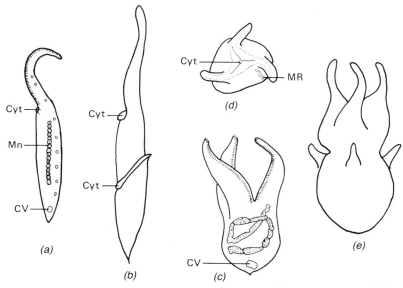

Fig. 5. Diagrams of (*a*) *Dileptus falciforme*, (*b*) *Dileptus anser*, and (*c*)–(*e*)
Teutophrys trisulca, all drawn to approximately the same scale (about 250 ×). (*a*),
(*c*), and (*d*) show non-dividing organisms, (*b*) and (*e*) dividing forms drawn in
outline only. (*a*), (*b*), (*c*), and (*e*) show lateral views, while (*d*) is an apical view
looking down upon the organism. Cyt = cytostome, CV = contractile vacuole,
Mn = macronucleus, and MR = marginal rows ('rangées marginales'). All diagrams are redrawn from original sources: (*a*) from Fig. 6A of Dragesco (1963),
(*b*) from Fig. 1 of Golinska & Doroszewski (1964), (*c*) and (*d*) from Figs. 2 and 1,
respectively, of Clement-Iftode & Versavel (1967), and (*e*) from Fig. 5 of Chatton
& de Beauchamp (1923).

Laboratory maintenance of cytotactically propagated deviant forms
with detectable manifestations in living cells, such as doublet
biotypes and cells with ciliary row inversions, requires frequent
artificial selection, and even this may not achieve indefinite propagation of the deviant form.

Despite these strictures, some observations are suggestive of the
possibility that cytotaxis might indeed have played some role in
ciliate evolution. To consider these observations, we will have to
digress temporarily from the general emphasis on experimentally
well-analyzed ciliates. There is one remarkable ciliate species,
Teutophrys trisulca, that possesses one trunk and three anterior
prosoces (Fig. 5*c*–*e*). Each of these three proboces resembles the
single proboscis of another ciliate genus, *Dileptus* (Fig. 5*a*, *b*). Two

circumstances suggest that *Teutophrys* is a distinct and well-established species rather than just an accidental monstrous *Dileptus*. First, although rare, *Teutophrys* has been reported about a dozen times in widely separated localities on three continents (Clément-Iftode & Versavel, 1967; Sonneborn, 1971; Corliss, 1979, p. 79). Secondly, *Teutophrys* has undergone substantial integration: although it has three well-defined probosces, it possesses only a single contractile vacuole (Fig. 5c), only a single cytostome at the conjoined bases of the probosces (Fig. 5d), and only a single set of certain specialized cilia (Fig. 5d; Clément-Iftode & Versavel, 1967). The various descriptions of *Teutophrys* are sufficiently consistent to suggest a well-defined, though rare, morphological species. Fauře-Fremiet (1954, p. 322) suggested that *Teutophrys* might have arisen as a consequence of a 'polymerization' event in an ancestral *Dileptus*-like organism; the 'polymerized' form might then have multiplied where it found conditions favorable for its continued stability. The only aspect of this otherwise plausible hypothesis that is somewhat surprising is the natural occurrence of a 'triplet' ciliate species with no naturally occurring 'doublet'; experimentally generated doublet ciliates are typically more stable than triplets (Fauré-Fremiet, 1945; Tartar, 1961, p. 208).

To find naturally occurring doublet forms, we have to leave the ciliates. One entire order of flagellates, the Diplomonadida, is characterized by cells that possess two nuclei and two complete flagellar systems. Although many of these organisms are superficially bilateral (see illustrations in Grassé, 1952), electron microscopy has revealed that they all express a two-fold rotational symmetry of their underlying patterns (e.g. *Octamitus*, Fig. 6) (Brugerolle, 1975b, 1975c). Brugerolle (1975a) has also described a flagellate genus, *Enteromonas*, in which each cell possesses a *single* ensemble of flagellar structures that resembles *each* of the sets that is doubled in diplomonads; *Enteromonas* also manifests transient doublet forms that resemble the stable free-living diplomonad species *Hexamita* in most respects (Brugerolle, 1975a). This, then, is the most highly suggestive example of a major change in cell organization that could have come about by an evolutionary stabilization of what might have begun as a developmental accident.

There is a major problem in interpreting events of evolutionary 'polymerization' such as we have considered in *Dileptus* –

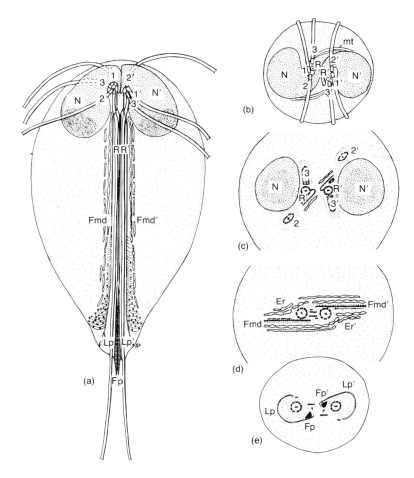

Fig. 6. Schematic diagram of the flagellar assemblies and associated structures of the diplomonad flagellate *Octamitus*. (*a*) Longitudinal section of axial structures, plus flagella; (*b*), (*c*), (*d*), and (*e*), cross-sections at progressively more posterior levels. The numbers within each diagram refer to the three free flagella of each set, R to the recurrent flagellum that courses parallel to the microtubular fibers (Fmd) and ordered layers of ergastoplasm (Er). N = nucleus, mt = prenuclear microtubules; Fp = terminal periodic fiber, Lp = layer with periodic structure. The nucleus and flagellar structures of the set on the viewer's left are unprimed, those of the set on the viewer's right primed. Note that the two sets are superimposable, and thus display a two-fold rotational symmetry despite the superficially bilateral form of the organism. (From Fig. 1 of Brugerolle *et al*. (1974) with permission of author and publisher.)

Teutophrys and in the diplomonad flagellates. Although a breeding analysis is not possible, we might suppose that *Teutophrys* would differ from *Dileptus*, and *Hexamita* from *Enteromonas*, not just in the states of 'polymerization' of patterns but also in genes that affect the stability of these states. If so, were the genic or the cytotactic differences the primary ones? At one extreme, if the 'polymerization' event were very rare and its cytotactic propagation very stable, then we could truly regard the morphogenetic 'accident' that generated the cytotactically propagated variant as the primary evolutionary event, and the probable subsequent shift of genotype that further stabilized this variant as secondary. At the other extreme, if the 'polymerization' process were relatively frequent but the cytotactic propagation of the ensuing variant were weak and error prone, then the decisive evolutionary event would be a shift in genotype to one that altered the 'stability center' in such a way that otherwise ephemeral pattern-variants would become more stable. Brugerolle's observations on *Enteromonas* suggest that the latter alternative is nearer the truth for the origin of the Diplomonadida, and I will review, below, a documented example of this kind of evolutionary change in the ciliate *Euplotes* (pp. 304–5). The manner of evolution of *Teutophrys* will doubtless remain forever enigmatic, but the general rarity of ciliate triplets suggests that an original sudden occurrence of a cytotactically propagated morphological change followed by subsequent genic stabilization is at least plausible.

3. New evolutionary directions might be initiated by developmental transitions brought about by single-gene mutations

The pseudomacrostome mutations in Tetrahymena

We will now consider whether any of the mutations of *Tetrahymena thermophila* that affect oral patterns (see pp. 290–1 and Fig. 4) have evolutionary potential. Some of the mutations, such as those that generate OAs with two or four membranelles, open evolutionary opportunities that, as far as we know, have not been exploited, since those ciliate species that have membranelles are remarkable in possessing either three or many (Corliss, 1979). Mutations of the *psm* (pseudomacrostome) type (Fig. 4*k*) are more promising in this regard. These mutations generate unusually large OAs. To understand how they do this, we must return to oral development in *Tetrahymena*. Figure 7 diagrammatically indicates the three major configurations of oral primordia found in *T. thermophila*. The first two, (*a*) and (*b*), are commonly manifested by wild-type cells. Formation of a mid-body oral primordium is an early step in preparation for cell division (Fig. 7*a*; cf. Fig. 2), while an anterior oral primordium typically develops in starved cells and forms a new OA that replaces the old one (Frankel, 1969; Nelsen, 1978). The anterior 'oral replacement' primordium is initially bipartite, being derived in part from a proliferative dispersion of the old UM, and in part from basal body formation adjacent to the extreme anterior portion of the right-postoral ciliary row (Frankel, 1969; Kaczanowski, 1976). Wild-type cells entering the stationary phase of culture growth in nutrient media also produce oral replacement primordia, which sometimes involve a more substantial portion of the right-postoral ciliary row (E. M. Nelsen, personal communication). Surprisingly often, 'pseudomacrostome' mutants are obtained that greatly increase the variability in latitude and length of the oral primordium, while not affecting its longitude (Frankel, 1979; J. Frankel & L. M. Jenkins, unpublished). A characteristic expression of these mutants is a very long oral replacement primordium that involves the UM plus the anterior $\frac{1}{2}$ to $\frac{3}{4}$ of the right-postoral ciliary row (Fig. 7*c*). This primordium typically develops into an unusually large OA that generally (though not invariably) possesses three membranelles, with the first two being much longer than the usual M1 and M2 (Fig. 4*k*). Such mutations have been found not only at the original '*psm*' locus (now *psmA*) (Frankel *et al.*, 1976; L. M. Jenkins, unpublished), but also at three other loci (L. M. Jenkins & J. Frankel, unpublished). Hence, recurring mutations, which doubtless sometimes occur in nature as well, can open a developmental window that allows formation of membranelles that are very much longer than the usual ones.

The 'pseudomacrostome' OAs of *T. thermophila* resemble the true macrostome OAs of *T. vorax* in two respects: first, M1 and M2 of 'pseudomacrostome' *T. thermophila* are often as long as the corresponding membranelles in the macrostome OA of *T. vorax*; secondly, the elongated oral primordia that generate the 'pseudomacrostome' OAs resemble, at least superficially, the oral primordia.

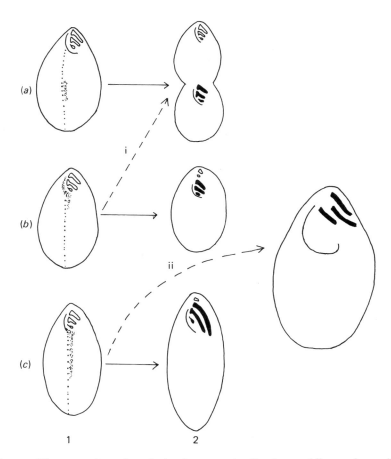

Fig. 7. Three modes of oral development in *T. thermophila*, and possible evolutionary transitions related to these modes. Column (1) indicates the as yet unorganized oral fields, column (2) a late stage near completion of oral development. Solid arrows indicate development in *T. thermophila*, dashed arrows development in other species. (*a*) Midbody oral development leading to cell division, (*b*) anterior oral development leading to oral replacement, (*c*) elongated oral replacement, leading to formation of a 'pseudomacrostome' OA. Sequence (i) represents formation of a *Tetrahymena*-like OA during cell division from an anterior primordium that then slips posteriorly (observed in a 'scuticociliate', *Dexiotricha media* (Peck, 1974) that anatomically greatly resembles *Tetrahymena*). Sequence (ii) indicates the formation of a macrostome OA from a very long oral replacement primordium, as is illustrated by Buhse (1966*b*) for *T. vorax*.

of *T. vorax* that bring about formation of the macrostome OA by oral replacement (Buhse, 1966*b*). Hence, it is not hard to imagine that the first step in the evolutionary transformation of a monomorphic bacteria-feeding *Tetrahymena* into a dimorphic form capable of producing a carnivorous macrostome OA might have been the occurrence of a mutant such as one of the *psm* series, followed by expression after inbreeding. This initial step might have been taken more than once, since analysis of major cortical proteins suggests that different dimorphic *Tetrahymena* species might be more closely related to other monomorphic species than to each other (Vaudaux *et al.*, 1977).

It is important to emphasize that even if this hypothesis were true, the occurrence of a 'pseudomacrostome' mutation would be only the first step in the evolution of a dimorphic *Tetrahymena* species. Despite the very large M_1 and M_2, the buccal design of 'pseudomacrostome' OAs is still basically bactivorous, with a small buccal cavity and a small and normally sculptured M_3 (Fig. 4*k*). None of the several *psm* mutations obtained thus far has gone further in the direction of a true macrostome OA. Thus, we might suspect that a mutation of this type occurring in nature would, at most, have allowed occasional sampling of larger prey, and thus promoted the selection of further modifications which, after an unknown number of steps, might have resulted in a true macrostome oral apparatus.*

Evolutionary transitions in marine Euplotes

While *Tetrahymena* provides an object lesson in evolutionary conservatism of cell-surface pattern plus some suggestive hints as to how pattern transitions might occur, it yields no documented example of actual evolutionary change in pattern. What we would like to have are pairs of incipient species that have undergone morphological divergence, in which the genetic bases of both the ongoing speciation process and the morphological change are understood. The only example in the Protozoa that comes close to

---

* This discussion presupposes the evolutionary primacy of the bactivorous microstome form. It is impossible to rule out the alternative possibility that the carnivorous macrostome form might have come first. In that event, a mutant of the 'pseudomacrostome' type in a monomorphic bacteria-feeding *Tetrahymena* would have to be regarded as an 'atavic' mutation, analogous to many of the mutations of the *bithorax* complex in *Drosophila melanogaster* (García-Bellido, 1977).

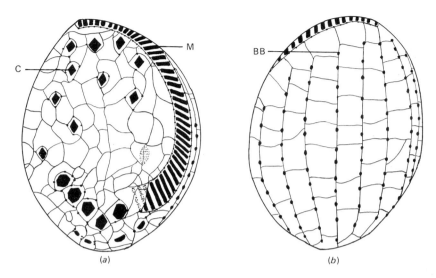

Fig. 8. Spatial organization of surface structures of silver-impregnated *Euplotes minuta*, from Heckmann & Frankel (1968). The ventral surface (*a*) possesses two major types of compound ciliary structures, membranelles (M) and cirri (C). The dorsal surface (*b*) has longitudinally oriented rows of 'bristles', which really are ciliated basal body (BB) pairs (Ruffolo, 1976). The lines indicate the silver-stained surface mesh, or 'argyrome', actually membrane folds (Ruffolo, 1976). Except for minor details such as the number of 'bristles' per row, the cell-surface organization of *E. vannus* and *E. crassus* is indistinguishable from that of *E. minuta*. (From Fig. 1 of Heckmann & Frankel (1968), with publisher's permission.)

meeting all of these requirements is encountered among certain very similar bottom-dwelling marine ciliates of the genus *Euplotes*, the *E. vannus–crassus–minuta* group. The surface anatomy of these ciliates is illustrated in Fig. 8. Although comprising at least three biological species (and quite probably more – cf. Machelon, 1978), the members of this group are similar in many ways: their ventral cirral patterns are indistinguishable (Gates, 1978) and their dorsal and caudal ciliary patterns are nearly so (Genermont *et al.*, 1976). The system of genic control of mating types is identical in all of these ciliates (Heckmann, 1963, 1964; Nobili, 1966). Further, they can pair with each other in interspecific matings, although interspecific gene exchange is aborted. Cells of one species can also induce members of another to conjugate intraclonally (Nobili *et al.* 1978). The species do differ in size (*E. minuta* is substantially smaller than either *E. crassus* or *E. vannus*) and shape (*E. crassus* and *E. minuta* ovoid, *E. vannus* nearly rectangular (Kahl, 1935, p. 636; Heckmann, 1963)). Very recent electrophoretic studies of total proteins indicate that the molecular distances among these species are large (Luporini & Seyfert, 1981).

Within this species-group at least two examples of incipient speciation have been analyzed in which there is information about how the divergence may have started and (in one case) about how an anatomical difference might subsequently have arisen. Since in both of these cases the probable initial event was a divergence in breeding systems, we need to consider this topic in more detail. The typical members of the *Euplotes vannus–crassus–minuta* group are all extreme outbreeders (Nobili *et al.*, 1978), with multiple genically determined mating types, long post-conjugation 'immaturity periods', and an absence of intraclonal conjugation except in aged cells of *E. crassus* heterozygous at the mating type locus (Heckmann, 1967). However, in two cosmopolitan members of this group, *E. minuta* and *E. crassus*, natural populations have been found that are capable of undergoing autogamy (Nobili, 1966; Siegel & Heckmann, 1966; Luporini, 1970; Luporini & Dini, 1977). Autogamy is a form of automictic parthenogenesis in which mitotic derivates of meiotic products fuse within a *single* cell instead of being interchanged between two conjugating cells. This process can occur without prior mixture of cells of different mating types. In the most familiar form of autogamy, encountered in the members of the '*Paramecium aurelia*' sibling-species swarm, the two nuclei that fuse are derived from mitosis of a *single* meiotic product, and the consequence is instant and total homozygosity (Sonneborn, 1947). In *Euplotes* the situation is more complicated, since in these organisms *two* meiotic products enter the final mitotic division, and then two of the four ensuing mitotic nuclei – not necessarily sister nuclei – survive. Surviving mitotic nuclei might thus be derived from different meiotic products (Heckmann, 1963; Nobili & Luporini, 1967). Results of crosses involving heterozygous cells of obligate outcrossing stocks indicate that these sequential nuclear selection events may be random (Heckmann, 1963, 1964) or may be biased either in the direction of excess homozygosity (Nobili & Luporini, 1967) or excess hetero-zygosity (Luporini & Dini, 1977); in contrast, stocks capable of autogamy always show excess heterozygosity even when conjugating (Nobili & Luporini, 1967), and do so to an extreme degree when going through autogamy (Nobili & Luporini, 1967; Heckmann & Frankel, 1968; Luporini & Dini, 1977), in a manner apparently independent of overall viability of the cross. These stocks were often heterozygous at the mating type locus when isolated from nature (Nobili & Luporini, 1967; Luporini & Dini, 1977).

Although the stocks capable of autogamy tend to maintain heterozygosity, there are three lines of evidence indicating that they are largely isolated in nature from stocks of the same 'species' incapable of autogamy. First, conjugation between cells of autogamous (A) and non-autogamous (NA) stocks is more sporadic (Luporini & Dini, 1977) and more delayed in onset (Nobili *et al.*, 1978) than is conjugation between cells of two NA stocks of complementary mating type. Secondly, substantial mortality is observed subsequent to crosses between A and NA stocks; this is especially pronounced in F_2s and backcrosses involving A and NA *E. minuta* (Siegel & Heckmann, 1966; Luporini & Nobili, 1967; Heckmann & Frankel, 1968). Thirdly, electrophoretic comparisons indicate that the differences between sympatric A stocks and NA stocks of *E. crassus* are greater than the differences among different A stocks or among different NA stocks, and are about the same as the difference between *E. crassus* (either A or NA) and *E. minuta* (NA) (Luporini & Seyfert, 1981; Dini & Giorgi, 1982). Autogamous stocks of *E. minuta* were not analyzed, but, since the mortality following crosses is greater between A and NA *E. minuta* than it is between A and NA *E. crassus*, we may presume that the divergence between A and NA stocks is at least as great within *E. minuta* as it is within *E. crassus*.

The capacity to undergo autogamy is dependent upon a dominant allele at a single gene locus both in *E. minuta* (Heckmann & Frankel, 1968) and in *E. crassus* (Dini & Luporini, 1980). Since the electrophoretic studies cited above indicate that the A and NA stocks must differ in many genes, it is likely that, as Luporini & Nobili (1967) first suggested, the A and NA stocks in fact represent different subspecies within *E. minuta* and *E. crassus*, and that, as Nobili *et al.* (1978, p. 611) state, 'The mutation of the *a* (autogamy) locus may have started the development of sexual isolation by the disruption of the breeding system. Afterwards, the autogamous and non-autogamous stocks (i.e. populations) may have come to be separated by multiple gene differences determining ethological and cytogenetic incompatibility.'

What about the morphological relationship between the auto-gamous and non-autogamous subspecies? Only *E. minuta* has been analyzed from this perspective. Starting from the observation that the autogamous stock A-23 of *E. minuta* 'looked different' from non-autogamous *E. minuta* stocks (K. Heckmann, personal communication), Heckmann & Frankel (1968) demonstrated that there is a difference between the maximum number of dorsal ciliary rows of A and NA stocks (ten in the former, nine in the latter), and that this difference is under genic control. Although the number of stocks analyzed in this study was very small, the results offer some support for the conclusion that the two subspecies of *E. minuta* differ morphologically as well as ethologically and cytogenetically.

4. Consolidation of evolutionary change in morphology probably involves many genes

In the previous section, we considered two examples of major developmental changes of potential evolutionary significance. In one of these examples, that of the 'pseudomacrostome' mutants in *Tetrahymena thermophila*, the insufficiency of the presumed primary change has already been pointed out (p. 300). However, we have no information about how many genetic steps might be required to derive a true carnivorous form from a putative 'pseudomacrostome' ancestor, and we know nothing about how such changes might have been integrated with speciation processes.

In the case of the divergence between autogamous and non-autogamous *Euplotes* subspecies, the mutation presumed responsible for initiating the divergence was a regulatory one that permitted cells to trigger meiosis and undergo self-fertilization in the absence of cells of another mating type. The morphological change probably came later. While the original study of Heckmann & Frankel (1968) demonstrated that control of the maximal ciliary-row number is genic, a follow-up study (Frankel, 1973) led to two further conclusions: (*a*) the difference in cortical pattern between the A and NA stocks is almost certainly controlled by alleles at more than one gene locus, and (*b*) what is controlled is not simply the capacity to express or not to express a specific digital phenotype (ten ciliary rows, in this case), but rather the *range* of phenotypes that may be stably perpetuated by cytotactic mechanisms. Even NA stocks could express ten ciliary rows, but in these stocks this cortical phenotype is extremely rare and almost certainly very unstable; in A stocks it is common and stable. What we may presume had happened during evolution was that the autogamous subspecies of *E. minuta*, following its initial establishment, underwent multiple genetic changes some of which had the consequence of conferring greater stability upon the ten-rowed cortical phenotypes that occasionally arose from nine-rowed progenitors. This change, although unspectacular, is still the only documented example of a morphological change that can be linked to a speciation event in a protozoan. Even this conclusion, however, is tentative until a larger number of A and NA stocks are suitably analyzed.

Conclusions

One proposed evolutionary mechanism that is making somewhat of a comeback in the literature is that of a major developmental switch generated by a single mutational step, bringing about formation of a 'hopeful monster' (Goldschmidt, 1940). In ciliates, we have identified two mechanisms by which such 'hopeful monsters' might originally arise. One is a genic mutation that generates a major change in some aspect of development, as in the length of membranelles in 'pseudomacrostome' *Tetrahymena*; the other is a developmental 'accident' giving rise to a new biotype that is cytotactically propagated, as in the postulated *Dileptus* triplet that might have been the ancestor of *Teutophrys*. In both cases, however, there is a large gap between the observed (or presumed) 'hopeful monster' and the new morphological species to which it might have given rise: pseudomacrostome *Tetrahymena thermophila* are still a long way from true macrostomes, and the hypothetical triplet *Dileptus*-like ancestor must have had three cytostomes and three contractile vacuoles, compared to one of each in the contemporary *Teutophrys*. Further, in the best-documented ciliate example of morphological change associated with ongoing speciation, that of the autogamous *Euplotes minuta*, the most likely sudden change was in the method of breeding; the morphological change is modest, probably polygenically based, and likely to have arisen secondarily.

Viewed from a developmental perspective, the great majority of the known mutations that alter pattern or form in ciliates tend to be rather limited in their effects. Even the 'pseudomacrostome' (*psm*) mutations in *T. thermophila* do not change the basic organization of membranelles but rather affect those features (especially membranelle length) that are known on independent grounds to be only slightly canalized. The one mutation that has a really far-reaching effect on geometrical organization, the *janus* mutation of *T. thermophila* (Frankel & Jenkins, 1979), which generates a new partial mirror-image organization in one half-cell (Jerka-Dziadosz & Frankel, 1979; Frankel & Nelsen, 1981), has helped to reveal the probable mechanism of action of a developmental constraint, rather than an evolutionary opportunity. A mutant of the *janus* type could be viewed as an early step in the evolution of bilateral organization in ciliates. There is, however, no ciliate species in which the organization of ciliary structures is truly bilateral: even where form is bilateral, ciliary organization remains asymmetric, as in the ciliate *Folliculina* in which one intrinsically asymmetrical membranelle band is spread over two wing-like structures whose form is bilaterally symmetrical (Fauré-Fremiet, 1932; Mulisch *et al.*, 1981). The reason why a truly bilateral organization is unlikely in ciliates is the inevitable conflict between any reversed higher-order pattern and the unchanging underlying asymmetry of the ciliary units themselves (Grimes *et al.*, 1980). This circumstance explains why surgically constructed mirror-image configurations are always incomplete and inviable (e.g. Fauré-Fremiet, 1945; Tartar, 1966), and similarly explains the incompleteness of the reversed oral structures of *janus* tetrahymenas (J. Frankel, unpublished), in which local ciliary asymmetry is known to remain unaltered (Jerka-Dziadosz, 1981) despite the global pattern reversal. This example illustrates the important point that even mutations that offer fascinating opportunities for analysis of underlying developmental mechanisms need not indicate feasible directions of evolutionary transformation.

If there is one highly tentative generalization that might best summarize our current fragmentary knowledge of the tempo and mode of ciliate evolution, it is that of a 'key mutation' (Simpson, 1953). This idea, to my knowledge first proposed for certain Eocene mammals by Bryan Patterson (1949), suggests that an initial mutation might bring about a relatively modest change in a single character that would enable its bearer to begin to exploit a new ecological niche; this initial exploitation would then subject the organism to new selective pressures that would elicit numerous and relatively gradual transformations in other characters. Applying this idea to ciliates, mutations such as those generating 'pseudo-macrostome' oral apparatuses in *T. thermophila*, or the capacity for autogamy in *E. minuta*, might have brought about the requisite 'key' changes. In ecological terms, a pseudomacrostome-type mutation might have promoted the occasional sampling of larger prey. A mutation that permits autogamy probably directly affected only breeding systems, but this in turn almost certainly initiated major physiological changes that are likely to be ecologically important (Dini, 1981).

A corollary of this hypothesis is that the secondary transformations will be intimately related to the initial 'key' change. Where the initial change is morphological and creates a mechanical challenge to the organism, such as that of 'swallowing' larger prey, a radical morphological transformation is inevitable. Where the initial basic change is physiological, subsequent morphological change might be minimal or non-existent.

How can we further improve our understanding of mechanisms of evolution of form and pattern in unicellular organisms? In my opinion, it is essential both to analyze mutational and cytotactic variations generated in the laboratory and to carry out quantitative morphological analysis of material from nature; the former shows what kinds of changes *can* occur, the latter examines which actually *do* occur in the 'real world'. The types of questions that need answering pertain to the relationship of evolutionary units to morphological variation and the genetic basis of both the breeding and the morphological divergence. It would be especially illuminating if this type of analysis could be carried out on morphologically differing populations obtained within a restricted geographical area. To give one illustrative example, Borror (1980) reported that neighboring populations of '*Diophrys scutum*-like hypotrichs' found in exposed beaches and tidal marsh pools differed greatly in number of dorso-lateral ciliary rows. The questions we would like to have answered, then, are as follows: Are these differences stable and inheritable in clonal culture? Are some or all of these populations capable of interbreeding? What are the breeding systems? If stable morphological differences are found among interbreeding populations, what are the genetic bases of these differences? If some coherent pattern of association of morphological differences, breeding patterns and barriers, and geographical or ecological differences do emerge, would analysis of proteins reveal the differences to be recently derived, or ancient? Questions of this kind, though not quickly and easily answered, are at least answerable. It is my personal suspicion that research of the above kind carried out on various sorts of ciliates will reveal that the extreme morphological conservation characteristic of the well-studied '*Tetrahymena pyriformis*' sibling species swarm might turn out to be less than universal, and that situations may be found in which recent or ongoing intracellular morphological change can be analyzed.

I am indebted to Drs Anne W. K. Frankel, E. Marlo Nelsen, Norman E. Williams, and the late T. M. Sonneborn for stimulating discussions relevant to the topic of this article, to Drs F. Dini, M. Jerka-Dziadosz, D. L. Nanney, H. E. Smith, and N. E. Williams for allowing me to read and cite their manuscripts prior to publication, to Drs Dini, Frankel, Nelsen, and Williams as well as Drs G. Grimes, K. Heckmann, J. Kaczanowska, P. Luporini, D. H. Lynn, R. Milkman, S. Ng, and C. Van Bell for their useful criticisms of earlier drafts of this manuscript, and to Mr L. M. Jenkins for having isolated and characterized the *Tetrahymena* mutants that made this paper possible. Dr Anne Frankel provided invaluable editorial assistance. The research in the author's laboratory that is reported here was supported by US National Science Foundation grant PCM 79–21405 and National Institutes of Health grant HD-08485 to J. Frankel, National Science Foundation grant PCM 80–21879 to E. M. Nelsen, and an International Exchange (Fulbright) fellowship to Dr J. Bakowska.

References

Alberch, P. (1980). Ontogenesis and morphological diversification. *Am. Zool.* **20**, 653–67.

Aufderheide, K. J., Frankel, J. & Williams, N. E. (1980). Formation and positioning of surface-related structures in Protozoa. *Microbiol. Rev.* **44**, 252–302.

Bakowska, J., Frankel, J. & Nelsen, E. M. (1982*a*). Regulation of the pattern of basal bodies within the oral apparatus of *Tetrahymena thermophila*. *J. Embryol. exp. Morph.* **69**, 83–105.

Bakowska, J., Nelsen, E. M. & Frankel, J. (1982*b*). Development of the ciliary pattern of the oral apparatus of *Tetrahymena thermophila*. *J. Protozool.* **29**, 366–82.

Beisson, J. & Sonneborn, T. M. (1965). Cytoplasmic inheritance of the organization of the cell cortex in *Paramecium aurelia*. *Proc. natn. Acad. Sci. USA* **53**, 275–82.

Borror, A. C. (1980). Spatial distribution of marine ciliates: micro-ecologic and biogeographic aspects of protozoan ecology. *J. Protozool.* **27**, 10–13.

Brugerolle, G. (1975a). Etude ultrastructurale du genre *Enteromonas* da Fonseca (Zoomastigophorea) et révision de l'Ordre des Diplomonadida Wenyon. *J. Protozool.* **22**, 468–75.

Brugerolle, G. (1975b). Contribution à l'étude cytologique et phylétique des Diplozoaires (Zoomastigophorea, Diplozoa, Dangeard 1910). II. Nouvelle interprétation de l'organisation cellulaire de *Giardia*. *Protistologica* **11**, 99–109.

Brugerolle, G. (1975c). Contribution à l'étude cytologique et phylétique des Diplozoaires (Zoomastigophorea, Diplozoa, Dangeard 1910). VI. Caractères généraux des Diplozoaires. *Protistologica* **11**, 111–18.

Brugerolle, G., Joyon, L. & Oktem, N. (1974). Contribution à l'étude cytologique et phylétique des Diplozoaires (Zoomastigophorea, Diplozoa Dangeard 1910). IV. Etude ultrastructurale du genre *Octomitus* (Prowazek, 1904). *Protistologica* **10**, 457–473.

Buhse, H. E. Jr (1966a). An analysis of macrostome production in *Tetrahymena pyriformis* strain V$_2$ type S. *J. Protozool.* **13**, 429–35.

Buhse, H. E. Jr (1966b). Oral morphogenesis during transformation from microstome to macrostome and macrostome to microstome in *Tetrahymena vorax* strain V$_2$ type S. *Trans. Am. micros. Soc.* **85**, 305–13.

Buhse, H. E. Jr & Williams, N. E. (1982). A comparison of cortical proteins in *Tetrahymena vorax* microstomes and macrostomes. *J. Protozool.* **29**, 222–6.

Chatton, E. & deBeauchamp, P. (1923). *Teutophrys trisulca* n.g. n. sp. Infusoire pélagique d'eau douce. *Arch. Zool. exp. gén.* **61** (Notes et Revue), 123–9.

Chatton, E. & Villaneuve, S. (1937). *Gregarella fabrearum* Chatton et Brachon, Protiste parasite du cilié *Fabrea salina* Henneguy. La notion de dépolarisation chez les Flagellés et la conception des apomastigines. *Arch. Zool. exp. gén.* **78**, 216–37.

Clément-Iftode, F. & Versavel, G. (1967). *Teutophrys trisulcata* (Chatton, de Beauchamp): Cilié planctonique rare. *Protistologica* **3**, 457–64.

Corliss, J. O. (1973). History, taxonomy, ecology, and evolution of species of *Tetrahymena*. In *The biology of* Tetrahymena, ed. A. M. Elliott, pp. 1–55. Dowden, Hutchinson, and Ross: Stroudsburg, PA.

Corliss, J. O. (1979). *The ciliated protozoa: characterization, classification, and guide to the literature*. Pergamon Press: Oxford.

de Puytorac, P. Savoie, A. & Roque, M. (1973). Observations cytologiques et biologiques sur le Cilié polymorphe *Glaucoma ferox* Savoie & de Puytorac 1971. *Protistologica* **9**, 45–63.

Dini, F. (1981). Relationship between breeding systems and resistance to mercury in *Euplotes crassus* (Ciliophora: Hypotrichida). *Mar. Ecol. Prog. Ser.* **4**, 195–202.

Dini, F. & Giorgi, F. (1982). Electrophoretic analysis of *Euplotes crassus* stocks from populations differing in their breeding systems. *Can. J. Zool.* **60**, 929–32.

Dini, F. & Luporini, P. (1980). Genic determination of the autogamy trait in the hypotrich ciliate, *Euplotes crassus*. *Genet. Res. Camb.* **35**, 107–19.

Dragesco, J. (1963). Révision du Genre Dileptus, Dujardin 1871 (Ciliata Holotricha). Systématique, cytologie, biologie. *Bull. biol. Fr. Belg.* **97**, 103–45.

Elliott, A. M. (1973). Life cycle and distribution of *Tetrahymena*. In *The biology of* Tetrahymena, ed A. M. Elliott, pp. 259–86. Dowden Hutchinson, & Ross: Stoudsburg, PA.

Fauré-Fremiet, E. (1932). Division et morphogenèse chez *Folliculina ampulla* O. F. Müller. *Bull. biol. Fr. Belg.* **66**, 78–110.

Fauré-Fremiet, E. (1945). Symetrié et polarité chez les Ciliés bi-ou-multicomposites. *Bull. biol. Fr. Belg.* **79**, 106–50.

Fauré-Fremiet, E. (1948). Doublets homopolaires et régulation morphogénétique chez le Cilié *Leucophrys patula*. *Arch. Anat. microsc. Morph. exp.* **37**, 183–203.

Fauré-Fremiet, E. (1950). Morphologie compareé et systématique des Ciliés. *Bull. Soc. zool. Fr.* **75**, 109–22.

Fauré-Fremiet, E. (1954). Les problèmes de la différenciation chez les Protistes. *Bull. Soc. zool. Fr.* **79**, 311–29.

Frankel, J. (1960). Morphogenesis in *Glaucoma chattoni*. *J. Protozool.* **7**, 362–76.

Frankel, J. (1969). Participation of the undulating membrane in the formation of the oral replacement primordium of *Tetrahymena pyriformis*. *J. Protozool.* **16**, 26–35.

Frankel, J. (1973). Dimensions of control of cortical patterns in *Euplotes:* the role of pre-existing structure, the clonal life cycle, and the genotype. *J. exp. Zool.* **183**, 71–94.

Frankel, J. (1974). Positional information in unicellular organisms. *J. theor. Biol.* **47**, 439–81.

Frankel, J. (1979). An analysis of cell surface patterning in *Tetrahymena*. in *Determinants of spatial organization*, eds. S. Subtelny and I. R. Konigsberg, pp. 215–46. Academic Press: New York.

Frankel, J. (1980). Propagation of cortical differences in *Tetrahymena*. *Genetics* **94**, 607–23.

Frankel, J. (1982). Global patterning in single cells. *J. theor. Biol.*, **99**, (in press).

Frankel, J. & Jenkins, L. M. (1979). A mutant of *Tetrahymena thermophila* with a partial mirror-image duplication of cell surface pattern. II. Nature of genic control. *J. Embryol. exp. Morph.* **49**, 203–27.

Frankel, J., Jenkins, L. M. & Debault, L. E. (1976). Causal relations among cell cycle processes in *Tetrahymena pyriformis*: an analysis employing temperature sensitive mutants. *J. Cell Biol.* **71**, 242–60.

Frankel, J. & Nelsen, E. M. (1981). Discontinuities and overlaps in patterning within single cells. *Phil. Trans. R. Soc. Lond. B* **295**, 525–38.

García-Bellido, A. (1977). Homoeotic and atavic mutations in insects. *Am. Zool.* **17**, 613–29.

Gates, M. A. (1978). Morphometric variation in the hypotrich ciliate genus *Euplotes*. *J. Protozool.* **25**, 338–50.

Gates, M. A. & Berger, J. (1976). Morphological stability in *Tetrahymena pyriformis*. *Trans. Am. microsc. Soc.* **95**, 11–22.

Genermont, J., Machelon, V. & Tuffrau, M. (1976). Donées expérimentales relatives au problème de l'espèce dans le genre *Euplotes* (Ciliés Hypotriches). *Protistologica* **12**, 239–48.

Goldschmidt, R. (1940). *The material basis of evolution.* Yale University Press: New Haven, Conn.

Golinska, K. & Doroszewski, M. (1964). The cell shape of *Dileptus* in the course of division and regeneration. *Acta protozool.* **2**, 59–67.

Gould, S. J. (1977). *Ontogeny and phylogeny.* Harvard University Press: Cambridge, Mass.

Gould, S. J. & Lewontin, R. C. (1979). The spandrels of San Marco and the Panglossian paradigm: a critique of the adaptationist programme. *Proc. R. Soc. Lond. B* **205**, 581–598.

Grain, J. (1969). Le cinétosome et ses dérivés chez les Ciliés. *Anneé biol.* **8**, 53–97.

Grassé, P. P. (1952). Ordre des Distomatinés ou Diplozoaires. *Traité de Zoologie*, ed. P. P. Grassé, **1**, (1) pp. 963–82. Masson: Paris.

Grimes, G. W., McKenna, M. E., Goldsmith-Spoegler, C. M. & Knaupp, E. A. (1980). Patterning and assembly of ciliature are independent processes in hypotrich ciliates. *Science* **209**, 281–3.

Hanson, E.D. (1977). *The origin and early evolution of animals.* Wesleyan University Press: Middletown, Penn.

Heckmann, K. (1963). Paarungssystem und genabhängige Paarungs-typdifferenzierung bei dem hypotrichen Ciliaten *Euplotes vannus* O. F. Müller. *Arch. Protistenk.* **106**, 393–421.

Heckmann, K. (1964). Experimentelle Untersuchungen an *Euplotes crassus.* I. Paarungssystem, Konjugation und Determination der Paarungstypen. *Z. Vererb. Lehre* **95**, 114–24.

Heckmann, K. (1967). Age-dependent intraclonal conjugation in *Euplotes crassus. J. exp. Zool.* **165**, 269–78.

Heckmann, K. & Frankel, J. (1968). Genic control of cortical pattern in *Euplotes. J. exp. Zool.* **168**, 11–38.

Jerka-Dziadosz, M. (1981). Patterning of ciliary structures in *janus* mutant of *Tetrahymena* with mirror-image cortical duplications. An ultrastructural study. *Acta protozool.* **20**, 337–56.

Jerka-Dziadosz, M. & Frankel, J. (1979). A mutant of *Tetrahymena thermophila* with a partial mirror image duplication of cell surface pattern. I. Analysis of the phenotype. *J. Embryol. exp. Morph.* **49**, 167–202.

Kaczanowski, A. (1975). A single gene dependent abnormality of adoral membranelles in syngen 1 of *Tetrahymena pyriformis. Genetics* **81**, 631–9.

Kaczanowski, A. (1976). An analysis of *mp* gene affected morphogenesis in *Tetrahymena pyriformis*, syngen 1. *J. exp. Zool.* **196**, 215–30.

Kahl, A. (1935). Urtiere oder Protozoa. I. Wimpertiere oder Ciliata (Infusoria). In *Die Tierwelt Deutschlands,* ed. F. Dahl. Gustav Fischer Verlag: Jena.

Kidder, G. W., Lilly, D. M. & Claff, C. L. (1940). Growth studies on ciliates. IV. The influence of food on the structure and growth of *Glaucoma vorax,* sp. nov. *Biol. Bull, mar. biol. Lab.,* Woods Hole, **78**, 9–23.

King, J. (1980). Regulation of structural protein interactions as revealed in phage morphogenesis. In *Biological regulation in development*, ed. R. F. Goldberger, vol. **2**, pp. 101–32. Plenum Press: New York.

King, M.-C. & Wilson, A. C. (1975). Evolution at two levels in humans and chimpanzees. *Science* **188**, 107–16.

Luporini, P. (1970). Life cycle of autogamous strains of *Euplotes minuta*. *J. Protozool.* **17**, 324–8.

Luporini, P. & Dini, F. (1977). The breeding system and the genetic relationship between autogamous and non-autogamous sympatric populations of *Euplotes crassus* (Dujardin) (Ciliata, Hypotrichida). *Monit. Zool. ital.* **11**, 119–54.

Luporini, P. & Nobili, R. (1967). New mating types and the problem of one or more syngens in *Euplotes minuta* Yocum (Ciliata, Hypotrichida). *Atti Ass. Genet. Ital.* **12**, 344–60.

Luporini, P. & Seyfert, H. M. (1981). Variations in the total protein patterns of *Euplotes* species with a single-type dargyrome. *Progress in protozoology*, Abstracts VI International Congr. Protozool. (Warsaw, 1981). p. 225.

Lwoff, M. (1950). *Problems of morphogenesis in ciliates.* Wiley: New York.

Lynn, D. (1981). The organization and evolution of microtubular organelles in ciliated protozoa. *Biol. Rev.* **56**, 243–92.

McCoy, J. W. (1974). New features of the tetrahymenid cortex revealed by protargol staining. *Acta protozool.* **8**, 155–9.

Machelon, V. (1978). Etude d'une troiseme espèce dans le complexe *Euplotes vannus* (Ciliés, Hypotriches). *Protistologica* **14**, 15–22.

Mulisch, M., Barthlott, W. & Hausmann, K. (1981). Struktur und Wachstum von *Eufolliculina* spec. Schwarmer und sessiles Stadium. *Protistologia* **17**, 285–312.

Nanney, D. L. (1958). Epigenetic control systems. *Proc. natn. Acad. Sci. USA* **44**, 712–17.

Nanney, D. L. (1966a). Cortical integration in *Tetrahymena*: an exercise in cytogeometry. *J. exp. Zool.* **161**, 307–17.

Nanney, D. L. (1966b). Corticotypes in *Tetrahymena pyriformis*. *Am. Nat.* **100**, 303–18.

Nanney, D. L. (1966c). Corticotype transmission in *Tetrahymena*. *Genetics* **54**, 955–68.

Nanney, D. L. (1977). Molecules and morphologies: the perpetuation of pattern in ciliated protozoa. *J. Protozool.* **24**, 27–35.

Nanney, D. L. (1980). *Experimental ciliatology.* Wiley-Interscience: New York.

Nanney, D. L. (1982a). Genes and phenes in *Tetrahymena*: the epigenetic paradox in protozoa. *BioScience* **32**, 783–88.

Nanney, D. L. (1982b). The molecular diversity and evolutionary antiquity of the *Tetrahymena pyriformis* species complex. *Acta protozool.* (in press).

Nanney, D. L., Chen, S. S. & Meyer, E. B. (1978). Scalar constraints in *Tetrahymena* evolution: quantitative basal body variations within and between species. *J. Cell Biol.* **79**, 727–36.

Nanney, D. L. & Chow, M. (1974). Basal body homeostasis in *Tetrahymena*. *Am. Nat.* **108**, 125–39.

Nanney, D. L., Chow, M. & Wozencraft, B. (1975). Considerations of symmetry in the cortical integration of *Tetrahymena* doublets. *J. exp. Zool.* **193**, 1–14.

Nanney, D. L. & McCoy, J. W. (1976). Characterization of the species of the *Tetrahymena pyriformis* complex. *Trans. Am. microsc. Soc.* **95**, 664–82.

Nanney, D. L., Nyberg, D., Chen, S. S. & Meyer, E. B. (1980). Cytogeometric constraints in *Tetrahymena* evolution: contractile vacuole positions in 19 species of the *T. pyriformis* complex. *Am. Nat.* **115**, 705–17.

Nelsen, E. M. (1978). Transformation in *Tetrahymena thermophila*: development of an inducible phenotype. *Devl Biol.* **66**, 17–31.

Nelsen, E. M. (1981). The undulating membrane of *Tetrahymena*: formation and reconstruction. *Trans. Am. Microsc. Soc.* **100**, 285–95.

Nelsen, E. M. & Frankel, J. (1979). Regulation of corticotype through kinety insertion in *Tetrahymena. J. exp. Zool.* **210**, 277–88.

Ng, S. F. & Frankel, J. (1977). 180°-rotation of ciliary rows and its morphogenetic implications in *Tetrahymena pyriformis. Proc. natn. Acad. Sci. USA* **74**, 1115–19.

Nobili, R. (1966). Mating types and mating type inheritance in *Euplotes minuta* Yocum (Ciliata, Hypotrichida). *J. Protozool.* **13**, 38–41.

Nobili, R. & Luporini, P. (1967). Maintenance of heterozygosity at the *mt* locus after autogamy in *Euplotes minuta* (Ciliata, Hypotrichida). *Genet. Res. Camb.* **10**, 35–43.

Nobili, R., Luporini, P. & Dini, F. (1978). Breeding systems, species relationships and evolutionary trends in some marine species of *Euplotidae* (Hypotrichida Ciliata). In *Marine organisms: genetics, ecology, and evolution*, eds. B. Battaglia and J. A. Beardmore, pp. 591–616. Plenum Press: New York.

Orias, E. (1981). Probable somatic DNA rearrangements in mating type determination in *Tetrahymena thermophila. Devl Genet.* **2**, 185–202.

Patterson, B. (1949). Rates of evolution in taeniodonts. In *Genetics, paleontology, and evolution*, eds. G. L. Jepsen, E. Mayr and G. G. Simpson, pp. 243–78. Princeton University Press: Princeton.

Peck, R. K. (1974). Morphology and morphogenesis of *Pseudomicrothorax*, *Glaucoma*, and *Dexiotricha*, with emphasis on the types of stomatogenesis in heterotrich ciliates. *Protistologica* **10**, 333–69.

Peck, R. K., Pelvat, B., Bolivar, I. & de Haller, G. (1975). Light and electron microscopic observations on the heterotrich ciliate *Climacostomum virens. J. Protozool.* **22**, 368–85.

Ruffolo, J. J. (1976). Fine structure of the dorsal bristle complex and pellicle of *Euplotes. J. Morph.* **148**, 469–87.

Seyfert, H. M. & Willis, J. H. (1981). Molecular polymorphism of ciliary proteins from different species of the ciliate *Tetrahymena. Biochem. Genet.* **19**, 385–96.

Shaw, R. R. & Williams, N. E. (1963). Physiological properties of *Tetrahymena vorax. J. Protozool.* **10**, 486–91.

Siegel, R. W. & Heckmann, K. (1966). Inheritance of autogamy and the Killer trait in *Euplotes minuta. J. Protozool.* **13**, 34–8.

Simpson, G. G. (1953). *The major features of evolution.* Columbia University Press: New York.

Smith, H. E. (1982a). Oral apparatus structure in the microstomal form of *Tetrahymena vorax. Trans. Am. microsc. Soc.* **101**, 36–58.

Smith, H. E. (1982b). Oral apparatus structure in the carnivorous macrostomal form of *Tetrahymena vorax. J. Protozool.* **29**, 616–26.

Sonneborn, T. M. (1947). Recent advances in the genetics of *Paramecium* and *Euplotes. Adv. Genet.* **1**, 264–368.

Sonneborn, T. M. (1957). Breeding systems, reproductive methods and species problems in protozoa. In *The species problem*, ed. E. Mayr, pp. 155–324. American Assn for the Advancement of Science: Washington, D.C.

Sonneborn, T. M. (1963). Does preformed cell structure play an essential role in cell heredity? In *The nature of biological diversity,* ed. J. M. Allen, pp. 165–221. McGraw-Hill: New York.

Sonneborn, T. M. (1964). The differentiation of cells. *Proc. natn. Acad. Sci. USA* **51**, 915–29.

Sonneborn, T. M. (1970). Gene action in development. *Proc. R. Soc. Lond.* **B, 176**, 347–66.

Sonneborn, T. M. (1971). Letter to J. Frankel, dated Aug. 25, 1971.

Sonneborn, T. M. (1977). Genetics of cellular differentiation: stable nuclear differentiation in eukaryotic unicells. *A. Rev. Genet.* **11**, 349–367.

Tartar, V. (1956). Pattern and substance in *Stentor.* In *Cellular mechanisms of differentiation and growth,* ed. D. Rudnick, pp. 73–100. Princeton University Press: Princeton.

Tartar, V. (1961). *The biology of Stentor.* Pergamon Press: Oxford.

Tartar, V. (1966). Stentors in dilemmas. *Z. allg. Mikrobiol.* **6**, 125–34.

Uhlig, G. (1960). Entwicklungsphysiologische Untersuchungen zur Morphogenese von *Stentor coeruleus* Ehrbg. *Arch. Protistenk.* **105**, 1–109.

Vaudaux, P. P. & Williams, N. E. (1979). Cytoskeletal proteins of the cell surface of *Tetrahymena.* II. Turnover of major proteins. *Exp. Cell Res.* **123**, 321–31.

Vaudaux, P. P., Williams, N. E., Frankel, J. & Vaudaux, C. (1977). Interstrain variability of structural proteins in *Tetrahymena. J. Protozool.* **24**, 453–8.

Waddington, C. H. (1957). *The strategy of the genes.* George Allen and Unwin: London.

Williams, G. C. (1966). *Adaptation and natural selection: a critique of some current evolutionary thought.* Princeton University Press: Princeton.

Williams, N. E. (1961). Polymorphism in *Tetrahymena vorax. J. Protozool.* **8**, 403–10.

Williams, N. E. & Bakowska, J. (1982). Scanning electron microscopy of the cytoskeletal elements in the oral apparatus of *Tetrahymena. J. Protozool.* **29**, 382–9.

Williams, N. E. & Frankel, J. (1973). Regulation of microtubules in *Tetrahymena.* I. Electron microscopy of oral replacement. *J. Cell Biol.* **56**, 441–57.

Williams, N. E., Van Bell, C. & Newlon, M. (1980). Variation in surface proteins of *Tetrahymena. J. Protozool.* **27**, 345–50.

Wilson, A. C., Carlson, S. S. & White, T. J. (1977). Biochemical evolution. *A. Rev. Biochem.* **46**, 573–639.

Embryological bases of evolution

T.J. HORDER

Department of Human Anatomy, University of Oxford, Oxford OX1 3QX, UK

 # 6 438

ONTOGENY
PHYLOGENETICS   EMB-MECH

'The eye to this day gives me a cold shudder.' (Darwin, 1887)

Our understanding of embryology is today in a state of quite remarkable confusion, which, as I shall attempt to show, is not the result of any scarcity of relevant information but must almost certainly be explained in terms of the vagaries of the scientific process: the presumptions contained in the survey which follows may perhaps be excused in the light of this confusion and justified by the importance of the implications for kindred subjects such as evolution. Embryology finds itself in a situation in many ways comparable to that of evolution immediately before Darwin, the requirement being not yet further data but a unifying conception of the appropriate generality to rationalize vast tracts of detailed but unconnected information. Therefore, if, as may well be the case, such a conceptual framework awaits us in embryology, we might anticipate that it will cut across traditional, entrenched subject barriers. It may be surprisingly simple; it will turn out to have been amply evident in the facts for some time and will probably have been unsuspectingly alluded to by many authors in the past; it will not require, and probably will not be amenable to, any proof in terms of any single, new experiments; indeed, the underlying molecular mechanisms may remain uncertain for some time to come or may even be quite incorrectly conceived, though with no serious effects on the particular formulation of the general issues, as with Darwin's misconceptions regarding mechanisms of inheritance.

I EMBRYOLOGY

Current theoretical issues and models

Mosaic development and the nature of preformed cytoplasmic factors in the egg

Preformationist views of embryology might well be thought to be of no more than historic interest but, to quote one recent influential textbook: 'Localization as used here is the specification of cell fate according to the section of egg cytoplasm inherited by an embryonic cell lineage . . .. The localization phenomenon is particularly interesting to contemporary students of development because it suggests that specific programmes of development are sequestered in the egg cytoplasm' (Davidson, 1976). The argument for cytoplasmic formative factors is based on the independent ('mosaic') differentiation of separated egg or early embryo fragments in accordance with their normal expected developmental fates. While the cells of all vertebrate embryos retain totipotentiality until the stage where the embryo consists of many hundreds or thousands of cells (as proved by the formation ('regulation') of whole embryos from embryo fragments (Fig. 1; Spemann, 1938)), many invertebrate embryos begin differentiation at a time when the total number of cells is much fewer and correspondingly their eggs are small. At the relevant pre-differentiative stages, tests of cellular potentialities by fragmentation and transplantation experiments comparable in detail to those done in vertebrates are difficult to perform. Yet the question of the degree of prior commitment of embryo fragments can only be judged by adequate testing. Possible initial multipotency can only be excluded if the earliest stages have been so tested. Mosaic behaviour as such may be explicable in different terms; isolated cells cleave in a predictable and characteristic way that may itself be a sufficient precondition for their future characteristic patterns of independent cellular specialization. In some cases (Conklin, 1905), experiments involved killing of selected blastomeres rather than separation; it is known that, even in regulative organisms such as amphibians, this can result in non-regulative development of the surviving blastomeres (Fig. 2), suggesting that the killed cells are still exercising a normal role (presumably here only a mechanical one) of restricting the behaviour of other cells.

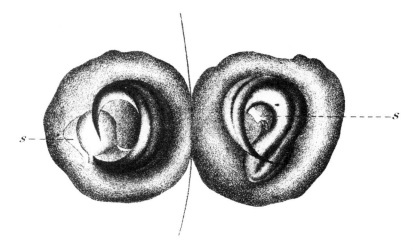

Fig. 1. 'Regulation' in the amphibian embryo: two normal larvae (S) derived from two initial blastomeres separated by a thread. (From Herlitzka, 1897.)

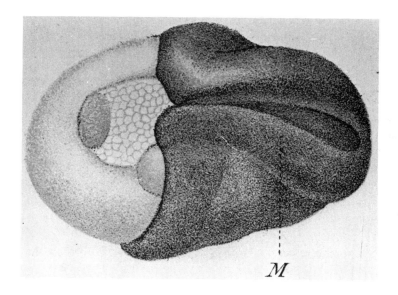

Fig. 2. 'Mosaic' development in the amphibian: after killing of one of the two initial blastomeres, the remaining cell goes on to form a rostral half-embryo (M). (From Roux, 1888.)

However, the most damaging evidence regarding preformed development is the fact that in the species which classically exemplify the phenomenon, such as ascidians or *Amphioxus* (Conklin, 1905, 1933), an entirely different, regulative, result can be obtained (see, for example, Wilson, 1893; Reverberi, 1971). Regeneration of complete organisms from adult components, which is a common potentiality even in species showing mosaic development (Berrill, 1961), is incompatible with the idea that cell fates are fixed by their descent. Furthermore, there is positive evidence in mosaic species that, far from being fully committed as a result of their pattern of derivation from specific regions of the egg, the development of some embryonic components can be made to vary according to the nature of neighbouring tissues (Reverberi, 1971). A good example is shown in Fig. 3: the blastopore lip was transplanted to the abdominal region in an *Amphioxus* embryo; the resulting induction of neural tissue was closely comparable to the results of the same experiments in amphibian or avian embryos.

I am therefore arguing that the time has come to lay the ghost of preformation. Many of the most cherished examples such as 'germ plasm', muscle or pigment cell precursors in ascidians and the amphibian 'cortical organizer' are now open to alternative interpretations (Nieuwkoop & Sutasurya, 1979; Gerhart *et al.*, 1981; Horder, 1981). There is no denying the theoretical possibility of intracellular spatial organization in egg cells; this reaches some complexity in protozoan cells while protozoan inheritance of 'cortical organization' (Beisson & Sonneborn, 1965) may be a mechanism available in controlling metazoan determinate cleavage. What is at issue is the question of the *nature* and *role* of such organization in the attainment of patterns of cell differentiation in multicellular organisms. The preconditions for development which must be programmed into the egg are the preconditions for the first formative event in embryogenesis, which, in vertebrates, is the morphogenetic process of gastrulation, and it happens that there is direct evidence, in the form of pseudogastrulation (Smith & Ecker, 1970), for corresponding morphogenetic preconditions in the egg, based presumably on the well-attested potential of eggs for complex cytoskeletal organization (Rappaport, 1970; Harris, 1979). Moreover, there is no reason to regard the evidence in species showing mosaic development in any fundamentally different way. The mosaic development of isolated early embryonic fragments

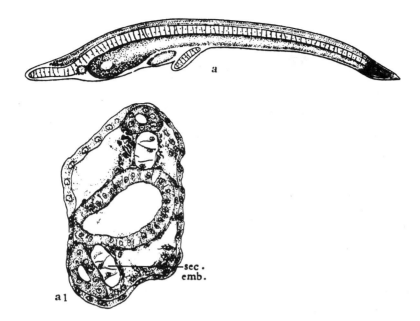

Fig. 3. Primary induction in *Amphioxus*: the blastopore lip of one early gastrula was transplanted to the blastocoele of another: *Top*: larva with transplant lying ventrally. *Bottom*: section showing notochord (sec. emb.) derived from transplant and neural tube induced by it from host abdominal ectoderm (Tung *et al.*, 1962).

may reflect no more than the continuance of a predisposition to certain patterns of cleavage, which, as the determinate cleavage of such species indicates, must be tightly controlled; it is not a test of the state of commitment of cells to their prospective patterns of differentiation at the time of isolation. It should be pointed out that, at best, such techniques reveal only low degrees of elaboration of preformed pattern: they cannot demonstrate a greater number of preformed features than the number of blastomeres or fragments that can be isolated and shown to behave differently.

Embryonic regulation; gradients; positional information

Having put preformation in perspective, we have to face up squarely to the central problem of development: on what basis can a pattern of discrete, differentiated, anatomical parts emerge from the starting conditions provided in the egg and early embryo, starting conditions which the phenomenon of regulation shows must be diffuse, totipotent and self-organizing? The idea of a gradient has an immediate intuitive appeal and it has surfaced on numerous occasions and in varying guises: e.g. 'fields', 'equipotential harmonious systems', 'polarity'. The existence of gradients has been posited in a wide but somewhat ill-assorted range of biological situations. Surprisingly, the most frequently quoted examples are not in embryonic development but in regeneration in invertebrates (Child, 1941) and the usual test of graded organization is 'susceptibility' of the formed organism to toxic agents. The choice of species is relevant because in organisms such as *Hydra* or planarians, with their relatively simple head to tail organization, the anatomical structure may well engender graded effects under certain experimental tests; whether this has any meaning as regards actual developmental controls is another matter. Segmented organisms such as annelids, on the other hand, may regenerate merely by repetition of their sequential polarized pattern of development. Slime mould aggregation is often put forward as a cast-iron case of a gradient controlling a pattern, but we are perhaps entitled to question the universality of mechanisms seen in highly specialized organisms facing quite unique 'developmental' problems. The widespread effect of a transplanted dorsal blastopore lip on a recipient amphibian embryo (Spemann, 1938) may be transmitted, not by a diffusing chemical, but by movement of cells during gastrulation.

The concept of positional information (Wolpert, 1969) encapsulates the problem of developmental pattern formation in a particularly clear-cut form. It is argued that embryos are provided with signalling systems which inform cells of their positions within the array of cells; integration of pattern and embryonic regulation reflect the operation and self-adjustment of this long-distance communication channel; and cells themselves differentiate autonomously on the basis of this information. The model was originally expressed in totally general terms without necessarily being linked with diffusion gradients or the like: two recent, quite independent and explicit formulations of positional information, based on particular classes of experimental evidence, will be considered here.

The evidence from imaginal discs; compartments; homoeotic mutations

The subspecialty of developmental biology concerned with pattern formation in the imaginal disc of larval *Drosophila* is of particular interest because it has offered a direct link between genes and development of complex patterns but it is important to examine the implications of the fact that it is based not only on a new phylum but on technology differing totally from that from which most embryological concepts were derived.

Genetics has been exploited to investigate clonal patterns of origin of cells in imaginal discs. Somatic crossing-over can be induced in single cells at chosen stages of development, and resulting expression of markers used to visualize the pattern of the clone of derived cells. Clones formed early are large and irregularly shaped, but later formed clones show ordered features such as elongation in the long axis of the formed wing (Bryant, 1970), most easily interpreted as the result of controlled orientation of mitoses (Bryant & Schneiderman, 1969; Vijverberg, 1974) together with stretching out of an initially small patch of cells during evagination of the disc. García-Bellido *et al.* (1973) have proposed that certain regions where clones never overlap may represent a hidden constraint to development (a 'compartment' boundary) actively involved in pattern control. In some situations where similar 'compartment' behaviour occurs, there are detectable anatomical boundary structures which could be associated with mitotic restriction. For example, clones early become restricted to the confines of single discs in conformity with the anatomical partitioning characteristics of these discrete epithelial structures: Lawrence & Green (1975) describe indications of a boundary between abdominal segments based on changing pigmentation patterns and cell shape parallel to epithelial folds; dorso-ventral clonal restriction in the wing corresponds precisely with the anterior and posterior margins of the wing (see Brower *et al.* (1982)), who also report indications of cell alignment (including peripodial membrane cells) parallel to the antero-posterior 'compartment boundary'). In such cases clonal restrictions correspond to eventual anatomical features because they share a common denominator (i.e. mitoses are orientated by disc folds which themselves anticipate adult morphology). In principle, there is no reason why mitoses should not be restricted

(Simpson, 1976), perhaps by patterns of tension in the disc epithelium, without there being any corresponding visible structural feature. The causal connection with the common denominator may not be one to one: antero-posterior wing boundary and vein IV may partially share an underlying morphological determinant even though their anatomical correspondence is not exact. This boundary is not a unique feature in the wing: clones elsewhere show a tendency not to cross veins (Bryant, 1970). There are situations in development (Lawrence & Morata, 1976) and in regeneration (Bryant, in Ashburner & Wright, 1978) where clonal boundaries shift *but* morphological patterning is not correspondingly distorted, thus indicating against a causal role of clonal restriction in pattern formation.

A second component of the compartment model is the correspondence between clonal restriction patterns and the spatial distributions of the effects of homoeotic mutations, which characteristically transform the pattern of differentiation of one disc into that of another. This raises the vexed question of the timing and early forms of developmental commitment of disc cells. It is frequently assumed that, because isolated fragments of discs fulfil their prospective fates accurately, all cells have their fates fixed but, as was the case in studies of embryos showing mosaic development, such a test is open to serious criticism. Specific patterns of development finally displayed by fragments may be the indirect result of varying specific patterns of growth, adhesiveness and shaping existing at the time of isolation (García-Bellido, 1967) or of later induction by patterns of underlying mesoderm (El Shatoury, 1955). It is not justifiable to infer that within each fragment (which is rarely smaller than one-tenth of the whole disc) single cells are individually committed. A second source of evidence for disc differentiation is that based on apparent selective reaggregation of randomly mixed disaggregated cells derived from marked disc or disc fragments, selective reaggregation being identified by the formation of 'mosaic' patterns, i.e. composite structural patterns of cuticular organelles showing integration across cells of two separate origins (Tobler, 1966). However, if cells are not yet developmentally committed, mosaic patterns could be the result of 'inductive' interactions across randomly associating clones. Poodry *et al.* (1971) were able to show that reaggregation was extremely limited: mosaics, which reached only limited anatomical complexity, were made up of clones of cells derived largely by mitosis from between one and six original cells. Therefore the final pattern of differentiation was not the result of ordering of already committed cells: each founding cell was the ancestor of a *number* of final differentiated structures. The situation is further complicated by the fact that surviving clones are probably derived from small undissociated cell clumps rather than single cells (Bryant, in Ashburner & Wright, 1978): preconditions for patterned development therefore exist regardless of reaggregation. Ultimately, however, the most telling evidence pointing to the labile state of commitment of imaginal discs is transdetermination (Hadorn, in Ashburner & Wright, 1978) and the heteromorphosis of discs induced by teratogens (Waddington, in Counce & Waddington, 1973; Bryant, in Ashburner & Wright, 1978).

The most positive evidence regarding the nature of pattern control comes from examples (reviewed by Waddington, in Counce & Waddington, 1973) in which the formation of one structure can be shown to depend on the presence of another: e.g. bracts (Tobler, 1966), veins (García-Bellido, 1977), siting of bristles (Stern, 1968), movement of sex-comb (Horder, 1976), growth stimulation by dominant zones (Lawrence & Morata, 1976), mitotic competition (Simpson, 1976), orientation of cuticular ridges (Lawrence, in Counce & Waddington, 1973). Induction is considered to be involved widely in insect development (Counce, in Counce & Waddington, 1973). It might be thought that there is counter-evidence in the form of the supposed position specificity of 'autonomously' developing clones of homoeotic cells (Postlethwait & Schneiderman, 1971) which fail to interact with surrounding wild-type tissue. However, there are difficulties in judging the appropriateness of the position of a given homoeotic structure in the context of a non-corresponding foreign structural background; in some cases (Postlethwait & Schneiderman, 1971) individual cuticular organelles are clearly abnormal, so perfect autonomy cannot apply. In studies such as these, clones may be large enough so that the relative normality of pattern within them may be accounted for by internal interactions. There is the further possibility (Horder, 1976) that, since discs are homologous, they can provide equivalent pattern-defining cues for homoeotic clones sufficient to allow fairly normal development.

The compartment model seeks to link the apparent decision-

making homoeotic genes with the control of integrated pattern
formation within specific developmental fields; but the available
data suggest pattern is built up gradually by multiple sequential
interactions between differentiated structures. The model is difficult to assess given the existing uncertainty as to the meaning and
significance of homoeotic mutations. Even *engrailed*, the prime
example of transformation of posterior wing into anterior, shows
unexplained anatomical abnormalities over and above those expected of a straight-forward anterior mirror duplication across the
antero-posterior compartment boundary, and here it is known that
pattern depends on local direct interactions between cells which can
actually distort pattern in the face of a largely intact posterior
compartment (Lawrence & Morata, 1976). Several considerations
point to the interpretation that homoeotic mutations owe many of
their effects to the organization of insect development rather than
any special feature of the wild-type genes as such. It is notable that
mutations with comparable effects are not prominent among vertebrates, whereas it is an obvious characteristic that insects are not
only organized segmentally but that the segments may be homologous and therefore share many aspects of their development
(Horder, 1976). Mutations blocking one divergent developmental
pathway may readily lead to reversion to another pattern of
homologous development. Heteromorphoses and teratological
phenocopies suggest the ease with which alternative pathways may
be selected. The restriction of the effects of homoeotic mutations
within the bounds of single discs may reflect the physical separation
and independence of disc structure.

The polar coordinate model; intercalary regeneration: distal transformation

This model (French *et al.*, 1976) is closely related to studies of
regeneration in *Drosophila* imaginal discs. After division of discs the
larger fragment regenerates a complete organ while the smaller
tends to form a set of duplicate parts (Bryant, in Ashburner &
Wright, 1978). This pattern applies regardless of the orientation of
the dividing line. The seeming equivalence of all points around the
circumference of the imaginal disc led to the proposal that organization is based on a continuous circumferential position-defining
mechanism. The 'shortest intercalation rule' would explain com-

plementary regeneration/duplication; for a large fragment, regenerative intercalation of the shortest sequence of circumferential values between its cut ends would result in a complete disc, but for a small fragment replacement of intervening positions by the shortest sequence will produce duplication. Complimentary regeneration/duplication patterns of regrowth of amphibian limbs are always explicable as 'distal transformation' (Bryant, 1975) which itself is a replay of the rule underlying the polarized pattern of normal development, namely that distal parts are sequentially laid down on previously formed proximal parts. In discs, there is no correspondingly marked epimorphic sequence in development and duplication does not fit an equivalent explanation because, depending on surrounding tissue context, one cut edge can duplicate *or* regenerate. For discs one alternative model might run along the following lines. Regeneration of a complete disc, which is associated with fragments which are large, may reflect the ability of that fragment to impose normal and integrated overall pattern on new tissues: duplication, associated with small fragments, may reflect absence of overall integration and domination of regionalities specific to the fragment. The polar coordinate model has difficulties explaining substitution of regeneration in place of duplication in long-term disc cultures (Schneiderman, 1979) and instances of spontaneous duplication (Wildermuth, 1968), as well as the evidence for direct interdependence of disc pattern elements already referred to.

The model has at the same time been applied to regeneration in adult amphibian and cockroach limbs, although certain differences are unexplained. Whereas the cockroach limb shows true intercalary regeneration when proximal limb parts are transplanted onto distal without rotation (Bohn, 1976), this does not occur (French *et al.*, 1976) in amphibians. This different pattern of behaviour may find an explanation in the different nature of the tissues involved; unlike the amphibian, the cockroach undergoes continual structural modification in the form of moults involving stretching out of an epithelium which gives rise to the intercalated tissue. Intercalation in the longitudinal axis leading to the formation of supernumerary limbs is subsumed by introducing a second rule (the 'complete circle rule'), whereby distal transformation in the proximo-distal axis of the limb can only occur if a complete set of circumferential values exist at the initiating proximo-distal level. Again there are paradoxical features to some of the data: if, as Bohn (1976) shows, segments of the cockroach limb are homologous (longitudinal intercalary regeneration occurs as if to complete each proximo-distally mismatched segment, regardless of whether mismatched tissue is from the same or different segments), then, after limb rotation, why does longitudinal supernumerary intercalary regeneration continue to include all distal segments? But it is undoubtedly with the vertebrate limb that the model becomes most obviously inappropriate and specific predictions of the model have not been borne out (Tank & Holder, 1981; Wallace, 1981) with the result that the model has been revised (Bryant *et al.*, 1981). It is difficult to see how such a model would ever explain findings such as those in which different components of limbs, such as skeleton, muscle and skin appendages, are independently patterned (Saunders *et al.*, 1976; Maden, 1980). Circumscribed wounding of the limb (involving a small fraction of the circumference) or flank (which has no circumference) (Fig. 5) is sufficient to initiate outgrowth of a supernumerary limb (Tank & Holder, 1981; Wallace, 1981). We know enough about the simplicity of the organization involved in initiation of a vertebrate limb (Figs. 5–7, and see later) for there to be much easier explanations for such a result than the strained and arbitrary modifications that are required of the polar coordinate model (Bryant *et al.*, 1981).

Experimentally established foundations for the analysis of development

A complete alternative to mechanisms based on the clonal history of a cell or on long-distance communication exists in the findings of classical experimental embryology which provide overwhelming, and undisputed, evidence for the decisive role, in the control of cell differentiation, of purely *local* and *progressively arising* circumstances.

Embryonic induction

Lens induction by the optic cup or neural plate induction by archenteron roof are facts beyond serious contention, but these stand for a considerable number of examples, so that we have effectively reached the point where there is evidence, scattered in

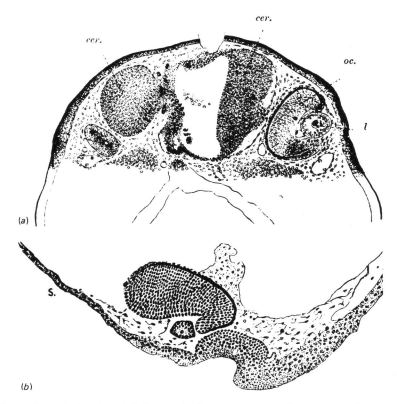

Fig. 4. Lens induction. (*a*) Removal of one eye primordium at neurula stage in *Rana fusca*: lens fails to form from prospective lens ectoderm on left side. *l*, normal lens; *cer*, brain; *oc*, normal eye. (From Spemann, 1901.)

(*b*) Transplanted optic vesicle of *R. palustris* can change developmental fate of trunk ectoderm of *R. sylvatica* to form lens (marked S): an example from the experiments of Lewis (1904).

the literature, to show that each tissue and organ in the vertebrate is dependent for its development on interaction with an appropriate inductor. In each case, the common feature is that an already specified tissue, such as the eye cup specified by previous events, including neural tube induction, comes to lie immediately adjacent to unspecified embryonic tissue: induction of a new specific cell type such as lens then results. This interaction is *necessary* (removal of the inductor prevents the appearance of the induced organ (Fig. 4*a*)) and *sufficient* (transplanted inductors can change the fate of competent tissue throughout the embryo (Fig. 4*b*)) for the formation of the induced structure.

To interpret inductive signals as forms of 'positional information' is a semantic device which undermines the principal claim of the positional information model (its integrating long-distance channel of communication), and leaves unexplained why the inductor acquires the properties of a source of positional information (this presumably requires introduction of new and arbitrary parameters into the originally simple model) and also why tissue whose fate is changed by an inductor transplanted to a foreign site does not respond to the more general positional information still available in the rest of the surrounding embryo.

Why induction has not found its rightful place as the very bedrock of all theories of development is not easy to explain, although a number of past contentious issues have presumably served to cloud the subject generally.

(*a*) It may be thought that the fact that, despite exhaustive investigations, no single inducing factor has been definitively isolated or chemically identified speaks against the reality of the phenomenon. Worse, in the case of a number of inductions (Saxen & Toivonen, 1962), a large number of heterologous agents, with no possible relation to biological reality, have been found to act in a fashion identical to that of normal inductors. For example, boiled salamander heart can induce a lens in urodele ectoderm (Holtfreter, 1934). To introduce concepts such as that of 'sublethal cytolysis' (the idea that induction of ectoderm into neural plate under tissue culture conditions at appropriate pH must be the result of release by the tissue itself of a suitable, normally masked, inductor molecule (Holtfreter, 1947)) is arbitrary, circular and unnecessary. The attempt to isolate an inductive molecule is founded on an assumption that induction is an instructive interaction. The effects of heterologous inductors become very much more intelligible if we place the burden of the interaction on the responding cell; if we were to suppose that it has receptors sufficiently selective to detect and distinguish between the relatively small number of inductors normally arising in the embryo (which themselves may differ by virtue of some aspect or product of their own already differentiated states), then it would be no surprise that many quite incidental extraneous agents could trigger the receptors and lead to apparently quite normal induced responses. In this view, no molecule would normally need to pass between inductor and induced tissue. The possession of appropriate receptors by induced cells does not imply prior developmental decisions in those cells; a complete battery of receptors may be an intrinsic property of all cells from the egg onwards.

(*b*) Confusion surrounds the phenomenon of 'double assurance'. The famous anomaly of *Rana esculenta*, in which, in marked contrast to most vertebrates, a lens develops despite early ablation of the eye cup (Spemann, 1938), gave rise to the convoluted argument that there must be a second potential inductor of the lens which is only of sufficient potency to act alone in certain species. It has subsequently been abundantly confirmed that a number of nearby tissues are contributory to typical inductions (Jacobson, 1958). However, it is not necessary to suppose that these tissues actually trigger lens differentiation or have anything more in common than that they contribute directly or indirectly to preconditions for induction such as the morphogenesis involved, for example, in invagination and cavitation of the lens primordium, without which lens formation cannot occur (Horder, 1981).

(*c*) One renowned and beautiful experiment (Spemann, 1938) – transplantation of urodele abdominal ectoderm onto anuran head with resulting induction of a correctly located balancer – poses intriguing problems. How could the anuran provide an inductor for an organ it itself cannot produce? At first sight this is strong evidence for 'universality' (with respect to species) of continuously distributed 'positional information', but not necessarily. The result may illustrate the retention through evolution of homologous structures and inductive mechanisms: the anuran has as an homologous organ, the structure which the urodele ectoderm normally regards as its inductor (Horder, 1976).

(*d*) It could be argued that the concept of embryonic induction provides no solution to the problem of development, because it does not explain regulation and it implies an infinite regress of pattern back into the egg: each inductor must be preceded by its own correspondingly positioned inductor indefinitely. Such an argument omits the essential involvement of morphogenetic cell movements!

Morphogenesis

If induction mediates the switching on of differentiated cells, it becomes essential to account for the control of the formation and spatial arrangement of inductors. An especially characteristic feature of embryos is morphogenetic cell movement, of which gastrulation is the earliest, largest-scale and most consistent example among species. In vertebrates, the massive cellular rearrangements of gastrulation lead to the formation of the archenteron, which induces the neural plate, which itself, by way of additional further morphogenesis, becomes further subdivided to produce a further generation of more localized inductors such as the optic vesicle. Starting from simple preconditions, morphogenetic rearrangements are therefore the mediators of step-wise localization and increasing diversification of inductors. Spemann (1938) showed that the entire anatomy is conditional on the dorsal lip of the blastopore, which mediates gastrulation, and there is no difficulty in envisaging the necessary morphogenetic preconditions for this event within the egg (Smith & Ecker, 1970; Freeman, 1981). Regulation, after halving an embryo or as shown in the spreading influence of a transplanted blastopore on the blastula, can be explained as the result of the infolding and redistribution of the invaginating archenteron, according to the extent of blastula tissue available.

Morphogenetic variables provide a basis for long-distance coordination of pattern formation in the embryo compatible with short-distance inductive interactions as the immediate determinants of specific differentiation in individual cells. Due consideration of morphogenetic forces alone can lead to ready explanations for the diversity with which often small numbers of apparently identical specialized cell types can become arranged into characteristic organ systems. By way of illustration the development of the limb will be discussed as a specific example.

At the heart of pattern formation in the limb is the condensation of initially homogeneous mesoderm (already 'induced' in the lateral plate) into precartilagenous rudiments. Thereafter, limb development follows a relatively predictable and autonomous (Chalmers & Ray, 1962), though complex, growth and patterning by means of processes such as joint formation, ossification, longitudinal epiphyseal growth and functional modelling. Precartilagenous condensation is a morphogenetic event which probably serves to space out rudiments competitively (Hinchliffe & Johnson, 1980) within the existing mesodermal field; their size and number is not absolutely fixed in any one species but appears to vary according to

Fig. 5. Supernumerary limb can be induced in flank of *Triton* larva by implant of celloidin. (From Balinsky, 1927.)

availability of mesoderm (Ede, 1971). Thus, it can be argued that the problem of control of skeletal pattern will reduce to the question of the control of the shape and dimensions of the early limb bud which in turn determines positions of condensations. Indeed, Slack (1977) describes how, after inducing enlarged limb outgrowths, the number of cartilage rudiments is proportional to limb bud size.

There is considerable evidence to show that the dimensions of limb buds and regenerating blastemas are actively controlled in each species: deliberate attempts to double or halve the bud (Harrison, 1921) or the blastema (Goss, 1961) usually (although results are complicated by regression or fusions of supernumerary elements) result in early adjustment to normal dimensions; early in the formation of supernumerary limbs (which originate from tissue originally destined to form a single limb) both buds are equal and normal in size (Harrison, 1921); and in transplantations of limbs to hosts of the same species but varying size, or to other species whose limbs are smaller or larger (Harrison, 1969), the transplant may catch up but never exceeds a growth rate characteristic of its own normal capacities. Because of the distortion of the blastema by the stump and the substantial contribution of covering epidermis, it is difficult to assess the dimensions of regeneration blastemas in relation to those of limb buds, but they may be comparable (Tank *et al.*, 1977).

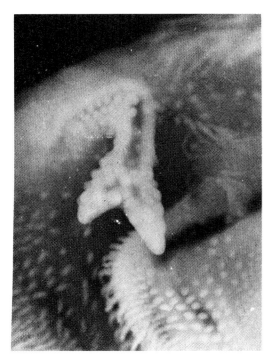

Fig. 6. Minimal spatial organization underlying limb outgrowth and digit formation: chick limb formed by placing disaggregated forelimb mesoderm within
forelimb ectodermal sheath. (From MacCabe *et al.*, 1973, reproduced with
permission.)

How is the specificity of limb bud shape controlled? It must be
inferred from the fact that new limbs can be caused to form
throughout a wide territory, including the flank (Fig. 5), that the
prerequisites for the formation of a limb are inherent as a diffuse
property in the mesoderm and ectoderm from which the limb
normally forms. Several lines of evidence confirm that the initial
impulse to bud outgrowth and the initial organization of the bud
involve minimal complexity in spatial terms (Fig. 6). Thus supernumerary limb outgrowth can be triggered by a wide variety of
extraneous agents and manipulations (Wallace, 1981; Tank &
Holder, 1981; Fig. 5) whose lowest common denominator appears
to be nothing more than non-specific wounding with delayed
healing. Formation of duplicate limb parts in the developing chick
limb can be induced by a diffusely organized zone of mesoderm

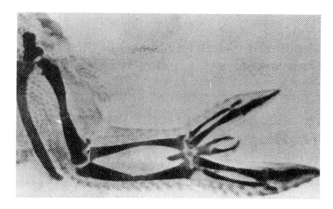

Fig. 7. Minimal spatial organization needed for control of digit number and type: chick forelimb duplicated distally by inserting disaggregated ZPA (zone of polarizing activity) cells into anterior limb bud. (From Tickle, 1981: reprinted by permission from *Nature* **289**, 295–8. © 1981 Macmillan Journals Ltd.)

(the 'zone of polarizing activity', ZPA) (Fig. 7) whose low level of complexity is indicated by the fact that it is not species or stage or limb type-specific and that it can be mimicked by various tissues from outside the limb (Saunders, 1977) and by vitamin A (Tickle *et al.*, 1982). It is hard to resist the conclusion that, just as a transplanted ZPA can induce new limb parts by interacting with non-ZPA limb tissue, normal limb outgrowth itself may be triggered in the first instance by interaction of the earliest ZPA rudiment with surrounding flank tissue. The same interaction explains limb duplication following amphibian limb bud rotation (Harrison, 1921). Although the initial trigger is therefore probably mesodermal, continued outgrowth is known to be the result of the interdependence of mesoderm and ectoderm: limb mesoderm induces the formation of the 'apical ectodermal ridge' (AER) and, after ZPA transplantation, delays its spontaneous regression (Saunders, 1977); if the AER is ablated outgrowth ceases, while transplantation of the AER shows that it mediates the shaping of the mesoderm (Saunders *et al.*, 1976). The AER is particularly prominent during the formation of the foot or hand plate and looks as though it is involved in the process of flattening out of the terminal bud, thereby leading to the characteristic shaping associated with the pattern of terminal skeletal parts.

At the present time it would be unwarranted to be more specific regarding mechanisms. The main points are, first, that the preconditions for outgrowth and eventual autonomous formation of a limb are diffuse and simple and, secondly, that all the important tissue components involved in continued outgrowth (AER, ZPA, apical mesoderm (Kieny, 1964)) retain the same potentialities at all stages. Therefore, from simple starting conditions, proximo-distal elaboration of pattern must arise in multiplicative fashion, each successively laid down distal limb region acquiring more complex features as the result of forces, such as AER shape changes, imposed by already formed more proximal tissues. There is no doubt that the distal limb tip can be influenced by proximal structures, e.g. by transplanted supernumerary skeletal elements (Goss, 1961), and this is reflected in the regulation of tips transplanted to older or younger limb stumps (Kieny, 1964).

With regard to the skeleton as a whole, we might expect, given the uniformity of cell types, frequent correlations in structure among regions of the adult skeleton of the kind described, for example, between fore- and hind-limbs, trunk and tail in lizards (Kramer, 1951). However, skeletal pattern achieves considerable regional diversity because it is arrived at by a variety of embryological routes, e.g. ossification may be membranous or cartilagenous and both forms of ossification are themselves conditional on a wide variety of prior morphogenetic permutations, such as those determining migration of neural crest and sclerotome. As the limb demonstrates, slight variations in starting conditions are capable of amplified and complex effects on later pattern. A major source of diversity and of detailed adaptive patterning and growth results

Fig. 8. (*a*) Ankylosed human knee showing formation of new trabeculae in lines of applied stress (inset *b*). (From Weiss, 1934.) (*b*) Transformation of pelvis towards form typical of bipedal species in congenitally bipedal goat: normal on left. (From Slijper, 1942.) (*c*) Adaptation of muscular attachments following experimental creation (on right) of a tibial pseudarthrosis in a dog. (From Wermel, 1935.) (*d*) Modelling of pattern of major vessels by blood flow: aortic arches of 4-day chick embryo were blocked with metal clips in various patterns shown above. (From Rychter, 1962.) (*e*) Functional control of gill growth: salamanders raised under 11% (*left*), 100% oxygen (*right*). (From Drastich, 1925.) (*f*) Functional control of growth of intestinal villi: 54–70% removal of the small intestine in a dog leg (*right*) to 30–50% taller cells and 4–5 × dilated crypts of Lieberkuhn. *Left*: control. (From Goss, 1964, after Flint, 1910, reproduced with permission.) (*g*) Control of melanocyte function by temperature (°C): thresholds for pigmentation on extremities in Himalayan rabbit. (From Schmalhausen, 1948.)

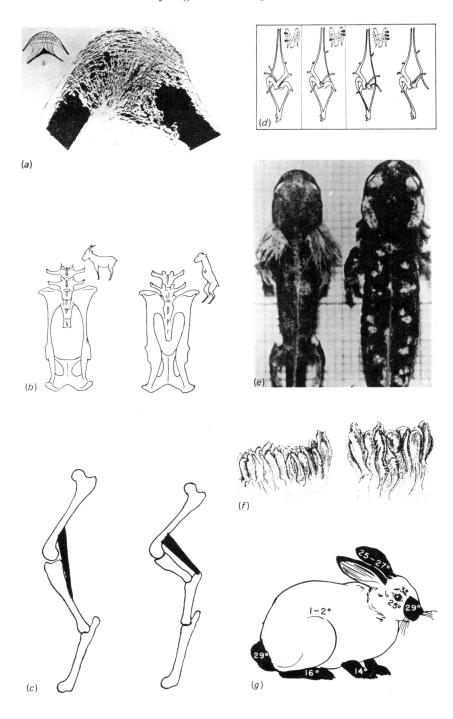

directly from the functional demands to which skeletal elements are subjected (Fig. 8*a*, *b*, *c*).

Summing up

Induction must once and for all be acknowledged as a central mechanism in embryogenesis. This mechanism is massively documented throughout the embryology of both vertebrates and invertebrates. The failure of present technology to substantiate induction in molecular terms is not important since the phenomenon can be unambiguously and fully defined operationally as described above while, for the purposes of explaining pattern formation, this level of description is both appropriate and adequate. Further, the concept of a battery of receptors lends itself readily to extension to include the role of other known developmental variables such as the effects of hormones or morphogenetic, mitotic and functional stimuli: any realistic definition of induction must embrace the integration of such multiple factors. The much neglected evidence for the role of 'function' in morphogenesis (Fig. 8) indicates the responsiveness of specific cells to particular features of their circumstances which, conceivably, could be the basis of their response to their inductor, of later maintenance of their differentiated state, and of growth or, equally, atrophy and degeneration in the case of organs no longer involved in 'functional demand' (Goss, 1964).

Morphogenetic mechanisms for the deployment of cell types are inextricably linked with induction. It is obvious from the few examples already mentioned that spatial integration and timing of pattern formation depend crucially on the scale, rate and form of morphogenetic events. For an explanation of how these embryonic phenomena are themselves acquired and inherited in various species, we turn to a consideration of evolution.

II EVOLUTION

It is ironic that it is Darwin's influence by way of Haeckel that has in one way or another dominated our thinking regarding the relation of evolution and embryology. De Beer (1958) demonstrated the way in which Haeckel's explanatory use of 'recapitulation' had overinterpreted the relationship, but, in his determination to refute that view, he came to concentrate on particular aspects of development in which the time relations in evolution had been changed ('heterochrony'), neoteny and the problem of larval forms being the classic examples. A closely related concept is allometry (Thompson, 1942; Huxley, 1972), by which evolutionary trends are analysed directly in terms of differential tissue growth rates controlled perhaps by 'rate genes' (Goldschmidt, 1940); again the emphasis is on temporal differentials in development. Such approaches establish the important point that evolutionary change could very plausibly be explained in terms of what are, in principle, simple embryological trends or transformations. However, they are essentially descriptive and beg the question of what controls expression of the hypothetical rate genes. As indicated by the multitude of factors affecting limb development, it is surely unrealistic to think in terms of single, organ-specific controllers of growth: the attainment of a predictable species-specific rate of growth (Harrison, 1969) is explicable rather as the result of the balancing of combinations of multiple positive and negative forces. One of the few laws of evolution of any generality that have emerged, 'Cope's law' (which states that there is a tendency for all established species to evolve into larger forms), has recently been shown to have explanations other than in terms of growth control (Stanley, 1973).

Approaches such as these share with Haeckel's a form of teleological thinking; conclusions were drawn about embryological processes by inference from the adult form, an approach which betrays a hidden assumption that the embryo is simply a miniature adult, whereas the embryological processes discussed above show the attainment of the adult to be confusingly indirect. Moreover, the starting point of these approaches is the adult form of ancestral species, since new species are considered as heterochronic or allometric variants of the original; as de Beer realized, this entirely leaves out of account any true evolutionary 'novelties'. In so far as the adults of ancestral species are the products of their own embryologies, then it follows that gradual evolutionary modification must entail modification at embryological stages. While it must be true that all processes in embryogenesis are potentially open to modification through natural selection, it must also be true that the selective advantages of changing certain processes will be very different from others. It is inconceivable that modification of vertebrate embryos could include any fundamental revision of the

basic processes of DNA replication and translation because so
much depends on them. For the same reason, the earlier the
embryological event the more dependent will be a greater number
of subsequent embryological processes and so the greater the
likelihood that it will be conserved in the evolution of new forms of
embryos. It follows that there will be a tendency for early
embryological features (such as early stages of limb development)
to be retained and shared in later-evolved species, with characteristically increasingly complex, sequential embryologies superimposed. It is therefore inevitable that a form of 'recapitulation'
should be seen in embryology. Von Baer's Laws, dating from well
before Darwin, express this fact free of any interpretation, evolutionary or otherwise.

Embryology and evolution are therefore fundamentally interdependent. Assuming the embryological mechanisms described
above, what potentialities and constraints do they impose for
evolutionary change? How have these mechanisms themselves
evolved and been perpetuated through generations. These matters
will be considered in the context of a specific example.

Speculations on the evolution of lens induction

In many respects, the development of the vertebrate eye has been
successfully charted out: many of the component tissues contribute
in patterns which parallel their deployment elsewhere throughout
the embryo, while shaping and growth are the end result of many,
often morphogenetic, interactions involving fluid pressures, scleral
skeletal elements, bony orbit and so on (Coulombre, 1965; Horder,
1981). But the unique and characteristic early event of lens
induction represents an obvious test case for any account of the
evolution of embryological mechanisms.

There is no doubt as to the strength of the evolutionary pressures
in favour of visual sensitivity. Salvini-Plawen & Mayr (1977)
estimate that photoreceptors have evolved independently in various
phylogenetic lines at least forty times. We can trace the vertebrate
eye reasonably securely from *Amphioxus* in which light receptors
occur diffusely throughout the cavity of the central nervous system;
as is true throughout evolution (Eakin, 1973; Salvini-Plawen &
Mayr, 1977), photoreceptivity is associated with cilia, which are
features of the ventricular cells.

Lateral and pineal eyes could then have evolved as gradual evaginations from a photoreceptor-rich region of the neural tube, which itself evolved into the diencephalon. Consistent with this scheme is the retention among vertebrates of light sensitivity in the diencephalon and varying degrees of anatomical overlap between diencephalon and the pineal (Hulsemann, 1967; Horder, 1981). This proposal satisfies an essential requirement which should be met by any hypothetical evolutionary sequence; a continuous sequence of morphogenetic events in an embryo is a repetition of a continuous sequence of morphological steps built up through the preceding evolving series of embryos, each stage of which must have been functionally advantageous in the transitional organism. This will be referred to as the continuity principle. In the case of the eyes the morphological continuum from the diencephalic origin means that functionally advantageous photoreceptors could have differentiated at each transitional stage in evolution, but how are we to apply the principle in the case of the lens of the lateral eye, which, given its separate origin from the ectoderm, is most unlikely to have evolved in morphological continuity with the lateral eye?

In spite of its simpler level of organization, the pineal eye (Fig. 9) should almost certainly not be regarded as a transitional form on the way to the evolution of the lateral eye. The two structures probably represent distinct morphological variants, founded on a common basic set of specialized cell types, which diverged from the start of their evolution as a consequence of their different points of emergence from the neural tube. In both, receptor cells face the lumen of the original ventricular cavity, but, whereas the pineal remains as a vesicle surrounding this cavity with receptors pointing forward, in the lateral eye infolding has resulted in orientation of receptors *away* from the lens. It has also resulted in a transformation of the interrelation of receptors with the pigment cells which are arranged between receptors in the case of the pineal (Studnicka, 1905; Eakin, 1973).

The structure of the pineal may, however, provide insights of possible relevance to the evolution of the lateral lens. The distal wall of the pineal optic vesicle is frequently specialized, reaching, in some species, degrees of elaboration which make it clearly recognizable as lens (Fig. 10). This identification is confirmed by the presence of crystallins (Eakin, 1973). This is of considerable interest because it demonstrates a possible sequence for lens

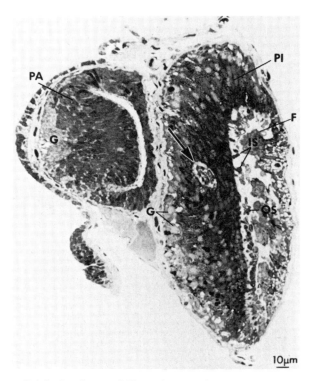

Fig. 9. Superficial pineal eye of *Entosphenus tridentatus* (PI) shows lens-type specialization distally (to right) where it is in contact with dermis, whereas there is no such specialization in the deeper parietal eye (PA). G = ganglion cells; IS and OS = inner and outer segments of receptor cells; F = fold in roof of pineal eye. (From Eakin, 1973; © 1973 by the Regents of the University of California, reprinted by permission of the University of California Press.)

evolution in which functional and morphological continuity are maintained. The structure of crystallins has so far offered few clues as to the evolutionary ancestry of lens cells (Bloemendal, 1981), but (echoing the diffuse arrangement of pigment cells between pineal receptors) they may well have originated from a cell type which was diffusely distributed in the neural tube and may initially have had quite different functions, perhaps supportive. Consistent with this supposition is the persistence of cilia in pineal lens cells (Eakin, 1973) and the widespread synthesis of lens crystallins in neural tissue in early embryos (Bloemendal, 1981). Parallelling pigment epithelial evolution from the stage seen in the pineal to that seen in the lateral eye, one can see, by comparing the pineal organs of

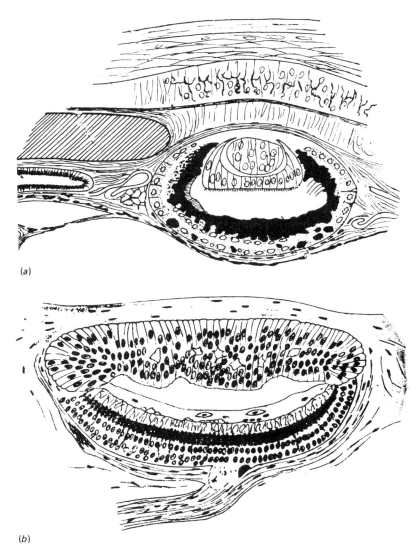

(a)

(b)

Fig. 10. (a) Pineal of *Anguis fragilis*. (From Studnicka, 1905, after Leydig, 1890.)
(b) Pineal of *Iguana tuberculata*. (From Studnicka, 1905, after Klinckowstroem,
1894.)

various species (Studnicka, 1905), a continuous progressive concentration of lens-type cells at the distal surface (Fig. 9), culminating in a tendency towards total segregation by self-organization
(Fig. 10). Of course pineal lens differentiation in extant vertebrates

may owe some of its features to properties evolved in the context of the lateral eye. Eakin (1973) has shown that pineal lens induction is not dependent on contact with specific surrounding tissues: the inductive stimuli responsible for lens differentiation are likely to be the result of interactions between cells in the vesicle itself (the vesicle surfaces are initially in cellular contact (Studnicka, 1905)) and an extension of the cell-by-cell stimuli that presumably induced the ancestor cells.

In the case of the lateral eye the initial evolutionary outgrowth of the optic vesicle need not have required provision for a lens; like the original ependymal photoreceptors it would have been functionally advantageous anyway provided it came sufficiently close to a sufficiently transparent body surface. A variety of circumstances can be envisaged by which it might have achieved its inductive recruitment of ectoderm for the formation of the lateral lens. This would include the possibility of gradual evolution *de novo* towards lens differentiation in an ectodermally derived organ of hitherto unconnected function which happened to occupy the site of the future lateral lens. However, an attractive alternative is suggested by the available facts taken as a whole. The eye-borne inductor and self-organizing properties of the lens cells may both have been available already as a result of evolution in the pineal eye. Then all that would be needed would be that the lateral eye recruits a source of uncommitted cells by, if necessary, an entirely non-specific morphogenetic stimulus to the ectoderm. The lineage of the cells is immaterial: cells merely have to be brought into the context of the lens inductor. This requires a coincidence, at that relevant stage of evolution, between the ability of the eye to recruit cells and/or the provision of cells by local ectoderm. Lateral ectoderm demonstrates an intrinsic disposition to delaminate and invaginate in that it gives rise to cranial placodes, the otocyst and the extensive lateral line system in early vertebrates. As suggested by the formation of neural tube and eyes by secondary cavitation from solid cores of cells in early vertebrates (Kerr, 1902) morphogenetic properties of early forms may have been quite different from those of current highly evolved embryos. The phylogenetically earliest examples of lens formation (Fig. 11) appear by delamination, consistent with having been recruited by a non-specific mitotic stimulus due to the approach of the optic vesicle. In higher vertebrates the morphogenetic stimulus has been further elaborated, involving for

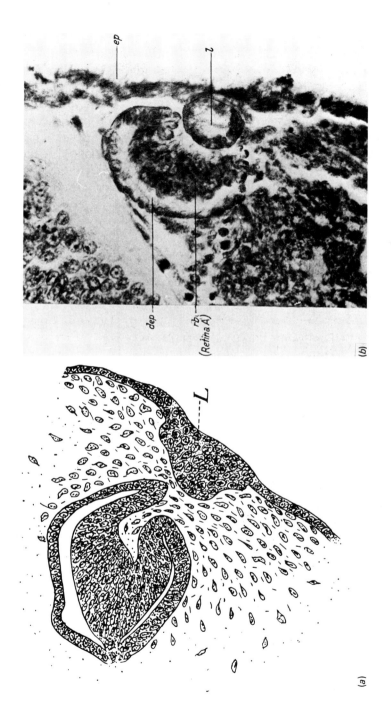

Fig. 11. Origin of lateral lens in cyclostome embryo. (a) Delamination of ectoderm (L) in *Bdellostoma stouti*. (From Stockard, 1907.) (b) Later lens vesicle (l) formation in *Petromyzon*: dorsal up (From Keibel, 1928.) dep = pigmented epithelium; ep = epidermis;

example local gut (Jacobson, 1958), permitting coordinated in-
vaginative folding. Whether lateral lens evolved independently, in
conjunction with, or consequent to the evolution of the pineal lens,
the foregoing discussion illustrates how an apparently discontinuous event like lateral lens induction can have evolved, perhaps
in parallel at different sites, in continuity.

In conclusion an attempt has been made to set out the sort of
procedures available, and some of the difficulties encountered, in
deducing evolutionary sequences using embryological evidence.
Naturally there is no reason to anticipate any final definitive
answers – that would require an ordered series of fossil embryos! –
but it is likely that convergence of evidence into a best fit pattern
may single out one scheme with overwhelming high probability.
The sort of evidence that can be brought to bear includes attention
to the plausibility of the required cellular and embryonic potentialities (these are likely to evolve relatively slowly and therefore to be
few and common between species), satisfying the continuity principle at all points, and use of any general rules for the elaboration of
new cell types and organ systems which may be found to apply
throughout evolution.

Some general rules for the embryological bases of evolution of organ-systems

(*a*) New cell types evolve on the basis of a diffusely arranged cell
by cell differentiation in an initially extensive and uniform cell
population (e.g. an epithelium) 'induced' perhaps by 'functional'
interdependencies. *Homogenous → heterogenous –*
(*b*) Later evolution primarily involves morphological re-location
and segregation into discrete organs of specific cell types which
themselves (together with their inductors) are relatively unchanging. Multiple morphological variants of the organ may occur (e.g.
pineal and lateral eyes): the tendency is for one to reach definitive
specialization while others change their function or atrophy in early
embryogenesis.

(*c*) An embryological morphogenetic sequence (such as that
leading from neural plate to eye) can only be understood as a
repetition of an evolutionary morphological progression, each stage
of which was functional. Evolutionary change does not anticipate
the final result: it is the result of selection and elaboration from the
inevitable morphogenetic variation occurring in ancestral embryos

at predifferentiation stages because of variables such as starting conditions in the egg, failure to mature of earlier evolved (vestigial) organs and minor changes in the properties of morphogenetically active cells prominent in early embryonic stages.

The inescapable gap between genes and evolution

Embryology's identity crisis, foreshadowed by schisms resulting from over-dogmatic theorizing such as surrounded recapitulation, intensified as the subject was overtaken by events in the form of the domination of biochemistry and genetics culminating in the triumphs of molecular biology. Should embryology use the Watson–Crick double helix as its model for a hoped-for 'breakthrough' or is it legitimate to operate at a less reductionist level of description equivalent to the framework created by Darwin for evolution?

Genetics has long appeared to represent the appropriately fundamental level for ultimate analysis of embryogenesis, as have mutations for the analysis of evolutionary change. If we knew the complete DNA sequence of a species, would we then necessarily completely understand embryology? In one important but poorly understood respect we would not: cell structure itself (including descent of organelles and 'maternal' factors) is inherited, independently of Mendelian considerations, by way of the template of the egg cell. Genetics has been disappointing in its contribution to the understanding of embryology: at the level of the final expression of differentiated states in cells, its relevance is obvious, and where pattern directly reflects properties of single differentiated cell types (as it does in situations such as cutaneous pigmentation patterns (Willier & Rawles, 1944; Fig. 8)) genetic variation can directly account for complex pattern changes. But, in general, known mutations do not fit well with the requirements one would expect of controllers of elements of spatial pattern; as surveys such as that by Kalter (1980) in the mouse show, single mutations very often have multiple (pleiotropic) effects at multiple locations and particular organs are disproportionately commonly involved. Disproportionately large groups of mutations share common teratological effects, suggesting that their effects are dictated by constraints resulting from embryological mechanisms. Homoeotic mutations in insects are exceptional. Their effects are clearly potentially valuable sources of segmental elaboration for insect evolution but, as discussed earlier, it is far from clear whether they involve a truly special class of pattern-controlling genes: they may be mutations of cellular expression whose effects are segmentally based as a result of the anatomical structure and homologies of insect segmentation.

Given the importance in embryogenesis of morphogenetic and functional interactions involving populations of cells, it is not surprising that the effects of many genes affecting pattern are buffered and transformed in a complex manner by embryonic processes (Grüneberg, 1963) which prevents resolution of their primary action as expressed in single cells. Integrated patterns of morphogenesis cannot be directly or explicitly programmed in the genome; they are the result of epigenetic interactions between differentiated embryonic cells progressively elaborated during embryogenesis and ultimately conditional on the organization of the egg cytoplasm.

Species differ most conspicuously in regard to differences in size and proportions, while cell types are, by contrast, few and stable in evolution. As we saw with the limb, it is the strategy for deployment of cells that generates anatomical complexity and variety: small changes to the initially simple starting conditions would be sufficient to cause major consequent anatomical modifications. Thus the gradual evolutionary modification of any particular embryonic feature, together with necessary harmonious changes elsewhere in the anatomy, would require small multidimensional genetic changes. Much may be achieved merely by rearrangement of genetic material (Wilson *et al.*, 1974). Embryology does not exclude the possibility of single mutations having global and relatively major ('macroevolutionary') effects which, in certain rare occasions, could be advantageous. While no aspect of embryogenesis can be entirely independent of the products of the genome, it may nonetheless be the case that control by genes is so indirect that analysis in such terms loses useful explanatory power. Pattern formation based on function (Fig. 8), in particular, enormously reduces the requirements for explicit genetic instructions, while also automatically ensuring adaptedness.

'A hen is only the egg's way of making another egg' (Butler, 1878)

The key to the eventual identification of the basic parameters for evolutionary change in body form is to see evolution from the point of view of the parameters of variation available in development, rather than drawing inferences from the adult end result. Concepts such as induction, morphogenetic strategy and functional demand provide unifying explanatory terms which refer to processes at a level likely to be actually selected for in evolution. Precise definition of cellular behaviour and potentialities operating in the embryo, which may be unexpectedly different from the familiar adult picture, may lead directly to identification of switching events underlying changes of genetic expression. Detailed study of individual organ systems such as the eye or the limb, from an embryological view-point, suggests plausible evolutionary sequences and mechanisms, avoiding the multidimensional complexity of the whole organism and biologists' concern with the species concept, and in spite of the current theoretical heart searching by taxonomists and palaeontologists. At the very least it is clear from the foregoing discussion that the particular embryological theory that is adopted will have a major bearing on how evolutionary events and the basis of their inheritance are conceptualized.

References

Ashburner, M. & Wright, T. R. F. (1978). *The genetics and biology of* Drosophila, vol. 2c. Academic Press: London.

Balinsky, B. I. (1927). Über experimentelle Induktion der Extremitatenanlage bei *Triton* mit besonderer Berucksichtigung der Innervation und Symmetrieverhaltnisse derselben. *Wilhelm Roux Arch. EntwMech. Org.* **110**, 71–88.

Beisson, H. & Sonneborn, T. M. (1965). Cytoplasmic inheritance of the organization of the cell cortex in *Paramecium aurelia. Proc. natn. Acad. Sci. USA* **53**, 275–82.

Berrill, N. J. (1961). *Growth, development, and pattern.* Freeman: San Francisco.

Bloemendal, H. (1981). *Molecular and cellular biology of the eye lens.* Wiley: New York.

Bohn, J. (1976). Tissue interactions in the regenerating cockroach leg. In *Insect development*, ed. P. A. Lawrence, Symposium of the Royal Entomological Society of London No. 8, pp. 170–85. Blackwell Scientific Publ.: Oxford.

Brower, D. L., Smith, R. J. & Wilcox, M. (1982). Cell shapes on the surface of the *Drosophila* wing imaginal disc. *J. Embryol. exp. Morph.* **67**, 137–51.

Bryant, P. J. (1970). Cell lineage relationships in the imaginal wing disc of *Drosophila melanogaster. Devl. Biol.* **22**, 389–411.

Bryant, P. J. (1975). Regeneration and duplication in imaginal discs. In *Cell patterning*, Ciba Foundation Symposium No. 29, pp. 71–93. Associated Scientific Publishers: Amsterdam.

Bryant, S. V., French, V. & Bryant, P. J. (1981). Distal regeneration and symmetry. *Science* **212**, 993–1002.

Bryant, P. J. & Schneiderman, H. A. (1969). Cell lineage, growth and determination in the imaginal leg discs of *Drosophila melanogaster*. *Devl. Biol.* **20**, 263–90.

Butler, S. (1878). *Life and habit*. Trübner: London.

Chalmers, J. & Ray, R. D. (1962). The growth of transplanted foetal bones in different immunological environments. *J. Bone Joint Surg.* **44B**, 149–64.

Child, C. M. (1941). *Patterns and problems of development*. University of Chicago Press: Chicago.

Conklin, E. G. (1905). Mosaic development in ascidian eggs. *J. exp. Zool.* **2**, 145–223.

Conklin, E. G. (1933). The development of isolated and partially separated blastomeres of *Amphioxus*. *J. exp. Zool.* **64**, 303–76.

Coulombre, A. J. (1965). The eye. In *Organogenesis*. eds. R. L. DeHaan and H. Ursprung, pp. 219–51. Holt, Rinehart & Winston: New York.

Counce, S. J. & Waddington, C. H. (1973). *Developmental systems: insects*, vol. 2. Academic Press: London.

Darwin, F. (1887). *The life and letters of Charles Darwin*, vol. 2, p. 273. Murray: London.

Davidson, E. H. (1976). *Gene activity in early development*, 2nd edn. Academic Press: New York.

de Beer, G. (1958). *Embryos and ancestors*, 3rd edn. Oxford University Press: Oxford.

Drastich, L. (1925). Uber das Leben der Salamandra-larven bei hohem und niedrigem Sauerstoffpartialdruck. *Z. vergl. Physiol.* **2**, 632–57.

Eakin, R. M. (1973). *The third eye*. University of California Press: Berkeley, Cal.

Ede, D. A. (1971). Control of form and pattern in the vertebrate limb. In *Control mechanisms of growth and differentiation*, eds. D. D. Davies and M. Balls, Symposium of the Society for Experimental Biology, No. 25, pp. 235–54. Cambridge University Press: Cambridge.

El Shatoury, H. H. (1955). Lethal no-differentiation and the development of the imaginal discs during the larval stage in *Drosophila*. *Wilhelm Roux Arch. EntwMech. Org.* **147**, 523–38.

Flint, J. M. (1910). Compensatory hypertrophy of the small intestine following resection of large portions of the jejunum and ileum. *Trans. Conn. State med. Soc.* 283–335.

Freeman, G. (1981). The cleavage initiation site establishes the posterior pole of the hydrozoan embryo. *Wilhelm Roux Arch. EntwMech. Org.* **190**, 123–5.

French, V., Bryant, P. J. & Bryant, S. V. (1976). Pattern regulation in epimorphic fields. *Science* **193**, 969–81.

García-Bellido, A. (1967). Histotypic reaggregation of dissociated imaginal disc cells of *Drosophila melanogaster* cultured *in vivo*. *Wilhelm Roux Arch. EntwMech. Org.* **158**, 212–17.

García-Bellido, A. (1977). Inductive mechanisms in the process of wing vein formation in *Drosophila*. *Wilhelm Roux Arch. EntwMech. Org.* **182**, 93–106.

García-Bellido, A., Ripoll, P. & Morata, G. (1973). Developmental compartmentalisation of the wing disk of *Drosophila*. *Nature New Biol.* **245**, 251–3.

Gerhart, J., Ubbels, G., Black, S., Hara, K. & Kirschner, M. (1981). A reinvestigation of the role of the grey crescent in axis formation in *Xenopus laevis. Nature* **292**, 511–16.

Goldschmidt, R. B. (1940). *The material basis of evolution.* Yale University Press: New Haven, Conn.

Goss, R. J. (1961). Regeneration of vertebrate appendages. *Adv. Morph.* **1**, 103–52.

Goss, R. J. (1964). *Adaptive growth.* Logos Press: London.

Gruneberg, H. (1963). *The pathology of development.* Wiley: New York.

Harris, P. (1979). A spiral cortical fiber system in fertilized sea urchin eggs. *Devl. Biol.* **68**, 525–32.

Harrison, R. G. (1921). On relations of symmetry in transplanted limbs. *J. Exp. Zool.* **32**, 1–136.

Harrison, R. G. (1969). *Organization and development of the embryo.* Yale University Press: New Haven, Conn.

Herlitzka, A. (1897). Sullo sviluppo di embrioni completi da blastomeri isolati di uova di tritone (*Molge cristata*). *Wilhelm Roux Arch. EntwMech. Org.* **4**, 624–58.

Hinchliffe, J. R. & Johnson, D. R. (1980). *The development of the vertebrate limb.* Clarendon Press: Oxford.

Holtfreter, J. (1934). Über die Verbreitung induzierender Substanzen und ihre Leistungen im *Triton-keim. Wilhelm Roux Arch. EntwMech. Org.* **132**, 307–83.

Holtfreter, J. (1947). Neural induction in explants which have passed through a sublethal cytolysis. *J. exp. Zool.* **106**, 197–222.

Horder, T. J. (1976). Pattern formation in animal development. In *The developmental biology of plants and animals*, eds. C. F. Graham and P. F. Wareing, pp. 169–97. Blackwell Scientific Publ.: Oxford.

Horder, T. J. (1981). On not throwing the baby out with the bath water. In *Evolution today*, eds. G. G. E. Scudder and J. L. Reveal, pp. 163–80. Hunt Institute for Botanical Documentation: Pittsburgh.

Hulsemann, M. (1967). Vergleichende histologische Untersuchungen uber das Vorkommen von Gliafasern in der Epiphysis cerebri von Saugetieren. *Acta anat.* **66**, 249–78.

Huxley, J. S.(1972). *Problems of relative growth*, 2nd edn. Dover Publ.: New York.

Jacobson, A. G. (1958). The roles of neural and non-neural tissues in lens induction. *J. exp. Zool.* **139**, 525–57.

Kalter, H. (1980). A compendium of the genetically induced congenital malformations of the house mouse. *Teratology* **21**, 397–429.

Keibel, F. (1928). Beitrage zur Anatomie, zur Entwicklungsgeschichte und zur Stammesgeschichte der Sehorgane der Cyklostomen. *Z. mikrosc.-anat. Forsch.* **12**, 391–456.

Kerr, J. G. (1902). The development of *Lepidosiren paradoxa. Q. J. microsc. Sci.* **46**, 417–59.

Kieny, M. (1964). Régulation des excédents et des déficiences du bourgeon d'aile de l'embryon de poulet. *Archs. Anat. micr. Morphol. exp.* **53**, 29–44.

Klinckowstroem, A. de (1894). Beitrage zur Kenntnis des Parietauges. *Zool. Jb. (Anat. Ont. Tiere)* **7**, 249–80.

Kramer, G. (1951). Body proportions of mainland and island lizards. *Evolution* **5**, 193–206.

Lawrence, P. A. & Green, S. M. (1975). The anatomy of a compartment border: the intersegmental boundry in *Oncopeltus*. *J. Cell Biol.* **65**, 373–82.

Lawrence, P. A. & Morata, G. (1976). Compartments in the wing of *Drosophila*: a study of the *engrailed* gene. *Devl Biol.* **50**, 321–37.

Lewis, W. H. (1904). Experimental studies on the development of the eye in Amphibia. I. On the origin of the lens. *Am. J. Anat.* **3**, 505–36.

MacCabe, J. A., Saunders, J. W. & Pickett, M. (1973). The control of the anteroposterior and dorsoventral axes in embryonic chick limbs constructed of dissociated and reaggregated limb-bud mesoderm. *Devl Biol.* **31**, 323–35.

Maden, M. (1980). Structure of supernumerary limbs. *Nature* **286**, 803–5.

Nieuwkoop, P. D. & Sutasurya, L. A. (1979). *Primordial germ cells in the chordates*. Cambridge University Press: Cambridge.

Poodry, C. A., Bryant, P. J., Schneiderman, H. A. (1971). The mechanism of pattern reconstruction by dissociated imaginal discs of *Drosophila melanogaster*. *Devl Biol.* **26**, 464–77.

Postlethwait, J. H. & Schneiderman, H. A. (1971). Pattern formation and determination in the antenna of the homoeotic mutant *antennapedia* of *Drosophila melanogaster*. *Devl Biol.* **25**, 606–40.

Rappaport, R. (1970). An experimental analysis of the role of cytoplasmic fountain streaming in furrow establishment. *Devl. Growth Diff.* **12**, 31–40.

Reverberi, G. (1971). *Experimental embryology of marine and fresh-water invertebrates*. North-Holland: Amsterdam.

Roux, W. (1888). Beitrage zur Entwicklungsmechanik des Embryo. V. Ueber die kunstliche Hervorbringung 'halber' Embryonen durch Zerstorung einer der beiden ersten Furchungszellen, sowie uber die Nachentwickelung (Postgeneration) der fehlenden Korperhalfte. *Arch. path. Anat.* **114**, 113–53.

Rychter, Z. (1962). Experimental morphology of the aortic arches and the heart loop in chick embryos. *Adv. Morph.* **2**, 333–71.

Salvini-Plawen, L. V. & Mayr, E. (1977). On the evolution of photoreceptors and eyes. *Evol. Biol.* **10**, 207–63.

Saunders, J. W. (1977). The experimental analysis of chick limb bud development. In *Vertebrate limb and somite morphogenesis*, eds. D. A. Ede, J. R. Hinchliffe and M. Balls, pp. 1–24, Cambridge University Press: Cambridge.

Saunders, J. W., Gasseling, M. T. & Errick, J. (1976). Inductive activity and enduring cellular constitution of a supernumerary apical ectodermal ridge grafted to the limb bud of the chick embryo. *Devl Biol.* **50**, 16–25.

Saxen, L. & Toivonen, S. (1962). *Primary embryonic induction*. Academic Press: London.

Schmalhausen, I. I. (1948). *Factors in evolution*. Blakiston: Philadelphia.

Schneiderman, H. A. (1979). Pattern formation and determination in insects. In *Mechanisms of cell change*, eds. J. D. Ebert and T. S. Okada, pp. 243–72. Wiley: New York.

Simpson, P. (1976). Analysis of the compartments of the wing of *Drosophila melanogaster* mosaic for a temperature-sensitive mutation that reduces mitotic rate. *Devl. Biol.* **54**, 100–15.

Slack, J. M. W. (1977). Control of anteroposterior pattern in the axolotl forelimb by a smoothly graded signal. *J. Embryol. exp. Morph.* **39**, 169–82.

Slijper, E. J. (1942). Biologic-anatomical investigations on the bipedal gait and upright posture in mammals, with special reference to a little goat, born without forelegs. II. *Proc. Ned. Akada. Wetensch. Amsterdam* **45**, 407–15.

Smith, L. D. & Ecker, R. E. (1970). Uterine suppression of biochemical and morphogenetic events in *Rana pipiens*. *Devl. Biol.* **22**, 622–37.

Spemann, H. (1901). Ueber Correlationen in der Entwickelung des Auges. *Verh. anat. Ges. Jena* 61–79.

Spemann, H. (1938). *Embryonic development and induction*. Yale University Press: New Haven, Conn.

Stanley, S. M. (1973). An explanation for Cope's rule. *Evolution* **27**, 1–26.

Stern, C. (1968). *Genetic mosaics and other essays*. Harvard University Press: Cambridge, Mass.

Stockard, C. R. (1907). The embryonic history of the lens in *Bdellostoma stonti* in relation to recent experiments. *Am. J. Anat.* **6**, 511–15.

Studnicka, F. K. (1905). *Die Parietalorgane, Lehrbuch der vergleichenden mikroskopischen Anatomie der Worbettiere*, ed. A. Oppel, Part V. Fischer: Jena.

Tank, P. W., Carlson, B. M. & Connelly, T. G. (1977). A scanning electron microscopic comparison of the development of embryonic and regenerating limbs in the axolotl. *J. exp. Zool.* **201**, 417–30.

Tank, P. W. & Holder, N. (1981). Pattern regulation in the regenerating limbs of urodele amphibians. *Q. Rev. Biol.* **56**, 113–42.

Thompson, D'A. W. (1942). *On growth and form*. Cambridge University Press: Cambridge.

Tickle, C. (1981). The number of polarizing region cells required to specify additional digits in the developing chick wing. *Nature* **289**, 295–8.

Tickle, C., Alberts, B., Wolpert, L. & Lee, J. (1982). Local application of retinoic acid to the limb bud mimics the action in the polarizing region. *Nature* **296**, 564–6.

Tobler, H. (1966). Zellspezifische Determination und Beziehung zwischen Proliferation und Transdetermination in Bein-und Flugelprimordien von *Drosophila melanogaster*. *J. Embryol. exp. Morph.* **16**, 609–33.

Tung, T. C., Wu, S. C. & Tung, Y. Y. F. (1962). Experimental studies on the neural induction in *Amphioxus*. *Scienta sin.* **11**, 805–20.

Vijverberg, A. J. (1974). A cytological study of the proliferation patterns in imaginal disks of *Calliphora erythrocephala* Meigen during larval and pupal development. *Neth. J. Zool.* **24**, 171–217.

Wallace, H. (1981). *Vertebrate limb regeneration*. Wiley: Chichester.

Weiss, P. (1939). *Principles of development*. Holt: New York.

Wermel, J. (1935). Untersuchungen uber die Kinetogenese und ihre Bedeutung in der onto- und phylogenetischen Entwicklung (Experimente und Vergleichungen an Wirbeltierextremitaten). VI. Mitteilung: Veranderungen der Muskulatur. *Morph. Jb.* **75**, 452–8.

Wildermuth, H. (1968). Autoradiographische Untersuchungen zum Vermehrungsmuster der Zellen in proliferierenden Russelprimordien von *Drosophila melanogaster*. *Devl. Biol.* **18**, 1–13.

Willier, B. H. & Rawles, M. E. (1944). Melanophore control of the sexual dimorphism of feather pigmentation pattern in the barred Plymouth Rock fowl. *Yale J. Biol. Med.* **17**, 319–40.

Wilson, A. C., Sarich, V. M. & Maxson, L. R. (1974). The importance of gene rearrangement in evolution: evidence from studies on rates of chromosomal, protein, and anatomical evolution. *Proc. natn. Acad. Sci. USA* **71**, 3028–30.

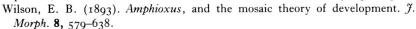 Wilson, E. B. (1893). *Amphioxus*, and the mosaic theory of development. *J. Morph.* **8**, 579–638.

Wolpert, L. (1969). Positional information and the spatial pattern of cellular differentiation. *J. theor. Biol.* **25**, 1–47.

Epigenetic control in development and evolution

BRIAN K. HALL*

Department of Biology, Medical and Biological Sciences Building, University of
Southampton, Southampton SO9 3TU, UK

#6439

'It has become increasingly clear from researches in embryology that the processes
whereby the structures are formed are as important as the structures themselves
from the point of view of evolutionary morphology and homology.' (de Beer, 1958,
p. 163.)

Introduction

Epigenetics, in the broader context of epigenesis as an alternative to
preformation, two major schema explaining development, goes
back to the time of Aristotle. Epigenetics has been variously
defined – all definitions having in common the concept of control
of gene expression by the micro-environment(s) encountered by
cells during development. Wilson (1925) gave *sole* control of
development to the epigenetic cytoplasmic environment: '. . .
heredity is effected by the transmission of a nuclear *preformation*
[my italics] which in the course of development finds expression in
a process of cytoplasmic epigenesis.' (*ibid.*, p. 1112). We now
realise that other factors also control development. Hormones,
tissue interactions, even physical factors such as mechanical stimuli, electrical and electromagnetic fields, can all elicit developmental processes such as differentiation, morphogenesis and
growth. I saw epigenetics as the 'causal analysis of development, in
particular . . . the mechanisms by which genes express their
phenotypic effects' (Hall, 1978a, p. 219). This definition is rather
broad but does capture the spirit of epigenesis as controlling gene
expression, whatever the nature of the epigenetic control.

* Permanent address. Department of Biology, Dalhousie University, Life Sciences Centre,
Halifax, Nova Scotia, Canada B3H 4J1

Gould (1977) has summarised the history of epigenesis and preformation as two major views of development, and emphasises (*ibid.*, p. 28) that the term *evolution* was coined by Haller in 1744 for the *developmental* process of preformation. Semantically, the duality of development and evolution has a long history. However, despite the early and major attempts of Schmalhausen (1949) and Waddington (1957) and more recently of Riedl (1978) to incorporate epigenetics into evolutionary theory, evolutionary biology has not embraced epigenetics as providing any necessary explanations for the mechanism(s) of evolutionary change. Very recently, this situation has begun to change. Epigenetic control has been seen as providing 'a developmental basis for producing major morphological shifts within but a few generations' (Alberch *et al.*, 1979, pp. 309–10). Alberch (1980, p. 653) has emphasised 'the role that developmental dynamics (epigenetics) plays in constraining the directionality of morphological evolutionary change'. Epigenetics provides 'the missing chapter of evolutionary biology' (Hamburger, 1980, p. 108); 'in a word, the key to macroevolutionary theory is epigenesis' (Wilson, 1981, p. 71); the basis for a structuralist approach to the evolution of biological form (Webster & Goodwin, 1981; Goodwin, this volume); and the increasingly more complex spatial component for cellular differentiation during both development and evolution (Ho & Saunders, 1979; Horder, this volume).

How is epigenetic control exercised during development and how has this fundamental developmental process been utilised during evolution? Epigenetic control over growth and morphogenesis has been extensively discussed and illustrated by Gould (1977), Alberch *et al.* (1979), Alberch (1980) and Alberch & Alberch (1981). To quote the last of these, '. . . any transformation in a complex morphological character must result from an epigenetic change in timing and rates of growth, differentiation, or morphogenesis' (p. 263). They have shown how the highly specialised morphology of the arboreal salamander, *Bolitoglossa occidentalis*, has been generated by the process of progenesis – earlier termination of development than occurs in related terrestrial species. Alberch (1980) has also used homoeotic mutants, skeletal mutants such as *wingless*, *polydactylous* and their phenocopies, and morphological polymorphisms in naturally occurring populations to illustrate epigenetic control over morphogenesis. However, little attention has been paid to the processes underlying such changes or to documenting whether the timing of cell differentiation (which is intimately linked to morphogenesis) is under similar epigenetic control. I want to illustrate aspects of epigenetic control with reference to the differentiation of cell types within the embryo. If epigenetic control and timing of development are important in evolution, we should be able to show that fundamental developmental processes such as embryonic induction and the initiation of cell differentiation are epigenetically controlled, and, more importantly, that the timing of these processes is also under epigenetic control.

Epigenetic control in development
The epigenetic landscape

Development can realistically be visualised as a hierarchical succession of epigenetic events, each one both eliciting a differentiative or morphogenetic response from pre-existing cells and creating a new microenvironment for subsequent generations of cells to undergo different and perhaps more complex interactions leading to further and more specialised differentiative and morphogenetic events. Horder and Maynard Smith in their chapters in this volume explore the epigenetic organisation of the whole embryo and the increasing spatial complexity which accompanies embryonic development.

Waddington (1940) introduced the analogy of the epigenetic landscape to visualise the progress of an embryo or one of its regions or cells during development. That he saw it as a useful analogy is exemplified by his inclusion of a frontispiece by John Piper to illustrate what he termed 'my somewhat romantic conceit, the epigenetic landscape' and his development of the concept in much of his later work (e.g. Waddington, 1957). The analogy is a simple one but its basis is profound and has yet to be fully translated into mechanisms of development and gene control, although the ontogenetic trajectories of Alberch *et al.* (1979) now provide a means of measuring and quantifying the end-products of pathways of development and Maynard Smith (this volume) has attempted a classification of the underlying developmental processes.

Waddington applied the epigenetic landscape to regions of early embryos competent to differentiate along more than one direction – the notion that cells during development come to forks where epigenetic factors send them down one pathway of differentiation rather than down another. Implicit in his scheme was the existence of bipotential cells able to exercise differentiative choices at one or various times during development. The earlier in development that these decisions occur, the broader and less developmentally restricted are the end-products. Furthermore, the end-products are clearly distinct from one another – intermediate states do not usually occur. Examples would be formation of inner cell mass or trophoblast in early mammalian development, neural or epidermal ectoderm but not neuroepidermis, cartilage or bone but not chondroid bone, etc. These patterns necessitated, for Waddington, thresholds of response to the epigenetic stimulus and a developmental mechanism to canalise development along each path or *creode* (the necessary path, as in the trajectory of a homing missile) as he termed them. Normal development is 'the most favoured path' with stabilising and canalising mechanisms operating to optimise the end-product. The evolutionary implications (*a*) of the existence of discrete end states, be they cells, tissues, organs or organisms, (*b*) of mechanisms to maintain only a small number of developmental pathways, and (*c*) of mechanisms to switch cells from one pathway to another without any continuous variation in morphology, are considerable (see the incisive discussion by Alberch, 1980).

Waddington saw the epigenetic landscape as equally applicable to the three ways of generating diversity during development, viz. segregation of cytoplasmic constituents, gradients of histogenetic substances (morphogens) and embryonic induction. I want to concentrate on the latter, viz. epigenetic control by inductive tissue interactions as a means of generating diversity of cell types and of spatially and temporally localising tissues within the embryo. Other workers (Saxén & Karkinen-Jaaskelainen, 1981) have also used the epigenetic landscape as the conceptual starting point for discussions of embryonic induction and tissue interactions. These authors emphasised that some developmental potential is lost as each choice is made (determination of the cells is more restricted) and that the responding cell may alter or have a reciprocal effect on its environment, e.g. a maintenance effect or render the inducer inactive, thereby preventing later, ectopic formations of the same cell type.

Embryonic induction

Induction is the evocative action of one cell or tissue upon another. Fertilisation itself can be viewed as the first of the many embryonic inductions which ensure that each developmental stage and the adult are coordinated and functional entities (we must view embryonic stages as entities in their own right and not just way-stations on the road to adulthood – a notion well recognised in studies on life history strategies but often forgotten by developmental biologists). After undergoing a complex series of interactions with the egg membrane(s) the sperm penetrates the egg. This activates the egg, thereby allowing cleavage, the first major developmental process, to be initiated. Without the inductive act of fertilisation, the egg remains impotent, its latent potential unexpressed.

The next wave of inductive interactions has to await the rearrangement of the major areas of the embryo which occurs during gastrulation and neurulation. It is the creation of these new associations which allows previously spatially separated parts of the embryo to come together, interact and trigger further associations as new cell types are differentiated. Thus, the presumptive notochordal tissue interacts with that portion of the presumptive ectoderm with which it comes into contact during gastrulation to evoke neural differentiation. Ectoderm which does not come into contact with notochord does not express neural differentiation (although it can, if transplanted into contact with notochord) but forms epidermal ectoderm instead (Fig. 1). Notochord is one of the forks or watersheds in Waddington's epigenetic landscape. Any ectoderm in embryos *at this stage*, poised at the fork in the landscape, can become neural ectoderm (the classic transplantation experiments of Spemann (1938)) but only that in contact with notochord has this potential released. Once this developmental stage is passed, epidermal ectoderm can no longer respond to notochord by forming neural tissues. The capacity for this epigenetic interaction is limited to a specific time and place during development. The choices are clear cut and non-overlapping – form neural ectoderm or form epidermal cells and organise those cells into nervous system or skin, respectively. Neither intermediate cell types nor tissues are found. The demarcation between the two end-products is unequivocal.

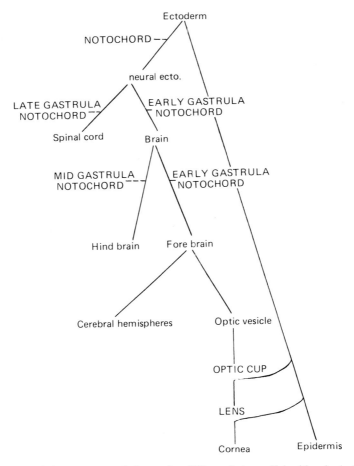

Fig. 1. Part of the sequence of alternative differentiations elicited by the inductive action of notochord upon ectoderm. Inducers. are in capitals, induced structures are not. Note that some structures (optic cup, lens) are both induced and inducers.

Differentiation of ectoderm into neural tissue in response to induction from notochord is not the end of this epigenetic interaction, for the neural tissue is organised into spinal cord and brain, and the brain into fore-, mid-, and hind-brain as part of the inductive process. Thus, notochord of early, middle and late gastrulae all share the common inductive ability of evoking neural differentiation from adjacent ectoderm but are fundamentally different in how they specify the spatial organisation of those neurones – early gastrula ectoderm specifying organisation into fore- and mid-brains, mid-gastrula into mid- and hind-brains and late gastrula into spinal cord (Fig. 1). Inducers, by acting *throughout* a developmental stage can elicit similar cytodifferentiated states but differing morphologies. This appears to represent at least a two-step induction, for although differentiative induction of neural ectoderm is not mediated by cellular contact between notochord and ectoderm the morphogenetic induction into regions is (Toivonen *et al.*, 1975; Toivonen & Wartiovaara, 1976). The differing morphogenetic organisation of the nervous system may in part be specified when differentiation is elicited. Thus, although similar in terms of their functional and cytological properties, neurones of fore-, mid-, and hind-brain may differ in those cell surface properties which specify cell-to-cell interactions and ultimately the three-dimensional organisation of the brain. What is unknown is whether the inducer provides one common signal for neuronal differentiation throughout gastrulation, with additional signals for morphogenesis, or whether the differentiative and morphogenetic signals are combined, one early and one late in gastrulation.

The action of the primary embryonic inducer also extends to the mesoderm. Mesoderm closest to the notochord becomes paraxial, mesoderm more laterally placed becomes intermediate, while mesoderm even further removed becomes lateral plate. This is a chain of epigenetic events whose character depends on position in relation to the source of the epigenetic control, the notochord of the dorsal lip. Further chains of epigenetic interactions are also initiated. Within the forebrain, one region becomes cerebral hemispheres, the other optic vesicle; the latter forms optic cup which differentiates along several different pathways (pigmented retinal epithelium, iris, ciliary zone, optic nerve, neural retina), while acting inductively on the overlying ectoderm to induce in it the lens, which in turn induces overlying ectoderm to diverge from its alternative pathway of forming epidermal ectoderm to form transparent cornea instead (Fig. 1). This last interaction continues throughout life, so that damage to the adult lens results, secondarily, in damage to the cornea (cataracts). A further chain of inductive interactions involves the hindbrain in the formation of the otic capsule (Hall, 1982*a*). Cranial ectoderm is induced to form an otic vesicle in two steps, first by cranial mesoderm during early to

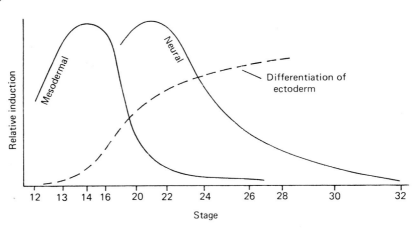

Fig. 2. In *Ambystoma*, ectoderm differentiates into otic vesicle as a result of two sequential inductions, one from the mesoderm during neurulation (stages 12–16) and one from the neural tube after neurulation. Stage 21 is the early tail-bud stage embryo; stage 32 the late tail-bud stage embryo. (Modified from Yntema, 1955.)

mid neurulation, and then by hindbrain from mid neurulation onwards (Fig. 2). The timing of these interactions is tightly controlled by the development of cellular contacts and gap junctions between the two layers only during the inductive process (Model *et al.*, 1981). Early neurula ectoderm will not respond to late neurula hindbrain, nor will post-neurulation ectoderm respond to neurula mesoderm. The differentiated otic vesicle then induces mesenchyme to chondrify as the otic capsule. The foot plate of the avian columella is induced to form only after epigenetic interaction with the otic capsule (the remainder of the columella forms without interaction with the capsule). The columella then acts back on part of the otic capsule to induce already differentiated chondrocytes to become fibroblastic and to form the annular ligament which anchors the foot plate to the capsule (Jaskoll & Maderson, 1978). Clearly, complex series of developmental events depend on the precise initiation of embryonic induction, both in time and in space.

Tissue interactions

A further example of epigenetic control during development (chosen non-arbitrarily from the many possible) is my own work on the differentiation of cartilage and bone in the craniofacial skeleton.

We have paid particular attention to the developing lower jaws in embryonic chicks and mice. The skeletal elements comprise a rod of Meckel's cartilage surrounded by membrane bones – five in the chick and one in the mouse. The mesenchyme from which these tissues develop has its origin in the embryonic neural crest. As we did for neural ectoderm, the eye and the otic capsule, we can follow the history of these cells and examine the various epigenetic events to which they respond.

Presumptive neural crest cells of the embryonic chick can first be detected in the anterior epiblast at the junction of neural and epidermal ectoderm (Rosenquist, 1981), a position which raises the question of whether primary embryonic induction exerts any control over their future. Is the tremendous differentiative potential of the neural crest (they form cartilage, bone, dentine, connective tissues, neurones, endocrine and pigment cells) causally related to their being exposed to combined inductions early in development? We do not know.

Neural crest can first be seen as a morphological entity as the embryo undergoes neurulation. These cells soon separate off and migrate away as the neural tube develops. Their fate has been mapped by grafting labelled neural crest into unlabelled hosts and following their differentiation (Johnston, 1966; LeLièvre, 1974; Noden, 1978). Thus, it is known that all except the myogenic mesenchyme of the mandibular arch has its origin in neural crest at the level of the midbrain and anterior hindbrain. Within the mandibular arch this neural crest-derived mesenchyme forms cartilage, bone, connective tissue and, in toothed vertebrates, dentine.

Clearly, it is a long journey from the epiblast to the neural tube, around the developing eye and into the future craniofacial region. If these cells respond to epigenetic activation then such interactions could occur (*a*) *before* migration, while the cells are in the epiblast or neural folds, (*b*) *during* migration, or (*c*) *after* migration when localised in the mandibular arch. For the embryonic chick, we know that very soon after migration commences, neural crest cells interact with lateral epidermal ectoderm and that this interaction is required for cartilage subsequently to form (Bee & Thorogood, 1980). Interactions with epithelia encountered later in development are not required (Hall, 1978*b*; Hall & Tremaine, 1979). We do not know whether all or only some neural crest cells interact with this cranial ectoderm. Similarly, we do not know whether those cells do undergo any further epigenetic interactions before actually differentiating as chondrocytes. Certainly, the interaction with cranial ectoderm is *sufficient* to allow chondrocytic differentiation, for it can be reproduced *in vitro* (*ibid.*). Interactions 'along the way' may be involved in the subsequent differentiation of connective tissue and bone. For membrane bones to form, the neural crest-derived cells must undergo an interaction with the epithelium of the developing mandibular arch, beginning soon after they reach the arch ($2\frac{1}{2}$ days of incubation) and continuing until 4 days of incubation, which is some 3 days before bone actually forms (Tyler & Hall, 1977; Hall, 1978*b*). We do not know what controls differentiation of connective tissue. It may be, like epidermal ectoderm, the cell type which is automatically produced unless the cells are specifically induced to form something else – neural tissue in the case of ectoderm and cartilage or bone in the case of neural crest.

Epigenetic interactions between neural crest-derived mesenchyme and epithelium also control initiation of chondrogenesis and osteogenesis in the lower jaw of the mouse. However, *both* cartilage and bone are induced by the epithelium of the lower jaw (Hall, 1980). In amphibians, an interaction between migrating neural crest-derived cells and the pharyngeal endoderm elicits differentiation of Meckel's cartilage (Horstadius, 1950; Hall, 1982*a*). Thus, the epigenetic events controlling differentiation of Meckel's cartilage from neural crest-derived mesenchyme can occur at the *beginning* of migration (birds), *during* migration (amphibians) or *after* migration (the mouse). The evolutionary implications of these variations in timing of the 'same' epigenetic event will be discussed below.

Other elements of the craniofacial skeleton are also controlled by similar epithelial–mesenchymal interactions – periocular mesenchyme and pigmented retinal epithelium or scleral papillae in the induction of scleral cartilage and bone; maxillary and palatal epithelia and mesenchyme in formation of the maxilla and palate; the brain and cranial ectoderm in induction of the skull, etc. (see Hall, 1982*b*, for a review).

These examples are but a few of the many which could be used to show that development is controlled epigenetically through tissue interactions. Cells, although they contain all the necessary genetic information for generating one or more clearly distinct differentiated states, must have that potential activated (and other potential pathways inactivated?) by epigenetic control, which, itself, is based on prior genetic activation of the inducer and the changing spatial relationships between responding and inducing tissues during development – the valleys, peaks and forks in Waddington's epigenetic landscape.

Epigenetic control in evolution

Intermediate tissues and epigenetic control of differentiation

Epigenetic control of development through embryonic inductions channels cells along one of (usually) only two alternative pathways: e.g. neural or epidermal ectoderm, cartilage or bone. Differentiated cells with intermediate characteristics are not usually found, but can be formed, as in the case of chondroid bone – a tissue with some characteristics of both cartilage and bone (Beresford, 1981).

The very earliest vertebrates, the Agnatha of the Ordovician, already contained skeletal tissues which are quite recognisable to the histologist as 'good' cartilage and bone (Ørvig, 1968; Hall, 1975). As epigenetic tissue interactions became established at this early stage in vertebrate evolution, we might expect to find instances of tissues with intermediate histology. In fact the Agnatha exhibit tissues intermediate between bone and dentine (osteodentine), between bone and cartilage (chondroid bone) and an acellular bone (aspidin), in addition to unequivocal cartilage, bone, dentine and enamel (Ørvig, 1951, 1968; Moss, 1977). The existence of these intermediate tissues shows (*a*) the close affinity between osteogenic (bone-forming), chondrogenic (cartilage-forming) and odontogenic (dentine-forming) cells, (*b*) that they can modulate their differentiated extracellular products to produce a variety of locally adapted skeletal tissues (see Hall, 1978*a*, for a full discussion) and (*c*) that such tissues must confer a selective advantage.

The existence of intermediate tissues implies that they serve local functions more effectively than do either of the alternative cell types, e.g. chondroid bone providing the resilience and high growth potential of cartilage with the strength and calcium storage potential of bone. Chondroid bone can be experimentally induced, e.g. by immobilisation of embryonic chicks whose secondary cartilage requires biomechanical stimuli for its initiation and maintenance (Hall, 1972), or by exposing such cartilage to vitamin C *in vitro* (Hall, 1981*a*). Interestingly, the former treatment alters both cells and extracellular matrix while the latter only affects the extracellular matrix, illustrating the potential of the same cells to form grades of intermediate tissues in response to particular external cues. The existence of modulation and metaplasia is a further indication of retention of the potential to modify pathways of development, either temporarily as in modulation, or permanently as in metaplasia (Beresford, 1981). As epigenetic control can select between such end-products, skeletal cells in our earliest vertebrate ancestors must have been capable of being directed by such control. The canalisation of these 'scleroblasts' into a more restricted number of end-products, characteristic of modern-day vertebrates (acellular bone is restricted to some teleost fishes; bone is not found in the endoskeleton of sharks or in cyclostomes; bone and dentine are now separate tissues, etc.), indicates that epigenetic control has narrowed the differentiative options during evolution, while retaining the potential to form other skeletal tissues if called upon to do so. Development of an intermediate tissue, be it chondroid bone or osteodentine, indicates that restriction into a particular creode has either not occurred or has been developmentally delayed. To move from osteodentine to bone and dentine involves the acquisition of a developmental mechanism to channel cells into one of the two alternative pathways *earlier in development*. The time of disappearance of these intermediate tissues from the fossil record would mark this shift in epigenetic control, a shift which probably occurred in the early Silurian as the Chondrichthyes and Osteichthyes, the cartilaginous and the bony fishes, arose. This was also the time at which the cartilaginous fishes stabilised the number of pairs of lateral fins at two (in earlier vertebrates the number and structure of fins was quite variable), perhaps also reflecting altered epigenetic control (Riedl, 1978, especially Fig. 45 and Table *E*; Hinchliffe & Johnson, 1980). Endochondral ossification also arose at this time (Hall, 1975, 1978*a*) – an important prerequisite for the development of the appendicular skeleton and for the progressive replacement of dermal by endochondral bone, which occurred in the subsequent evolution of the pectoral and pelvic girdles. The bony fishes retained both cartilage and bone as differentiated end products, the cartilaginous fishes only cartilage. The bony fishes replace cartilage by mineralised bone to gain skeletal support, the cartilaginous fishes heavily calcify their cartilage instead (both bone and this calcified cartilage contain around 1 g of mineral per cm³ of extracellular matrix), giving them the strength but not the structure of bone. In so doing, the cartilaginous fishes evolved some parallels to bone not seen in other groups – the chondrocytes remain viable and fail to hypertrophy when the matrix mineralises – behaviour typical of osteocytes in other vertebrate groups (Moss, 1977). This movement away from the osteogenic pathway in the Chondrichthyes did not involve loss of the ability to form bone, for acellular bone is found at the base of the teeth and in the dermal denticles in both recent and fossil sharks (Moss, 1977). Whether osteogenesis has been permanently turned off in the endoskeleton or could be epigenetically reactivated remains to be determined although a very recent report by Peignoux-Deville *et al.* (1982) claims that the neural arches of dogfish vertebrae are capped with bone, containing both osteoblasts and osteocytes and that this perichondral bone is induced to form by the same epigenetic mechanisms found in other vertebrates (see Scott-Savage & Hall, 1980, for a discussion of those mechanisms, and Hall, 1982*f*, for bone in sharks).

Modern-day cyclostomes have an extensive mucocartilage throughout the larval head which is replaced by, or transformed into, hyaline cartilage at metamorphosis (Hardisty, 1979). They seem to have retained both an intermediate tissue (mucocartilage) and true cartilage but lost, or never had, bone as a developmental option.

An examination of other modern groups also sheds some light on epigenetic control and the retention or loss of intermediate tissues. Modern-day teleosts display cellular and acellular bone and osteodentine (continuing to link the dental and skeletal tissues), in addition to cartilage, dentine, enamel and enameloid, the latter a mineralised tissue which caps teleost teeth (Moss, 1964). Dentine and enamel are normally formed by mesenchymal and epithelial cells, respectively, after a complex series of epithelial–mesenchymal interactions. However, in some fishes, an enameloid is formed for which mesenchymal cells (odontoblasts) provide the collagen, but epithelially derived ameloblasts (the normal enamel-producing cells) provide the other proteins (Shellis & Miles, 1974; Moss, 1977). Enameloid is also found as a cap over the teeth of larval urodele amphibians, whereas ameloblasts deposit enamel in adult urodeles (Meredith-Smith & Miles, 1971). Epigenetic control over dental development can therefore alter the products produced at different phases of the life cycle. An epigenetic basis for the formation of intermediate tissues such as osteodentine can be found in the close association of the dental sac (which produces the periodontal ligament, cementum and alveolar bone) with the dental mesenchyme, and the fact that, early in development, dental sac mesenchyme is as effective an inducer of dental epithelium into ameloblasts as is dental mesenchyme (Yoshikawa & Kollar, 1981). A slight shift in position of the dental sac towards the dental mesenchyme and enamel organ could establish the spatial relationship required to form a tissue such as osteodentine.

The mucocartilage–cartilage transition at metamorphosis in lampreys, or the cartilage–bone transition in anuran amphibians as they go from tadpole to adult, may be similarly controlled, although the critical experiments have yet to be done.

A dramatic example of the ability of epigenetic control to alter the pathway of cytodifferentiation comes from the recombination of mesenchyme from the developing first molar of the embryonic mouse with ectoderm from the chick limb bud. In recombination with dental ectoderm the mesenchyme forms dentine but in recombination with limb ectoderm it forms cartilage (Hata & Slavkin, 1978), a tissue which occasionally appears ectopically within developing teeth (Hall, 1971), presumably also because of disrupted epigenetic control. Experimental evidence for altered morphogenesis arising as a result of the *responding* cells acquiring a new position during development comes from the congenital anomalies which follow defective migration of neural crest cells (Poswillo, 1976) or the misplaced cartilage and bone which forms when neural crest cells destined for the mandible are redirected to the maxilla after vitamin A treatment of pregnant rats (Morriss & Thorogood, 1978). The extensive literature on the induction of ectopic tissues such as cartilage and bone provides further evidence for epigenetic evocation of a previously unexpressed pathway of differentiation. Thus, implanted demineralised bone matrix in muscle interacts with local mesenchymal cells to induce them to form cartilage which is then replaced by bone (Urist, 1982). Neither chondrogenesis nor osteogenesis is seen in the absence of the implant.

In situations such as the lower jaw of the embryonic mice, the same epigenetic stimulus (mandibular epithelium) is required for the initiation of two different cell types, cartilage and bone. Is the epithelium providing two different inductive messages, e.g. one cell contact mediated and one not, as in primary embryonic induction, or is the same inductive message being responded to differently by different mesenchymal cells with the developing mandible? To answer such questions we need clonal cell analysis to be sure that the epigenetic control is switching single cells down a particular pathway, rather than selecting predetermined cells from a mixed population. In either case, the primacy of epigenetic control in initiating differentiation remains.

Can epigenetic control activate pathways of differentiation which have been inactive for a long time? The occurrence of atavisms provides one class of evidence in the affirmative. I plan to discuss them in a subsequent paper. Experimental evidence comes from the still to be confirmed demonstration that mouse dental mesenchyme, when recombined with chick mandibular arch epithelium, can elicit the formation of enamel by these epithelial cells (Kollar & Fisher, 1980). Clearly, birds do not make enamel and have not done so since their separation from the reptiles. Nevertheless, they have retained the ability to form this dental tissue, only requiring suitable epigenetic conditions to elicit it. Clearly such studies tell us much about the conservation of basic developmental pathways during evolution and of the adequacy of epigenetic control to reactivate them. Although dramatic, this is not the only example of induction of a character not normally seen in a species. Balancers, organs found in some urodele amphibian larvae but not in frogs, can be induced from frog mesenchyme in response to transplanted urodele epithelium. Chimaeric urodele teeth can be similarly formed in frogs (see de Beer, 1958, and Riedl, 1978, for examples and literature).

Thus, epigenetic control is a means of stabilising the selection of a particular cell type during both development and evolution. The combination of (*a*) populations of cells capable of differentiating in more than one direction, (*b*) an epigenetic process to call forth a particular differentiation, coupled with (*c*) changing tissue relationships during development, provides a powerful means of controlling not only what will form but when and where it will form in the embryo.

Epigenetic control of morphogenesis

A further indication of the ancient origin of epigenetic control comes from the commonality of the mechanism of primary embryonic induction within the vertebrates. Although it was originally· thought that the notochord did not induce the neural tube in ascidians (Tung, 1934; de Beer, 1958, p. 150), it is now known from more recent experiments isolating, transplanting and recombining ascidian blastomeres, that notochord along with endoderm induces the neural tube to form (Reverberi *et al.*, 1960). The notochord of the cephalochordate, *Amphioxus*, induces neural tube formation (Tung *et al.*, 1962) as do all other vertebrate notochords tested. Schaeffer & Thomson (1980) use this commonality as one of the major pieces of evidence for regarding the cephalochordates 'as the sister group of the vertebrates' and use homology of epigenetic tissue interactions to propose groupings of the agnathan and gnathostome vertebrates – clearly a valuable use of epigenetic information. (But see de Beer (1958) for a cautionary discussion on homology and developmental processes: homologous structures can arise by quite different developmental mechanisms and vice versa.) That all vertebrates use the notochord as an inducer of neural ectoderm, coupled with the fact that notochord from members of one class can elicit neural differentiation from members of another (amphibian neural tube induced by fish or avian notochord, Hatt (1933), Oppenheimer (1936), rabbit neural tube induced by avian notochord, Waddington (1934)), and that cyclostome notochord can induce amphibian ectoderm to form neural tube (Bytinski-Salz, 1937), indicates how ancient in vertebrate evolution is the origin of this epigenetic control. Riedl (1978) has provided an extensive argument for why and how such mechanisms have been retained, even though the notochord has lost its original *adult* function as a supportive tissue.

Development of Meckel's cartilage and the associated membrane bone(s) has already been introduced. The data most relevant to the evolution of epigenetic control are (*a*) the non-specificity of the epithelium in the induction of bone (Hall, 1978*b*, 1981*b*), (*b*) the importance of epithelial age (Hall, 1978*b*, and unpublished data), (*c*) that timing of differentiation is set by the epithelium rather than by the responding mesenchyme (Hall, 1980), (*d*) that epithelium loses inductive ability when the induction is complete (Hall,

Table 1. *The timing of the epigenetic initiation of chondrogenesis in forming Meckel's cartilage*[a]

Group	Site of inductively active epithelium	Relationship to neural crest cell migration
Cyclostomes[b]		
Lampetra fluviatilis	Branchial ectoderm	
	Branchial endoderm	Mid migration
Urodele amphibians		
Triturus alpestris		
Ambystoma mexicanum	Pharyngeal endoderm	Mid migration
Pleurodeles waltlii	Pharyngeal endoderm	
	Dorsal mesoderm	Mid migration
Anuran amphibians		
Discoglossus pictus	Pharyngeal endoderm	Mid migration
Bombina spp.		
Birds		
Gallus domesticus	Cranial ectoderm	Early migration
Mammals		
Mus musculus	Mandibular epithelium	After migration

[a]Based on data summarised in Hall (1982a).
[b]Although not homologous to Meckel's cartilage, the branchial cartilage of the lamprey, *Lampetra fluviatilis*, is included for completeness.

1978b), and (e) that these interactions are neither species nor class specific (Hall, 1980, 1982e).

All vertebrates examined derive their mandibular skeletons from the neural crest and use an epithelium to elicit skeletal differentiation. I have summarised the timing of the epithelial–mesenchymal interactions which initiate chondrogenesis in Table 1. Detailed literature is available in Hall (1982a, b). In the lamprey, urodele and anuran amphibians, the chondrogenic interaction is with pharyngeal endoderm and therefore *during* neural crest cell migration. In birds, the interaction is with lateral cranial ectoderm *early* in migration, while in the mouse, the interaction is with mandibular epithelium and therefore *after* migration. These interactions are not just important because they allow Meckel's or branchial cartilages to form. They set in motion a succession of epigenetic tissue interactions which, at least in amphibians, have been shown to result in formation of bone, teeth, and in fact the whole mouth cavity (Cusamano *et al.*, 1962; Cassin & Capuron, 1979).

It is clear from the data in Table 1 that the timing of epigenetic events which trigger chondrogenesis and set in motion the development of the oral region has changed during vertebrate evolution. Evolutionary implications of shifting epithelial–mesenchymal boundaries have been noted before (Hall, 1975; Maderson, 1975). It is difficult to imagine the sequence of events which produced the ossicles of the mammalian middle ear from the elements of the reptile-like mammalian lower jaw occurring in a group in which epithelial control of chondrogenesis occurred early in neural crest cell migration. Evolution of these ossicles required a shift in timing so that the mandibular epithelium could allow some cells to form Meckel's cartilage while others, freed from interaction with mandibular epithelium by their new position, could shift their site of chondrogenesis to the developing middle ear. Mandibular epithelial control seen in modern mammals must then have arisen in the transition from reptile to reptile-like mammal. An examination of epigenetic control over ossicle initiation (are these cells still induced by an epithelium, which epithelium, and when in development?), along with a study of the tissue interactions operating in the reptilian lower jaw, would determine whether changing epigenetic control is a plausible mechanism for these changes. That such tissue interactions are neither species or class specific (mouse mandibular epithelium will induce chick mandibular mesenchyme and vice versa, Hall, 1980) highlights their common basis.

Can the timing of these epigenetic tissue interactions be correlated with growth of particular regions? Does the early induction of mesenchyme for chondrogenesis in the chick mean that the chick has more chondrogenic cells available to make Meckel's cartilage than the mouse where the interaction occurs later? If so, and the experiments have yet to be done, then developmental processes which have retarded or accelerated growth of a character, or allowed it to arise earlier or later in development (see Gould, 1977, and Alberch *et al.*, 1979) could have their basis in altered timing of epigenetic tissue interactions. It is clear that the *site* of the tissue is under epigenetic control – it is fixed by the 'stationary' epithelium, not by the migrating mesenchyme. It is also clear that the *timing* of these interactions is under epithelial control. Recombination of different-aged epithelia and mesenchyme shows that, in the mandibular arch of the mouse, epithelium becomes inductively active one day later (10 days of gestation) than the time when the mesenchyme is competent to respond to it (Hall, 1980, 1982*a*, *c*). Even a slight change in (*a*) the time when the epithelium becomes or loses inductive activity, (*b*) when the mesenchymal cells position themselves near the epithelium, or (*c*) in the duration and closeness of that contact, could profoundly affect morphology. This would be especially true if the induced tissue occupied, as Meckel's and sense capsule cartilages do, a key position in a functional unit, or initiated subsequent inductions. Timing, position and degree of contact all play their parts in the development of the avian mandible (Hall, 1982*c*, *e*). Perhaps the search for the regulation of gene expression in development and evolution should, in addition to looking for regulator genes, also look for epigenetic associations and how they activate the genomes of adjacent cells.

Duration of these interactions may also be important for the generation of pattern. Hinchliffe & Johnson (1980) have noted that there is only a short inductive interaction between apical ectodermal ridge and fin mesenchyme in fish where only a single row of proximal skeletal elements is formed but a much more prolonged interaction in amniotes where several rows of elements are specified. Similarly, within the limb of a single species 'short' regions containing many elements (wrist) require a similar duration of tissue interactions as 'longer' regions which contain fewer elements (forearm, metacarpals), in each case about half a day in the embryonic chick (Summerbell & Lewis, 1975). Timing relates to pattern, a key element in Wolpertian pattern formation (Wolpert, this volume). The longer inductive interaction between epithelium and mesenchyme in the chick as compared with the mouse lower jaw may be related to the induction of five membrane bones in the chick but only one in the mouse. Again, the experiments have yet to be done but this view would see timing of epigenetic control as fundamental to the many evolutionary changes involving alteration in the number of skeletal elements.

Analysis of the time over which inducers remain active also provides useful information with evolutionary implications. I will compare timing of four inductive interactions in the embryonic chick in terms of how long the inducer remains active *after* the responding tissue has been induced. The four are (i) mandibular epithelial induction of bone, (ii) spinal cord induction of vertebral cartilage, (iii) otic vesicle epithelial induction of otic capsular cartilage and (iv) pigmented retinal epithelial induction of scleral cartilage.

Detailed discussion of these systems may be found in Hall (1978*a*, 1982*a*, *b*).

The time when the induction is completed, based on the earliest age that the mesenchyme can be removed from the inducer and differentiate, is shown in Table 2. By recombining young, uninduced mesenchyme with inducers taken from embryos *after* the induction has been completed *in ovo*, we can determine whether the inducer retains any activity beyond the normal inductive period. These data are summarised in Table 2 as the last age at which the inducer is active (I have only used examples where older tissues were shown to *lack* inductive activity, e.g. mandibular epithelia from 6-day-old embryos and pigmented retinal epithelia from 8-day-old embryos are inactive, Newsome, 1976; Hall, 1978*b*). Comparison with timing *in ovo* shows whether the last inductive age coincides with, or is later than, the duration of induction *in ovo*. For mandibular membrane bones, the two ages are the same – four days of incubation. Comparable data for Meckel's cartilage are not available. For the inducers of otic capsular, somitic and scleral cartilage, inductive activity is prolonged beyond the time required to induce these cartilages *in ovo*, by 3 days for otic vesicle epithelium, 2 days for pigmented retinal epithelium, and 17 days for ventral spinal cord. Why these disparities between the four systems ? Serendipity or fundamental process?

On the one hand, we can see the utility of the synchronous timing seen in the mandibular mesenchyme. Having induced mesenchyme to form bone by 4 days of incubation, mandibular epithelium loses the ability to induce any further bone, thus avoiding the possibility of accidental induction of ectopic bones or even ectopic lower jaws later in development (Hall, 1978*b*). Evolution has synchronised inducer and responding tissue, a synchrony also seen in primary embryonic induction and in induction of the eye, presumably for the same reasons. Why do not all inductions conform to this pattern? At the other extreme we can see why ventral spinal cord should retain inductive ability for a prolonged period after completion of induction. The spinal cord is inducing serially repeated vertebrae. In some vertebrates, notably fish, amphibians and mice, vertebral number can be increased or decreased in response to environmental temperature. Retaining an ability to add vertebrae by retaining prolonged inductive ability gives tremendous plasticity to evolution of body length. Inducing 'ectopic' vertebrae

Table 2. *Duration of induction in relation to the timing of induction* in ovo[a]

Tissue induced	Inducer	Age when induction completed *in ovo* (days)	Last age at which inducer is active[b] (days)	Age when tissue first differentiates (days)	Reference
Membrane bones of lower jaw	Mandibular epithelium	4	4	7	Hall, 1978b
Otic capsule cartilage	Otic vesicle	5	8	$5\frac{1}{2}$	Benoit & Schowing, 1970
Vertebral cartilage	Ventral spinal cord	3	20	4	Tremaine & Hall, 1979
Scleral cartilage	Pigmented retinal epithelium	3	5	$7\frac{1}{2}$	Newsome, 1972, 1976

[a]Based on data from studies on the embryonic chick, discussed in detail in Hall (1982a, b).

[b]Based on recombinations of various ages of inducer with uninduced mesenchyme from younger embryos.

would not be the developmental disaster that would accompany production of a second set of jaws.

Otic vesicle and pigmented retinal epithelium retain inductive activity for some days after induction is completed. Perhaps this has evolved as a correlate to the fact that growth and morphogenesis of the eye and ear are so strongly influenced by the developing brain and adjacent regions of the chondrocranium. Retaining an 'opting-in clause' of a somewhat prolonged period of inductive activity would enable otic or scleral mesenchyme to compensate for changes in the timing or rate of development of these associated and dependent tissues. Such a mechanism could control all four parameters of the ontogenetic trajectories formulated by Alberch *et al.* (1979), viz. (i) onset age of growth, (ii) offset signal for growth, (iii) growth rate, and (iv) initial size. Alberch and colleagues have applied these trajectories to limb development; I have discussed them in the context of the cellular basis of growth (Hall, 1982*d*). The challenge for developmental biologists is to document the timing of epigenetic tissue interactions (not an easy task in a complex region like the otic capsule), to compare them in closely related species and groups, and to understand how they control development. The challenge for the evolutionary biologists is to integrate epigenetic control into evolutionary theory so that its role in generating, while at the same time limiting, diversity may be clarified.

Conclusions

Development is controlled epigenetically. Evolution acts by altering development. Therefore, epigenetic control must have played an important role in the evolutionary process. Evidence from the commonality among modern-day vertebrates of particular epigenetic processes (cell migration, primary embryonic induction, epithelial induction of skeletal tissues, etc.) can be used to argue that epigenetic control via tissue interactions arose very early in vertebrate evolution and that it has been conserved with amazing fidelity. Changing elements of epigenetic control such as position of epithelia and mesenchyme within the embryo, timing of cell migration, onset and termination of induction, length of time inducers retain inductive activity, etc., without any necessary change in the nature of the inductive signals or in the competence of the responding cells (the evolutionarily long-term stability of skeletal cell and tissue types) provides a fundamental mechanism for generating evolutionary change.

I thank NSERC of Canada for funding my research programme, the Nuffield Foundation for the award of a Nuffield Fellowship, the University of Southampton for providing a Visiting Fellowship, Professor Michael Sleigh for making space and facilities available at Southampton, Dr James Hanken and Dr Peter Thorogood for stimulating discussions on this and other topics and for their incisive comments on the first draft of this chapter.

References

Alberch, P. (1980). Ontogenesis and morphological diversification. *Am. Zool.* **20**, 653–67.

Alberch, P. & Alberch, J. (1981). Heterochronic mechanisms of morphological diversification and evolutionary change in the neotropical salamander, *Bolitoglossa occidentalis* (Amphibia: Plethodontidae). *J. Morph.* **167**, 249–64.

Alberch, P., Gould, S. J., Oster, G. F. & Wake, D. B. (1979). Size and shape in ontogeny and phylogeny. *Paleobiology* **5**, 296–317.

Bee, J. & Thorogood, P. V. (1980). The role of tissue interactions in the skeletogenic differentiation of avian neural crest cells. *Devl Biol.* **78**, 47–60.

de Beer, G. R. (1958). *Embryos and ancestors*. Oxford University Press: Oxford.

Benoit, J. A. A. & Schowing, J. (1970). Morphogenesis of the neurocranium. In *Tissue interactions during organogenesis*, ed. Et. Wolff, pp. 105–40. Gordon & Breach: New York.

Beresford, W. A. (1981). *Chondroid bone, secondary cartilage and metaplasia*. Urban & Schwarzenberg: Munich & Baltimore.

Bytinski-Salz, H. (1937). Trapianti di 'organizzatore' nelle uova di Lampreda. *Arch. ital. Anat. Embryol.* **39**, 177–228.

Cassin, C. & Capuron, A. (1979). Buccal organogenesis in *Pleurodeles waltlii* Michah (Urodele, Amphibian). Study by intrablastocoelic transplantation and *in vitro* culture. *J. Biol. Bucc.* **7**, 61–76.

Cusamano, T., Fagone, A. & Reverberi, G. (1962). On the origin of the larval mouth in the anurans. *Acta embryol. morph. exp.* **5**, 82–103.

Gould, S. J. (1977). *Ontogeny and phylogeny*. Belknap Press: Cambridge, Mass.

Hall, B. K. (1971). Histogenesis and morphogenesis of bone. *Clin. Orthopaed. rel. Res.* **74**, 249–68.

Hall, B. K. (1972). Immobilisation and cartilage transformation into bone in the embryonic chick. *Anat. Rec.* **173**, 391–404.

Hall, B. K. (1975). Evolutionary consequences of skeletal differentiation. *Am. Zool.* **15**, 329–50.

Hall, B. K. (1978a). *Developmental and cellular skeletal biology*. Academic Press: New York.

Hall, B. K. (1978*b*). Initiation of osteogenesis by mandibular mesenchyme of the embryonic chick in response to mandibular and non-mandibular epithelia. *Arch. oral Biol.* **23**, 1157–61.

Hall, B. K. (1980). Tissue interactions and the initiation of osteogenesis and chondrogenesis in the neural crest-derived mandibular skeleton of the embryonic mouse as seen in isolated murine tissues and in recombinations of murine and avian tissues. *J. Embryol. exp. Morph.* **58**, 251–64.

Hall, B. K. (1981*a*). Modulation of chondrocytic activity *in vitro* in response to ascorbic acid. *Acta anat.* **109**, 51–63.

Hall, B. K. (1981*b*). The induction of neural crest-derived cartilage and bone by embryonic epithelia: an analysis of the mode of action of an epithelial–mesenchymal interaction. *J. Embryol. exp. Morph.* **64**, 305–30.

Hall, B. K. (1982*a*). Tissue interactions and chondrogenesis. In *Cartilage*, vol. 2, *Development, differentiation and growth*, ed. B. K. Hall, pp. 187–222, Academic Press: New York.

Hall, B. K. (1982*b*). Epithelial–mesenchymal interactions in cartilage and bone development. In *Epithelial–mesenchymal interactions*, eds. R. H. Sawyer and J. F. Fallon. Praeger Press: New York (in press).

Hall, B. K. (1982*c*). Distribution of osteo- and chondrogenic neural crest cells and of osteogenically inductive epithelia in mandibular arches of embryonic chicks. *J. Embryol. exp. Morph.* **68**, 127–36.

Hall, B. K. (1982*d*). The role of tissue interactions in the growth of bone. In *Factors and mechanisms influencing bone growth*, eds. A. D. Dixon and B. G. Sarnat, Alan R. Liss Inc.: New York (in press).

Hall, B. K. (1982*e*). How is mandibular growth controlled during development and evolution? *J. craniofac. Genet. devl Biol.* **2**, 45–9.

Hall, B. K. (1982*f*). Bone in the cartilaginous fishes. *Nature* **298**, 324.

Hall, B. K. & Tremaine, R. (1979). Ability of neural crest cells from the embryonic chick to differentiate into cartilage before their migration away from the neural tube. *Anat. Rec.* **194**, 469–76.

Hamburger, V. (1980). Embryology and the modern synthesis in evolutionary theory. In *The evolutionary synthesis. Perspectives on the unification of biology*, eds. E. Mayr and W. B. Provine, pp. 97–112. Harvard University Press: Cambridge, Mass.

Hardisty, M. W. (1979). *Biology of the cyclostomes*. Chapman & Hall: London.

Hata, R. I. & Slavkin, H. C. (1978). *De novo* induction of a gene product during heterologous epithelial–mesenchymal interactions *in vitro*. *Proc. natn. Acad. Sci. USA* **75**, 2790–4.

Hatt, P. (1933). L'induction d'une plaque mèdullaire secondaire chez le *Triton* par implantation d'un morceau de ligne primitive de poulet. *C. r. Séanc. Soc. Biol.* **113**, 246–8.

Hinchliffe, J. R. & Johnson, D. R. (1980). *The development of the vertebrate limb. An approach through experiment, genetics and evolution*. Clarendon Press: Oxford.

Ho, M. W. & Saunders, P. T. (1979). Beyond neo-Darwinism – an epigenetic approach to evolution. *J. theor. Biol.* **78**, 573–92.

Horstadius, S. (1950). *The neural crest: its properties and derivatives in the light of experimental research.* Oxford University Press: Oxford.

Jaskoll, T. F. & Maderson, P. F. A. (1978). A histological study of the development of the avian middle ear and tympanum. *Anat. Rec.* **190**, 177–200.

Johnston, M. C. (1966). A radioautographic study of the migration and fate of cranial neural crest cells in the chick embryo. *Anat. Rec.* **156**, 143–56.

Kollar, E. J. & Fisher, C. (1980). Tooth induction in chick epithelium: expression of quiescent genes of enamel synthesis. *Science* **207**, 993–5.

LeLièvre, C. (1974). Rôle des cellules mésectodermiques des crêtes neurales céphaliques dans la formation des arcs branchiaux et du squelette viscéral. *J. Embryol. exp. Morph.* **31**, 453–77.

Maderson, P. F. A. (1975). Embryonic tissue interactions as the basis for morphological change in evolution. *Am. Zool.* **15**, 315–28.

Meredith-Smith, M. & Miles, A. E. W. (1971). The ultrastructure of odontogenesis in larval and adult urodeles; differentiation of the dental epithelium. *Z. Zellforsch. mikrosk. Anat.* **121**, 470–98.

Model, P. G., Jarrett, L. S. & Bonazzoli, R. (1981). Cellular contacts between hind-brain and prospective ear during inductive interaction in the axolotl embryo. *J. Embryol. exp. Morph.* **66**, 27–41.

Morriss, G. M. & Thorogood, P. V. (1978). An approach to cranial neural crest cell migration and differentiation in mammalian embryos. In *Development of Mammals*, vol. 3. ed. M. H. Jonhson, pp. 363–412. North-Holland: Amsterdam.

Moss, M. L. (1964). The phylogeny of mineralised tissues. *Int. Rev. gen. exp. Zool.* **1**, 297–331.

Moss, M. L. (1977). Skeletal tissues in sharks. *Am. Zool.* **17**, 335–42.

Newsome, D. A. (1972). Cartilage induction by retinal pigmented epithelium of chick embryos. *Devl. Biol.* **27**, 575–9.

Newsome, D. A. (1976). *In vitro* stimulation of cartilage in embryonic chick neural crest cells by products of retinal pigmented epithelium. *Devl. Biol.* **49**, 507.

Noden, D. M. (1978). The control of avian cephalic neural crest cytodifferentiation. 1. Skeletal and connective tissues. *Devl. Biol.* **67**, 296–312.

Oppenheimer, J. (1936). Structures developed in amphibians by implantation of living fish organizer. *Proc. Soc. exp. Biol. Med.* **34**, 461–3.

Ørvig, T. (1951). Histological studies of placoderms and fossil elasmobranchs. 1. The endoskeleton with remarks on the hard tissues of lower vertebrates in general. *Ark. Zool.* **2**, 321–456.

Ørvig, T. (1968). *Current problems of lower vertebrate phylogeny.* Nobel Symposium No. 4. Almqvist & Wiksell: Stockholm.

Peignoux-Deville, J., Lallier, F. & Vidal, B. (1982). Evidence for the presence of osseous tissue in dogfish vertebrae. *Cell Tiss. Res.* **222**, 605–14.

Poswillo, D. (1976). Mechanisms and pathogenesis of malformation. *Brit. med. Bull.* **32**, 59–64.

Reverberi, G., Ortolani, G. & Farinella-Ferruzza, N. (1960). The causal formation of the brain in the ascidian larva. *Acta embryol. morph. exp.* **3**, 296–336.

Riedl, R. (1978). *Order in living organisms*. John Wiley & Sons: New York.

Rosenquist, G. C. (1981). Epiblast origin and early migration of neural crest cells in the chick embryo. *Devl. Biol.* **87**, 201–11.

Saxén, L. & Karkinen-Jaaskelainen, M. (1981). Biology and pathology of embryonic induction. In *Morphogenesis and pattern formation*, eds. T. G. Connelly, L. L. Brinkley and B. M. Carson, pp. 21–48. Raven Press: New York.

Schaeffer, B. & Thomson, K. S. (1980). Reflections on agnathan–gnathostome relationships. In *Aspects of vertebrate history*, ed. L. L. Jacobs, pp. 19–33. Museum of Northern Arizona Press: Arizona.

Schmalhausen, I. I. (1949). *Factors of evolution*. Blakiston: Philadelphia.

Scott-Savage, P. & Hall, B. K. (1980). Differentiative ability of the tibial periosteum from the embryonic chick. *Acta anat.* **106**, 129–40.

Shellis, R. P. & Miles, A. E. W. (1974). Autoradiographic study of the formation of enameloid and dentine matrices in teleost fishes using tritiated amino acids. *Proc. R. Soc. Ser. B*, **185**, 51–72.

Spemann, H. (1938). *Embryonic development and induction*. Yale University Press: New Haven.

Summerbell, D. & Lewis, J. H. (1975). Time, place and positional value in the chick limb bud. *J. Embryol. exp. Morph.* **33**, 631–43.

Toivonen, S., Tarin, D., Saxén, L., Tarin, P. J. & Wartiovaara, J. (1975). Transfilter studies on neural induction in the newt. *Differentiation* **4**, 1–7.

Toivonen, S. & Wartiovaara, J. (1976). Mechanisms of cell interaction during primary embryonic induction studied in transfilter experiments. *Differentiation* **5**, 61–6.

Tremaine, R. & Hall, B. K. (1979). Retention during embryonic life of the ability of avian spinal cord to induce somitic chondrogenesis *in vitro*. *Acta anat.* **105**, 78–85.

Tung, T. C. (1934). L'organisation de l'oeuf fécondé d'*Ascidiella scabra* au début de la segmentation. *C. r. Séanc. Soc. Biol.* **115**, 1375–8.

Tung, T. C., Wu, S. C. & Tung, T. T. F. (1962). Experimental studies on the neural induction in *Amphioxus*. *Scient. sin.* **11**, 805–20.

Tyler, M. S. & Hall, B. K. (1977). Epithelial influences on skeletogenesis in the mandible of the embryonic chick. *Anat. Rec.* **188**, 229–40.

Urist, M. R. (1982). Origin of cartilage cells. In *Cartilage*, vol. 2 *Development, differentiation and growth*, ed. B. K. Hall, pp. 2–81. Academic Press: New York.

Waddington, C. H. (1934). Experiments on embryonic induction. III. A note on inductions by chick primitive streak transplanted to the rabbit embryo. *J. exp. Biol.* **11**, 224–6.

Waddington, C. H. (1940). *Organisers and genes*. Cambridge University Press: Cambridge.

Waddington, C. H. (1957). *The strategy of the genes. A discussion of some aspects of theoretical biology*. George Allen & Unwin: London.

Webster, G. & Goodwin, B. C. (1981). History and structure in biology. *Perspect. Biol. Med.* **25**, 39–62.

Wilson, E. B. (1925). *The cell in development and heredity*, 3rd edn. Macmillan: New York.

Wilson, E. O. (1981). Epigenesis and the evolution of social systems. *J. Hered.* **72**, 70–7.

Yntema, C. L. (1955). Ear and nose. In *Analysis of development*, eds. B. H. Willier, P. A. Weiss and V. Hamburger, pp. 415–28. W. B. Saunders: Philadelphia.

Yoshikawa, D. K. & Kollar, E. J. (1981). Recombination experiments on the odontogenic roles of mouse dental papilla and dental sac tissues in ocular grafts. *Arch. Oral Biol.* **26**, 303–8.

Axial organisation in developing and regenerating vertebrate limbs

M. MADEN, M. C. GRIBBIN AND
D. SUMMERBELL

Division of Developmental Biology, National Institute for Medical Research,
Mill Hill, London NW7 1AA, UK

6440

Introduction

The vertebrate limb was at the centre of the debate leading up to the publication of the *Origin of species* because it had long been recognised that there seemed to be a common plan to the structure of all vertebrate limbs. The term coined to describe such similarities was 'homology'. Much was made of the homology of vertebrate limbs by Darwin and the supposed explanation of the phenomenon was that the limbs of each species had descended from a common ancestor whose archetypal limb pattern had subsequently undergone modification. Despite the fact that these conclusions were derived from studies of the anatomy of adult limbs, as many others have been since then, the common ancestor theory had implications for embryology. First, it became accepted that all vertebrate limbs passed through this common ancestor stage in their development, without any embryological evidence to support it. Only recently have detailed studies been performed to discredit this view (Hinchliffe, 1977) and it is now clear that the prechondrogenic condensation pattern of the limb bud mirrors, in miniature, the pattern of the adult form. The limb of each species therefore develops uniquely in terms of the prechondrogenic condensation pattern, a conclusion which is reinforced below.

Secondly, in evolutionary terms, development is viewed as a highly ordered sequence of hierarchical processes. This means that the earliest events in the development of the limb should be common to all vertebrates, with adaptation only working on the latest stages to provide the variety of adult forms we see. In this hierarchical scheme, one of the earliest processes to occur in limb development must surely be the establishment of the limb axes, without which organised growth could not take place. All vertebrate limbs have an antero-posterior, a dorso-ventral and a proximo-distal axis, and, according to the implications of the above evolutionary thinking, each axis should be organised by the same mechanisms throughout the vertebrates.

Unfortunately, we do not yet know the precise details of how the axes of developing limbs are organised. The only techniques presently available to us to gain this knowledge are the observation of normal development and experimental interference in various ways. By performing these experiments on several species of embryo we can test the above hypothesis. Clearly, the implications are that, if all vertebrate limbs organise their axes by the same mechanisms, the effects of experimental interference should be the same. There have up to now been almost no studies of what might be called experimental homology, since most experimenters prefer to concentrate their efforts on the embryos of only one species. The beginnings of this study were reported recently (Maden, 1981) and the work described here continues this effort. Four species of vertebrate limb are compared – the developing limbs of *Ambystoma mexicanum* (urodele amphibian), *Rana temporaria, Xenopus laevis* (anuran amphibian) and *Gallus domesticus* (Aves) as well as data on the regenerating limbs of *Ambystoma mexicanum*. We can therefore look for similarities not only between the developing limbs of amphibians and birds, but also between developing and regenerating limbs.

Normal development

We begin by examining the normal developmental sequence in the species under study here. Limb buds at various stages of development can be stained with Alcian green to observe the pattern of emergence of prechondrogenic elements. We find that this pattern reflects faithfully the cartilages in the adult form. There are no similarities between the stages of developing limbs of axolotl, *Rana, Xenopus* or chick, indeed the elements do not even develop in the same antero-posterior sequence. For example, in axolotl limb buds the digits appear in approximately an anterior to posterior sequence – in the hindlimb the sequence is digits 2, 1, 3, 4, 5 but in *Rana* and *Xenopus* the sequence is from posterior to anterior –

digits 4, 3, 5, 2, 1. In the chick, where the forelimb bud is routinely used, the sequence is 4, 3, 2. Furthermore, the chick limb bud condensation pattern does not pass through a primitive amphibian stage as the common ancestor theory would have us believe (Hinchliffe, 1977).

A more detailed examination of the size of elements when they first appear in the developing limb has also revealed differences between species. In chicks, the size of the elements in each of the four segments (humerus, radius and ulna, wrist, digits) when they appear in the proximo-distal sequence is approximately 300 μm (Summerbell, 1976). We might expect such a fundamental mechanism to be common to all vertebrates, but again this seems not to be so – each species has its own characteristic element size. In axolotls, the figure is approximately 250 μm, in *Xenopus* 150 μm and in *Rana* 130 μm.

Experiments on the proximo-distal axis

These have involved two types of operation. In the first, the tip of the limb bud or regeneration blastema is simply cut off, leaving about 50% of the original amount of tissue. Can the remaining tissue regulate to form a complete limb? The result of this operation on either developing or regenerating limbs of axolotls is the production of perfect limbs in 100% of the cases (Maden & Goodwin, 1980). In *Rana* limb buds, a few limbs with defects begin to appear – 95% of limbs regulate (Maden, 1981). In *Xenopus*, only 38% of limbs are perfect and, in chicks, all limbs are truncated at the level of the cut. Thus a clear decrease in the regulative ability is apparent, as it is in the second type of experiment.

Here, the distal tip of the limb bud or regeneration blastema is cut off and a second cut made at a more proximal level. This slice of tissue is removed and the distal tip replaced on the proximal stump. We can then ask whether the intercalary defect is regulated. Again, in axolotl limb buds and regeneration blastemas, perfect regulation takes place and 100% of the limbs are normal (Stocum, 1975; Maden & Goodwin, 1980; Maden, 1980*a*). In *Rana* limb buds, this operation only results in 42% of perfect limbs (Fig. 1), and in *Xenopus* (Fig. 2) and chicks all limbs have deficiencies of one kind or another, although the results are highly stage dependent (Sum-

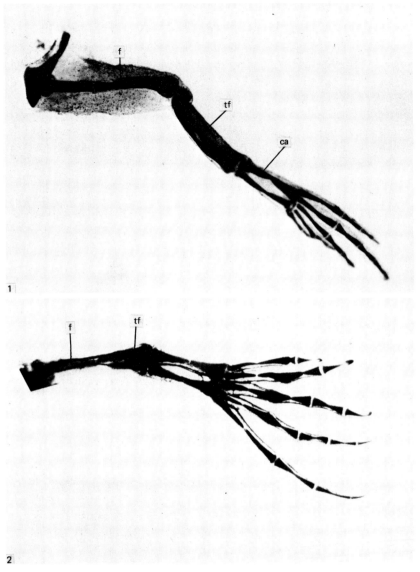

Figs. 1 and 2. Typical limbs resulting from the grafting of a distal hindlimb bud
tip into the proximal stump. Victoria blue staining. Fig. 1. A normal *Rana*
hindlimb in which the missing tissue has been intercalated. Compare with Fig. 2 in
which the deleted parts have not been replaced. f = femur, tf = tibia and fibula,
ca = calcaneum and astragalus. Mag. × 25. Fig. 2. A *Xenopus* hindlimb which
did not replace the missing tissue. Here, the proximal end of the femur (f) joins
directly into the distal end of the tibia and fibula (tf). Mag. × 18.

Table 1. *Frequency of supernumerary limb induction after three types of operation on the developing limb buds of* Rana, Xenopus *and chick and the regenerating limb blastemas of axolotl. (Regeneration data from Tank, 1978, 1981; Maden, 1980b; Turner, 1981; Maden & Mustafa, 1982)*

Animal	Operation	Number of cases	Number of limbs with supers (%)	Number of supernumeraries		
				Single (%)	Double (%)	Triple (%)
Rana	AP inversion	26	54	21	79	0
Xenopus	AP inversion	37	30	90	10	0
Chick	AP inversion	32	31	90	10	0
Axolotl	AP inversion	143	57	81	19	0
Rana	DV inversion	28	61	12	88	0
Xenopus	DV inversion	34	26	89	11	0
Chick	DV inversion	18	0	0	0	0
Axolotl	DV inversion	37	81	81	19	0
Rana	AP/DV inversion	58	67	46	41	13
Xenopus	AP/DV inversion	43	47	65	30	5
Chick	AP/DV inversion	31	68	76	24	0
Axolotl	AP/DV inversion	83	94	57	40	3

AP = antero-posterior; DV = dorso-ventral.

merbell, 1977). Thus again a decreasing regulative ability is apparent throughout these vertebrate species.

Experiments on the transverse axes

Investigations of the organisation of the transverse axes (antero-posterior and dorso-ventral axes) have involved the grafting of limb buds or regeneration blastemas to invert either one or both axes. By grafting a left limb bud to a right stump (or vice versa) each axis can be inverted independently, and by grafting a limb bud back onto its own stump after 180° rotation both transverse axes are inverted at once.

The antero-posterior axis

Inversion of the antero-posterior (AP) axis results in the production of supernumerary limbs, their frequency varying according to the species concerned (Table 1). They arise at the points of maximum

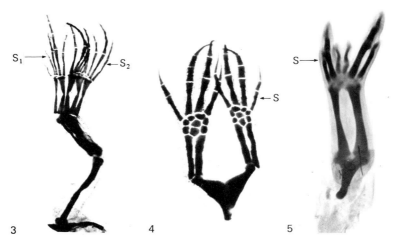

Figs. 3–5. Supernumerary limbs produced by contralateral grafting of limb buds or regeneration blastemas to invert the antero-posterior axis. Victoria blue or Alcian green staining. Fig. 3. *Rana* hindlimb with two supernumeraries S_1 and S_2 at the anterior and posterior poles of the limb. In between is the grafted contralateral limb. The digital sequence from left to right is 1 2 3 4 5 (S_1), 5 4 3 2 1 (graft), 1 2 3 4 5 (S_2). Note the supernumeraries are in mirror-image relation to the grafted limb. Mag. × 18. Fig. 4. A single supernumerary from an axolotl regeneration blastema graft. On the left is the grafted limb with a digital sequence of 4 3 2 1 and on the right is the supernumerary (S) with a digital sequence of 1 2 3 4 in mirror-image relationship. Mag. × 10. Fig. 5. A single supernumerary (S) produced by grafting a chick limb bud. The similar mirror-image digital sequence of 4 3 2 2 3 4 is apparent here too. Mag. × 14.

incongruity, i.e. at the anterior and/or posterior poles, and can be either single or double (Figs. 3–5). The relatively low incidence of supernumerary limbs produced by AP inversion in chick limbs should be contrasted with the effect of grafting the organising region from the posterior margin of the limb bud to the anterior margin. Using equivalent stage embryos, supernumeraries can be typically produced in more than 90% of the cases.

Analysis of cleared whole mounts such as those in Figs. 3–5 reveals entirely normal sequences of digits, carpals etc. That is, the AP axes of the supernumeraries follow a well-defined pattern, even if a full complement of digits is not present. In Fig. 3, for instance, the sequence of digits is 1, 2, 3, 4, 5 (supernumerary) 5, 4, 3, 2, 1 (graft) 1, 2, 3, 4, 5 (supernumerary), and in Fig. 5 the sequence is

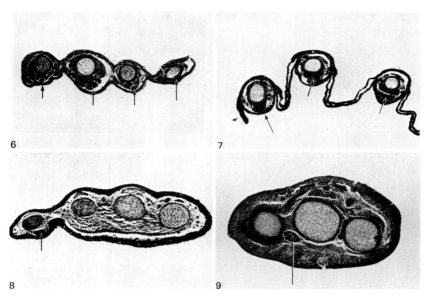

Figs. 6–9. Sections through the digits of normal limbs showing the dorso-ventral asymmetry of muscle patterns. Haematoxylin and eosin staining. Fig. 6. *Rana* limb showing the absence of muscle on the dorsal (upper) surface of the digits and the presence of ventral muscle and prominent ventral tendons (arrows) on the lower surface. Mag. × 130. Fig. 7. *Xenopus* limb with the same characteristics as *Rana*. Ventral tendons arrowed. Mag. × 82. Fig. 8. Axolotl regenerated limb. The digit on the left shows the same characteristics (arrow) as the previous sections. The rest of the hand has more muscle on the ventral surface. Mag. × 56. Fig. 9. Chick limb. Note the prominent adductus secundus muscle (arrow) between digits 2 and 3 which provide a good ventral marker. Mag. × 106.

4, 3, 2 (supernumerary) 2, 3, 4 (graft). To determine the organisation of the dorso-ventral (DV) axis, AP supernumeraries have been serially sectioned to study their muscle patterns and together these two methods of analysis allow one to determine their handedness. The normal muscle structure for each species is depicted in Figs. 6–9. It is characterised in the amphibians by the early disappearance of dorsal muscles and the persistence of ventral muscles to the tips of the digits. The muscles of the chick, too, have a characteristic dorsoventrality. Analysis of a total of 38 AP supernumeraries revealed that in every case their muscle patterns were entirely normal (Table 2). From this we can conclude that, after inversion of the AP axis in limb buds or regeneration blastemas, the

Table 2. *Cartilage structure, muscle structure and handedness of those AP and DV supernumeraries which have been sectioned in the four species*

Note that chick limbs do not form supernumeraries after DV inversion. *Rana, Xenopus* and chick results are from operations on developing limb buds, the axolotl results are from operations on regenerating blastemas.

Animal	Type of supernumerary	Number analysed	Structure of cartilage	Structure of muscle	Handedness of supernumeraries
Rana	AP	11	Normal	Normal	Same as stump
Xenopus	AP	6	Normal	Normal	Same as stump
Chick	AP	10	Normal	Normal	Same as stump
Axolotl	AP	11	Normal	Normal	Same as stump
Rana	DV	11	Normal	Normal	Same as stump
Xenopus	DV	6	Normal	Normal	Same as stump
Axolotl	DV	10	Normal	Normal	Same as stump

AP = antero-posterior; DV = dorso-ventral.

supernumerary limbs which are produced are structurally normal and show *stump handedness*. Thus in this series there is considerable uniformity within the species studied.

The dorso-ventral axis

Inversion of the DV axis results in the production of supernumeraries in the amphibian species, but not in the chick (Table 1). In the former, one to two supernumerary limbs are produced at varying frequencies, usually at the points of maximum incongruity which, this time, are the dorsal and ventral poles of the host limb (Figs. 10 & 11). Analysis of the muscle patterns in *Rana, Xenopus* and axolotl regenerating limbs revealed that, as in the previous case, all supernumeraries were normal in their DV organisation as well as in their AP sequence of digits and were of *stump handedness* (Table 2).

Inversion of both axes

After 180° rotation of limb buds or regeneration blastemas, a high frequency of supernumerary limbs is produced in both amphibians and birds (Table 1). It is again of interest to note that the incidence

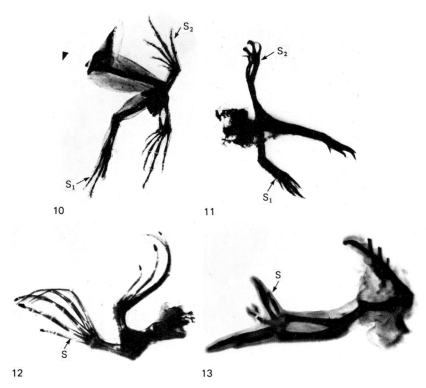

Fig. 10. Supernumerary limbs (S_1 and S_2) produced in a *Rana* limb after DV inversion of the grafted (middle) limb bud. The supernumeraries arose at the dorsal and ventral poles of the limb. Victoria blue staining. Mag. × 18. Fig. 11. DV inversion of an axolotl regeneration blastema showing the same type of supernumeraries (S_1 and S_2) as the previous figure. Victoria blue staining. Mag. × 10. Fig. 12. 180° rotation (AP/DV inversion) of a *Xenopus* limb bud which resulted in a single five-digit supernumerary (S). Victoria blue staining. Mag. × 12. Fig. 13. AP/DV inversion of a chick limb bud which produced a single supernumerary. Alcian green staining. Mag. × 14.

in chick limb buds is lower after this type of operation than after organising region grafts.

Several differences begin to emerge between the effects of uniaxial and biaxial inversion. Instead of just single or double supernumeraries, triples appear in a small, but consistent number of cases in *Rana, Xenopus* and axolotl. The position of origin of supernumeraries in *Rana, Xenopus* and axolotl is unpredictable

Table 3. *Dorso-ventral organisation of supernumeraries generated by inversion of both transverse axes, deduced from studying the muscle patterns in serial sections*

Examples of each type are – Figs. 6–9 normal; Fig. 17 double ventral (2V); Fig. 18 double dorsal (2D); Fig. 19 part normal/part mirror-imaged; Fig. 20 part normal/part inverted. *Rana, Xenopus* and chick results are from operations on developing limb buds, the axolotl results are from operations on regenerating blastemas.

Animal	Number of super-numeraries analysed	Numbers of each type of muscle pattern in AP/DV supernumeraries			
		Normal	2D or 2V	Part normal/ part mirror-imaged	Part normal/ part inverted
Rana	18	5	3	4	6
Xenopus	12	9	1	1	1
Chick	17	17	0	0	0
Axolotl	100	22	22	32	24

rather than being at defined poles, although in the chick they only appear at the posterior and occasionally at the anterior pole as well. Analysis of their AP organisation by cartilage staining reveals surprising differences between the species. In *Xenopus* and chick developing limbs, all supernumeraries had a normal sequence of digits (Figs. 12 & 13), i.e. the AP axis was as expected. However, in *Rana* developing limbs, a new category of supernumeraries appeared for 28% (11 out of 39) were double posterior in structure, the remainder being normal (Fig. 14). These double posterior supernumeraries were composed of either three (Fig. 15), four (Fig. 16) or five digits.

Analysis of their DV organisation by muscle patterns revealed further suprises. In axolotl regenerates (Maden, 1982a; Maden & Mustafa, 1982) and *Rana* limb buds, there were approximately equal numbers of four types of muscle patterns (Table 3). First, there were normal supernumeraries as had been found before (Figs. 6–9). Secondly, there were perfectly mirror-imaged limbs, either double ventral (Fig. 17) or double dorsal (Fig. 18). Thirdly, there were limbs of mixed symmetry and asymmetry, that is partly double dorsal or double ventral and partly normal (Fig. 19). Fourthly, there were limbs which were partly normal and partly inverted (Fig. 20). The double posterior *Rana* supernumer-

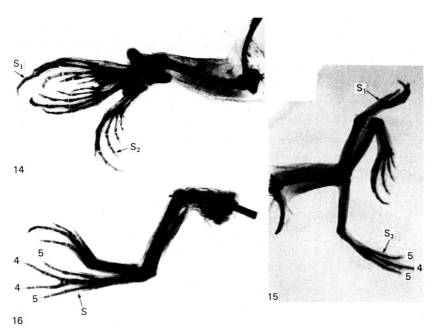

Figs. 14–16. Typical results of AP/DV inversion of *Rana* hindlimb buds. Fig. 14. Here two normal 5 digit supernumeraries (S_1 and S_2) have been produced. The grafted limb bud is in the middle. Victoria blue staining. Mag. × 20. Fig. 15. Here a five-digit normal supernumerary (S_1) and a three-digit double posterior supernumerary (S_2) have been produced. The digital formula of the double posterior limb is 5 4 5. Mag. × 18, Fig. 16. Here a single four-digit double posterior supernumerary (S) with a digital formula of 5 4 4 5 has been produced. Above the supernumerary is the grafted limb. Mag. × 21.

aries (Figs. 15 & 16) were all found to be of normal muscle structure. Thus only one transverse axis can be mirror-imaged at a time.

In *Xenopus* the majority of supernumeraries (9 out of 12) were normal in their muscle patterns. Only one of each of the other three classes of structure was found. Supernumeraries produced in chick limbs often had a reduced complement of muscles, although those that could be identified were always arranged normally. Some effect on the DV axis may be involved here, since AP supernumeraries had well-formed muscles.

Thus, after AP/DV inversion, there seems to be a gradation in the structure of supernumerary limbs generated from 100% normal

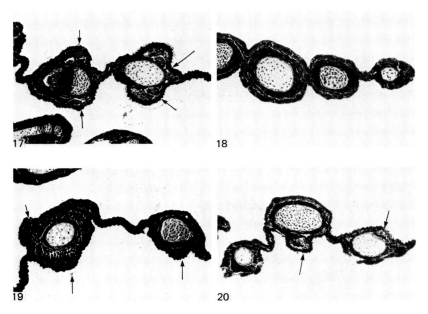

Figs. 17–20. Sections through the digits of *Rana* supernumeraries resulting from 180° rotation of limb buds to show abnormal muscle patterns. All these figures should be compared with Fig. 6 which shows the normal arrangement of muscles and tendons in *Rana* digits. Haematoxylin and eosin staining. Fig. 17. A double ventral (2V) supernumerary with ventral tendons (arrows) and muscles on both surfaces of the digits rather than just one. Mag. × 250. Fig. 18. A double dorsal (2D) supernumerary with no ventral tendons present at all. Mag. × 250. Fig. 19. Here the digit on the left is double ventral with ventral tendons (arrows) on both surfaces of the limb. The digit on the right is normal. Thus this supernumerary is part double ventral/part normal. Mag. × 250. Fig. 20. Here the two digits on the left are in normal orientation with their dorsal surfaces above and ventral (arrow) below. These are adjacent to a digit with its ventral surface uppermost (arrow) and dorsal surface below. This limb is part normal/part inverted. Mag. × 250.

(chicks) to various frequencies of four (*Xenopus* and axolotl) and ultimately five different classes of structure in *Rana*, including double posteriors.

The effect of retinoids

One further method of analysing axial organisation which has recently come to light is the effect of vitamin A and its derivatives, the retinoids, on developing and regenerating limbs.

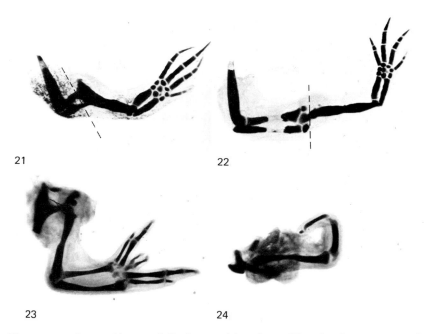

21

22

23

24

Figs. 21 and 22. Abnormal limbs resulting from Vitamin A treatment of regenerating axolotl limbs. Victoria blue staining. Fig. 21. This limb was amputated through the mid-radius and ulna (dashed line) and instead of regenerating just the distal radius and ulna and hand, a complete new limb regrew. Mag. × 15. Fig. 22. After amputating through the carpals (dotted line) again, a complete new limb has regenerated. Note the normal antero-posterior organisation of the digits in these figures. Mag. × 12.

Figs. 23 and 24. Abnormal limbs resulting from Vitamin A treatment of chick limb buds. Alcian green staining. Fig. 23. After implanting a retinoic acid soaked paper graft into the anterior margin of the limb bud a duplication of digital formula 4 3 2 2 3 4 can be produced. Note the similarity between this result and AP or AP/DV inversions of limb buds (Figs. 5 & 13). Note also that the AP axis is affected here rather than the proximo-distal axis as in Figs. 21 and 22. Mag. × 16. Fig. 24. Some limbs also have defective growth after retinoic acid treatment. Mag. × 16.

If the amputated limbs of axolotls are immersed in a solution of retinol palmitate, then, instead of replacing those elements removed by amputation, extra tissue in the proximo-distal sequence is regenerated (Maden, 1982*b*). For example, after amputating through the mid-radius and ulna, an extra radius and ulna can be produced in tandem as well as the hand. If the concentration of vitamin A is increased, a whole limb can be regenerated from the cut stump (Fig. 21). The same phenomenon occurs more dramatically after amputation through the hand (Fig. 22). It is important to note that, in both these figures, the AP sequence of digits is normal, as is the DV organisation of muscles when such limbs are sectioned. The effects of various retinoids have been tested and retinoic acid was found to be the most potent (Maden, 1982*b*).

In the developing chick limb bud, the effect of Vitamin A analogues seems to be very different. We have attempted to produce homologous effects on the organisation of the proximo-distal axis to those that we have shown in the Amphibia. However, ten different experiments extending over a dose range of six orders of magnitude and using over 1400 chick embryos have so far failed to produce any cases showing unequivocal additional skeletal elements along the proximo-distal axis. Instead, Vitamin A causes modifications of the AP axis including the formation of mirror-image reduplicate supernumerary limbs (Fig. 23) or deficiencies (Fig. 24) in which parts of the limb are deleted. In particular, local application of retinoic acid to the anterior margin of the limb bud gives a similar result to AP or AP/DV reversals. It therefore seems that in some ways it may mimic the effect of the zone of polarising activity (ZPA). Naïvely, we could consider retinoic acid to be an analogue of the signalling molecule from the ZPA, thereby causing the formation of supernumerary structures. The deficiency results can be explained by assuming that the extra vitamin A systemically raised the concentration of this molecule to a background level incompatible with organised pattern. Alternatively, it could have side effects such as causing cell death, the inhibition of proliferation or the inhibition of cell differentiation.

Discussion

It is clear from the work described above that the four species of vertebrate limbs used here (three amphibians and one bird) behave differently in some aspects of development and similarly in others. The similarities, far fewer in number, seem to be only two. First, all limbs develop in a proximal to distal sequence of elements (stylopodium, zeugopodium, autopodium) rather than, for example, distal to proximal or various combinations of the two. Secondly, after inversion of the AP axis limbs all seem to develop supernumeraries of normal cartilage and muscle patterns with the same handedness as the stump.

On the other hand, the differences are many. During normal development the digits appear in an anterior to posterior sequence in the axolotl limb bud and in a posterior to anterior sequence in anurans and birds. The size of the elements when they first appear is characteristic of the species, as is the number and disposition of the elements. Experiments revealed that there is an increasing mosaicism in the proximo-distal axis from the axolotl to the chick which results in the loss of the ability to replace missing tissue. After inversion of the DV axis, amphibians produce supernumerary limbs of stump handedness, but chicks do not. After inversion of both transverse axes there are five types of supernumerary limb structure that appear in *Rana* limb buds, four in *Xenopus* and axolotl, and only one (normal) in chicks. Thus the normal pattern seems to be most stable in the higher vertebrate class. Other differences between these limbs have also been observed such as the presence of an apical ectodermal ridge (AER), which plays an essential role in distal outgrowth, and its lack of regeneration in chicks (Saunders, 1948). In anurans an AER is present and it is presumably capable of regeneration (Maden, 1981), but in urodeles there is *no* AER at all (Sturdee & Connock, 1975; Maden & Goodwin, 1980).

From this array of developmental behaviour in vertebrate limbs, we can draw one important conclusion. There seems to be more than one way of developing a limb – it can involve highly regulative mechanisms such as those utilised by axolotl limb buds and regeneration blastemas, or it can involve mosaic mechanisms such as those in chick limb buds. The degree of mix of regulative and mosaic mechanisms seems to determine the developmental behaviour of the limbs of any one species. This conclusion is clearly contrary to the evolutionary view of the development of limbs, where such fundamental mechanisms as the organisation of the axes should be common to all vertebrates. Neither does there seem to be a common ancestral stage in development through which all limbs pass. Therefore not only have evolutionary concepts been of no value in advancing our understanding of developmental mechanisms, they have been positively misleading. There is surely no reason *a priori* why developmental mechanisms should not evolve and we concur with Sander (this volume) that homology does not imply conservation of mechanisms.

Finally, it is interesting to consider the relation between development and regeneration. Only urodeles are capable of limb regeneration in adults, and it is surely no coincidence that urodeles have the most highly regulative form of limb development. *Rana* can regenerate for a short period after limb development and have a less regulative form of limb development. *Xenopus* loses regenerative ability before limb development has terminated and has a relatively mosaic type of development. Chicks, which can never regenerate, are the most mosaic of all. It is possible therefore, that the reason why the limbs of higher vertebrates have not evolved regenerative ability is because their mosaic type of development precludes it.

References

Hinchliffe, J. R. (1977). The chondrogenic pattern in chick limb morphogenesis: a problem of development and evolution. In *Vertebrate limb and somite morphogenesis*, eds. D. A. Ede, J. R. Hinchliffe and M. Balls, pp. 293–309. Cambridge University Press: Cambridge.

Maden, M. (1980*a*). Intercalary regeneration in the amphibian limb and the rule of distal transformation. *J. Embryol. exp. Morph.* **56**, 201–9.

Maden, M. (1980*b*). Structure of supernumerary limbs. *Nature* **287**, 803–5.

Maden, M. (1981). Experiments on Anuran limb buds and their significance for principles of vertebrate limb development. *J. Embryol. exp. Morph.* **63**, 243–65.

Maden, M. (1982*a*). Supernumerary limbs in amphibians. *Am. Zool.* **22**, 131–42.

Maden, M. (1982*b*). Vitamin A and pattern formation in the regenerating limb. *Nature* **295**, 672–5.

Maden, M. & Goodwin, B. C. (1980). Experiments on developing limb buds of the axolotl *Ambystoma mexicanum*. *J. Embryol. exp. Morph.* **57**, 177–87.

Maden, M. & Mustafa, K. (1982). The structure of 180° supernumerary limbs and a hypothesis of their formation. *Devl Biol.* **93**, 257–65.

Saunders, J. W. (1948). The proximo-distal sequence of origin of the parts of the chick wing and the role of the ectoderm. *J. exp. Zool.* **108**, 363–403.

Stocum, D. L. (1975). Regulation after proximal or distal transposition of limb regeneration blastemas and determination of the proximal boundary of the regenerate. *Devl Biol.* **45**, 112–36.

Sturdee, A. & Connock, M. (1975). The embryonic limb bud of the Urodele: morphological studies of the apex. *Differentiation* **3**, 43–9.

Summerbell, D. (1976). A descriptive study of the rate of elongation and differentiation of the skeleton of the developing chick wing. *J. Embryol. exp. Morph.* **35**, 241–60.

Summerbell, D. (1977). Regulation of deficiencies along the proximal distal axis of the chick wing-bud: a quantitative analysis. *J. Embryol. exp. Morph.* **41**, 137–59.

Tank, P. W. (1978). The occurrence of supernumerary limbs following blastemal transplantation in the regenerating forelimb of the axolotl, *Ambystoma mexicanum*. *Devl Biol.* **62**, 143–61.

Tank, P. W. (1981). Pattern formation following 180° rotation of regeneration blastemas in the axolotl, *Ambystoma mexicanum. J. exp. Zool.* **217**, 377–87.

Turner, R. N. (1981). Probability aspects of supernumerary production in the regenerating limbs of the axolotl, *Ambystoma mexicanum. J. Embryol. exp. Morph.* **65**, 119–26.

The vertebrate limb: patterns and constraints in development and evolution

NIGEL HOLDER

Department of Anatomy, King's College, Strand, London WC2R 2LS, UK

#6441

Introduction

The tetrapod vertebrates are represented by four classes; the amphibians, reptiles, birds and mammals; and the first of these groups to evolve, the amphibians, did so about 200 million years ago. It is a striking fact, as others in this volume have pointed out, that the overall structure of the limbs of the tetrapods has remained remarkably consistent throughout this time. The limb of a human differs only in minor detail from that of a salamander (Fig. 1). At the same time, however, many diverse specialisations of specific parts of the limb have appeared. These specialisations often involve specific tissues of the limb, such as the feathers and scales of the skin, or occur in localised limb regions, such as the extraordinary growth of the central digits of the forelimb of the bat (see Fig. 11). In addition to these more striking developments, many subtle anatomical variations have emerged which are only apparent upon close scrutiny.

The intriguing challenge posed by the vertebrate limb is to establish how such overall conservatism of basic form through evolution can be reconciled with the many subtle structural changes which characterise the limbs of individual species. An obvious starting point for the answer to this challenge is to examine how the vertebrate limb is built during development and to assess the similarities in developmental mechanism between developing limbs of animals in different classes. If a common framework can be identified, the developmental constraints upon anatomical modification maybe much easier to describe.

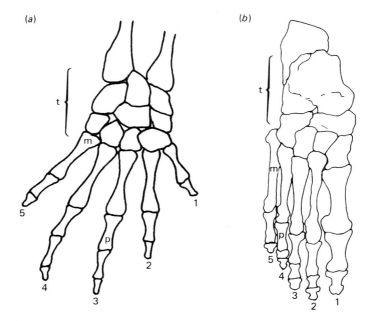

Fig. 1. Drawings depicting the bones of the foot of (*a*) the salamander and (*b*) the human. About two hundred million years separate the emergence of these animals yet the basic pattern of the skeletons is very similar. Numbers 1 to 5 denote traditional identification of digits from medial (1) to lateral (5). t, tarsals; m, metatarsals; p, phalanges.

Developmental constraints

In this essay, the term developmental constraint applies to a specific developmental mechanism and the anatomical alterations expected if the mechanism concerned is altered in a certain way (see Kauffman, this volume). The aim is to examine a series of developmental mechanisms and to establish the anatomical constraints which each has in developmental terms. This kind of argument yields a series of possible structural possibilities. Once the developmental possibilities for each particular mechanism have been established, the record of evolutionary change of tetrapod limbs will be discussed and the likely links between developmental constraint and evolutionary possibility will emerge. The developmental mechanisms selected for analysis are discussed within the framework of positional information (Wolpert, 1969, 1971, and see this volume).

Common anatomical features of limb development

All vertebrate limbs first appear during ontogeny as small, tongue-shaped outpushings of mesoderm which are covered in a thin layer of ectoderm. Although the exact position of origin of the limb on the embryo and the relative timing of limb bud outgrowth may vary, the overall appearance and initial growth of the limb of different vertebrates is remarkably similar. All of the limbs of the amniote vertebrates (reptiles to mammals) and the anuran amphibians, have a specialisation of the distal ectoderm called the apical ectodermal ridge or AER (Saunders, 1948; O'Rahilley *et al.*, 1956; Milaire, 1957, 1965; Goel & Mathur, 1977; Tschumi, 1957; Jurand, 1965; Yntema, 1968; Stébler, 1973; Tarin & Sturdee, 1973; Fallon & Kelley, 1977). This AER is not externally evident in either the developing or regenerating limbs of urodelean amphibians, but a distinct apical thickening of the ectoderm called the apical cap, is present (Thornton, 1965; Tank *et al.*, 1977).

The limbs of all tetrapods, including the urodeles, show a clear proximal to distal sequence in the differentiation of the limb parts (see Saunders, 1948; Tschumi, 1957; Milaire, 1965; Summerbell, 1976). The control of this sequential appearance of parts and its implications for pattern formation mechanisms will be discussed below.

The embryological origin of the limb bud mesoderm also seems to be comparable in different animal groups. In the most extensively studied system, the chick limb bud, it has been experimentally demonstrated that the muscle cell lineage derives from the somitic or para-axial mesoderm, whereas the connective tissues of the limb, the cartilage, tendons and fibrocyte tissues which invest the muscles, all derive from the lateral plate region of the embryo (Christ *et al.*, 1974, 1977; Chevalier *et al.*, 1976, 1977). This dual origin of the limb bud mesoderm has also been suggested from histological analyses in the reptiles (see review by Raynaud, 1977).

The common anatomical features of the development of at least all of the amniote vertebrate limbs and those of the anuran amphibians, strongly implies that the developmental mechanisms underlying their formation also share many common features. In the following section this possibility will be discussed in some detail.

Common developmental mechanisms underlying limb formation

Permissive versus instructive interactions

The developmental mechanisms underlying limb development can conveniently be divided into two types: those concerned with the control of limb outgrowth and those concerned with the spatial arrangement of the cell types which characterise the limb mesoderm. This distinction has important implications for the search for developmental constraints which affect the evolution of limb form and structure.

The control of limb outgrowth is effected by a series of epithelial–mesenchymal interactions involving the limb mesoderm and the apical ectodermal ridge (see reviews by Stocum, 1975; Saunders, 1977, and Hinchliffe & Johnson, 1981). These interactions are permissive in that they are not responsible for establishing the spatial patterning of the anatomy of the limb mesoderm cells directly, but are responsible for creating a population of cells in which the anatomy of the limb is manifested by a separate, albeit intimately related, series of mechanisms.

This point is well illustrated by examining the role of the AER in limb development. The initial formation of the AER results from the inductive property of the limb mesoderm prior to limb outgrowth, and it is maintained only by the continuing influence of the limb mesoderm (Zwilling & Hansborough, 1956; see review by Saunders, 1977). In turn, the AER is itself indispensible to the developing limb, because, if it is removed surgically in the chick limb, outgrowth ceases (Saunders, 1948; Summerbell, 1974). However, the AER of a young staged limb bud is equivalent to that of an older staged limb bud, even though the pattern, and therefore the character, of the underlying limb mesoderm, is changing with time (Summerbell *et al.*, 1973). This point was clearly demonstrated by an experiment in which the AER and the mesoderm of different staged limb buds were exchanged (Rubin & Saunders, 1972), and the limb structures which resulted were exclusively of the form determined by the age of the mesoderm. In this sense, therefore, the interaction of the AER with the mesoderm is permissive.

A different mechanism must exist in the limb mesoderm which governs the formation of the spatial pattern of cell types. This mechanism, or series of mechanisms, are instructive in the sense that they arrange the mesodermal cells into a defined anatomy. The nature of these instructive interactions is as yet unclear, but some well-developed theoretical schemes have been put forward.

The majority of schemes have been built within the general framework of pattern formation which was clarified by Wolpert (1969, 1971), whereby a cell is made aware of its position within a bounded field of cells, in this case the limb field, and the cell then interprets this positional information with reference to both its genome and its developmental history (Lewis & Wolpert, 1976). In this framework, a mechanism must exist whereby the first part of the process, the specification of cell position, can be achieved.

The most extensively developed theory for the specification of position in the limb field is the view that a gradient of a diffusible morphogen is produced across the anterior to posterior limb axis by a specialised, localised region of cells on the posterior limb margin, which is called the zone of polarising activity or ZPA (Tickle *et al.*, 1975; Summerbell & Tickle, 1977; Summerbell, 1979, 1981; Smith, 1980; Wolpert & Hornbruch, 1981; Tickle, 1981). This idea is based on the original demonstration that a small piece of posterior chick wing bud tissue will induce a complete new set of chick wing structures if it is grafted to the anterior wing margin during limb outgrowth (Saunders & Gasseling, 1968; see also review by Saunders, 1972). In the ZPA model, the mesoderm cells infer their position along the posterior to anterior limb axis with reference to the exact amount of morphogen which they see. The more anterior the cell is the lower this value will be.

This global signalling model involving a localised source region producing a diffusible morphogen is just one way of specifying position. An alternative mechanism has been put forward to explain patterning in epimorphically regenerating systems such as the cockroach and amphibian limb and the *Drosophila* imaginal disc (French *et al.*, 1976; Bryant *et al.*, 1981), and this scheme has recently been adapted to attempt to explain pattern regulation in the developing limb field (Iten & Murphy, 1980; Javois & Iten, 1981; Javois *et al.*, 1981; Iten, 1982). This model is based not on long range signalling but on short-range cell–cell interaction and intercalation of new positional values by an averaging process involving cell division. In this case, following the grafting of a piece of tissue from the posterior limb bud, the extra structures which result are assumed to arise as the result of cell proliferation caused by the positional mismatch of the graft and host tissue.

As mentioned earlier, the purpose of this paper is not to discuss the developmental merits of either of these models but rather to examine the nature of the constraints that they place upon the production of novel limb structures throughout the course of evolution. Prior to attempting this kind of analysis with regard to instructive interactions, the evidence that demonstrates the application of these ideas to limbs other than those of the chick will be presented.

The possibility of universal mechanisms controlling limb development

The very fact that the AER is found on the limbs of nearly all vertebrate groups indicates that they use similar methods for controlling limb bud outgrowth. This assumption has been upheld by the demonstration that the AER of developing rodent limbs and the AER of several species of birds will support the outgrowth of the chick wing (Zwilling, 1959; Pautou, 1968; Jorquera & Pugin, 1971; Saunders, 1977).

In terms of instructive mechanisms, whatever the theoretical mechanism that is preferred, it is also clear that the basic principles governing pattern formation in the amniote limb bud are similar. Post-axial limb bud tissue from numerous mammalian species, including human, and from reptiles and birds, will cause certain degrees of anterior–posterior duplication of limb parts when grafted into the pre-axial margin of the chick limb (MacCabe & Parker, 1976; Tickle *et al.*, 1976; Fallon & Crosby, 1977). It is also evident that the same universal principle governs patterning in the different limbs of individual animals. The same post-axial graft of tissue from the chick leg bud to the wing causes duplication of wing parts (Summerbell & Tickle, 1977).

Thus it is clear that both the permissive interactions of epithelium and mesoderm governing limb bud outgrowth, and the cellular interactions within the mesoderm which control pattern formation, are likely to be based on similar principles, at least within the amniotes. This observation makes a discussion of developmental constraints and evolutionary possibility more meaningful. If this universal applicability were not evident and many totally different developmental schemes were employed by vertebrate embryos, the task of clarifying the relationship between development and evolution would be made considerably more difficult.

Alterations in permissive and instructional interactions and the structural consequences

In the preceding sections, evidence has been presented for the universal nature of the developmental mechanisms governing limb bud development. These mechanisms have been divided into two basic sets, permissive and instructive interactions. This section deals with what may happen to limb structure if these mechanisms are interfered with, and attempts therefore to identify the likely constraints that a particular developmental mechanism places upon possible evolutionary change.

Permissive interactions

In all amniote vertebrates, the AER is a transient structure that degenerates when the most distal part of the limb pattern has been created. In the chick wing, this is the third phalange of the largest digit. It was mentioned above that the surgical removal of the AER before this stage truncates the proximo-distal pattern. In fact, a clear temporal relationship exists between the limb stage from which the AER is removed and the eventual proximo-distal level of truncation (Summerbell *et al.*, 1973; Summerbell, 1974; Summerbell & Lewis, 1975).

In evolutionary terms, it is straightforward to see that any change that brings about the early degeneration of the AER will have major anatomical consequences. If the AER degenerates very early, only the most proximal tissues will form. This is the exact anatomy that is seen in many of the limbless vertebrates, such as the snakes. It is also the case that the AER borne on the limb buds which begin outgrowth in these reptiles degenerates very prematurely (Rah-

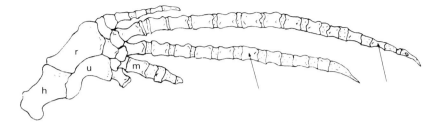

Fig. 2. Drawings of a dorsal view of a right forelimb of a round-headed dolphin. Note the many phalanges (arrows). h, humerus; r, radius; u, ulna; m, metacarpal. (After Flower, 1885.)

mani, 1974; Gans, 1975; Renous, 1977; Lande, 1978). In the limbless forms of reptiles, the early loss of the AER occurs in parallel with large scale necrosis of the limb mesoderm. Whether this is brought about by the early cessation of AER function or vice versa is unclear. Limb loss is also produced by specific mutations in the chick (Zwilling, 1949, 1956; Hinchliffe & Ede, 1973), and large scale mesodermal necrosis is again associated with early death of the AER in these cases.

It seems that the death of the AER at different developmental stages has a major effect on limb organisation, such as early death and the creation of the limbless phenotype. However, because of the permissive nature of the function of the AER, any addition of proximal to distal parts at the distal end of the limb must entail not only an extended life for the AER but also a concomitant change in the pattern formation mechanisms. An extension of the life of the AER alone will merely allow the creation of more limb cells but a pattern formation mechanism is needed to turn these cells into a useful pattern. Such an increase in the distal limb pattern is feasible, however, because many marine vertebrates, such as the dolphins, have many rows of phalanges in their paddle like forelimbs (see Fig. 2 and Felts, 1966).

Instructive interactions

It was pointed out above that the framework of positional information emphasises a dualistic mode of action in the creation of patterns; the specification of cell position within a field, and the subsequent interpretation of this information by the cells then leads to an appropriate differentiation of the cells in the field, and to a characteristic anatomy.

Alteration of each stage of this two-part process will have specific anatomical repercussions. The nature of these repercussions depends not only on which specific mode is being discussed, i.e. positional specification or cellular interpretation, but also upon which particular mechanism for establishing positional information is preferred. In the following discussion, specific ways of altering two positional information schemes will be outlined. The two schemes are the global signalling zone of polarising activity model and the short-range cell–cell interaction polar coordinate model, both of which were introduced earlier in the discussion.

1. A global signalling scheme

The formal presentation of the ZPA hypothesis (Tickle *et al.*, 1975) suggested that the source of the diffusible morphogen was localised to a point on the posterior limb bud margin. The concentration profile of the morphogen in the one-dimensional antero-posterior limb axis is given by the diffusion equation,

$$C = C_o \, e^{-\lambda x}$$

where C is the concentration of the morphogen given the initial morphogen concentration at the source (C_o) and the distance of the diffusion path x. The diffusion constant (λ) is arbitrarily chosen and in the diffusion model presented by Tickle *et al.* (1975) was given a value to allow the diffusion profile to drop from $C_o = 100$ to $C = 10$ over a distance of 1 mm, which is the approximate width of the young chick wing bud (Fig. 3).

If the values given to C_o and λ in this initial formulation are realistic, then the developmental constraints contained within this global signalling theory can be examined by altering the initial source concentration C_o and the diffusion constant λ independently of one another. In order to make the effects of these alterations clear in structural terms, the approximate thresholds for the formation of the radius and ulna (the bones of the forearm region of the chick wing) have been measured using the original morphogen concentration profile of Tickle *et al.* (1975).

The approximate positions of these cartilagenous rudiments across the antero-posterior axis were measured from Alcian green

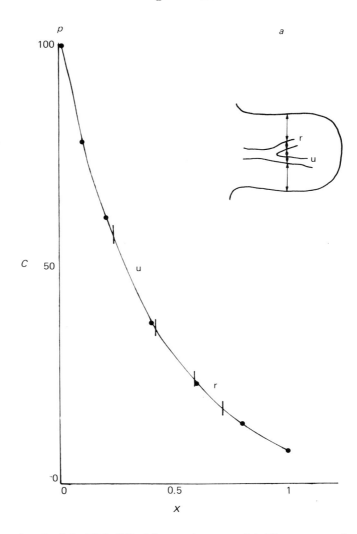

Fig. 3. Graph of the ZPA diffusible morphogen model. The concentration (C) of the substance falls from the source on the posterior wing margin (p) to a low value at the anterior wing margin (a). The distance (x) between these points is given as 1 mm. A drawing of the cartilagenous skeleton of a stage-24 wing bud is also shown to demonstrate the measurements used to establish the thresholds for radius (r) and ulna (u) specification upon the gradient (see text).

stained whole mounts of stage 24 chick wings (Fig. 3 and see Summerbell, 1976). Stage 24 is the first stage in which the radius and ulna become clear in a whole-mount preparation and, although this stage is later than the time that specification of the elements occurred (see Summerbell & Lewis, 1975), the increase in width of the wing bud during these early stages is very slight and the distances measured are likely to be accurate representations of the relative positions of these elements at this earlier time.

As shown in Fig. 3, the radius is seen to be 0.12 mm wide, with its anterior edge situated 0.29 mm from the anterior wing margin and its posterior edge 0.59 mm from the posterior wing margin. The ulna is 0.18 mm wide, with its anterior edge 0.59 mm from the anterior wing margin and its posterior edge 0.23 mm from the posterior wing margin. When these measurements are fitted into the original gradient depicted in Fig. 3, the radius has thresholds (C values) of 17 anteriorly and 23.5 posteriorly and the ulna has thresholds of 36.5 anteriorly and 56 posteriorly.

Having obtained these values, the new positions of the radius and ulna can be calculated with varying concentrations of C_0 and varying values of λ. The results of these calculations for C_0, varying between 200 and 50 with λ set at 2.5 and for λ varying from 2.0 to 3.5 with C_0 set at 100, are shown in Fig. 4a, b. The position of the radius and ulna with their thresholds constant at 17–23.5 and 36.5–56, respectively, are shown both on these graphs and diagrammatically in idealised limb segments in Fig. 4.

This simple mathematical exercise reveals nothing startling to the theoretical biologist but yields specific anatomical variations which define the developmental constraints to the evolution of the antero-posterior pattern of this segment of the limb, given a long-range signalling mechanism as the starting point. Varying the original output concentration from the ZPA alone causes the movement in parallel of the radius and ulna from anterior ($C_0 = 100$) to posterior ($C_0 = 50$) (Fig. 4). Varying the diffusion constant alone with C_0 set at 100 causes changes in proportion of limb space occupied by the skeletal elements in a precise way (Fig. 4).

The results of this analysis reveal two clear points. The most obvious of these is that any particular alteration in the properties of the source of the diffusion of the molecule produced by the ZPA will affect the whole field in a predictable and coordinated way.

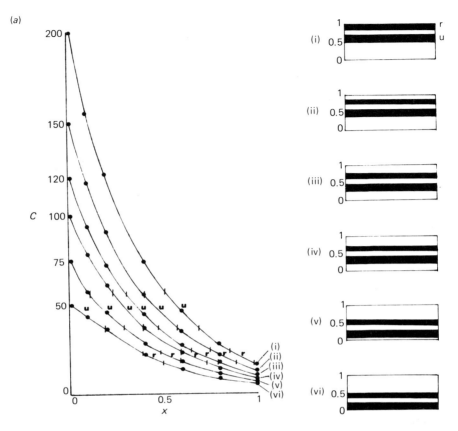

Fig. 4. Graphic and diagrammatic representations of threshold alterations for specification of the radius (r) and ulna (u) for independently variable values of C_o (a) and λ (b). In each case, the adjacent set of diagrams represent the same thresholds as the corresponding curves.

The constraint of this system in evolutionary terms is that no single part of the pattern can be altered independently from another. If the radius shifts position in the limb, so does the ulna. The second point is that the type of analysis outlined above is quantifiable. The position of each skeletal element can be measured in whole mounts at an early stage, and the variations in these positions within limbs of different species could reveal such a predictable pattern of variation. Furthermore, because of the different specific alterations produced by altering C_o and λ independently, the part of the

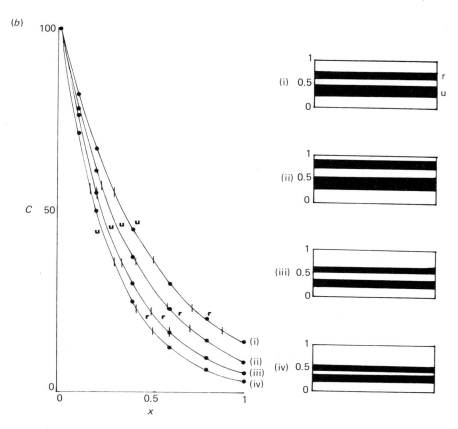

signalling system which has undergone a change in evolution may be identifiable.

2. A short-range cell–cell interaction scheme

Pattern specification and regulation can also be thought of as occurring as the result of localised interactions between cells in growing limb buds or regenerating limb blastemas. An example of such a system is the polar coordinate model (see for example Holder, 1981; Bryant *et al.*, 1981, 1982; Tank & Holder, 1981), which relies on local interactions and a specific set of rules which govern the result of the production of new positional values following contact between normally non-adjacent cells and requires that positional values are spatially arranged in a two-dimensional

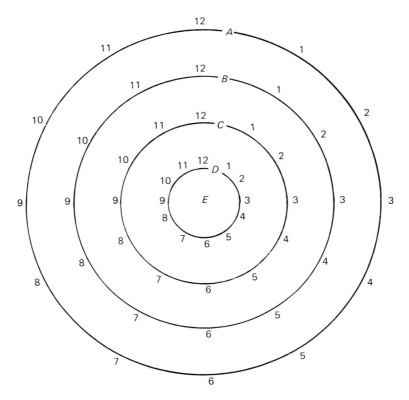

Fig. 5. The polar coordinate model can be formally represented as a two-dimensional array of positional values. One set of values lies around the circumference (numbers) the other set is radial (letters) with *A* being proximal (shoulder) and *E* distal (digits).

polar coordinate array (Fig. 5). The constraints placed upon the likely creation of form and pattern during evolution if this model is tinkered with are strikingly different from those of the long-range diffusion model discussed above. The important difference is that, because the cell–cell interactions are localised in the polar coordinate model, the alterations in pattern resulting from specific disturbances can also be localised. This point is clearly illustrated by a recent experiment performed by Stock & Bryant (1981).

The system used in this experiment was the regenerating hind-limb of the newt *Notophthalmus viridescens* (Fig. 6). Normally, following amputation of any region of the limbs of the urodelean amphibians, a faithful copy of the limb tissues removed regenerates

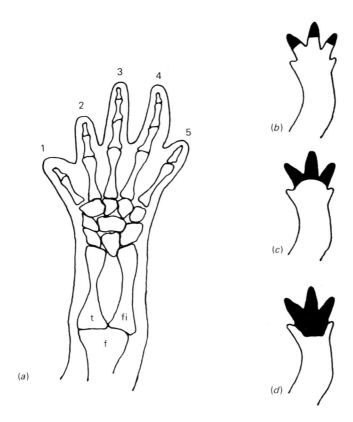

Fig. 6. Representation of the digit amputation performed by Stock & Bryant (1981). (*a*) A drawing of the skeleton of a normal hindlimb of *Notophthalamus viridescens*. (*b–d*) Limbs where the actual tissues amputated (shaded) differ. They are distal to the webbing level, at the webbing level and proximal to the webbing level respectively. t = tibia, fi = fibula, f = femur.

(see Tank & Holder, 1981). However, Stock & Bryant demonstrated that amputation at one specific region of the proximo-distal axis led to highly abnormal regeneration. This specific location was the point where the foot split into separate toes during regeneration. For example, if the centre three toes were removed at the point of separation, as many as five toes reformed in their place, leaving a foot with seven toes (Fig. 7). Amputation at levels distal or proximal to the level of digital separation led to faithful replacement of the parts that were removed. How can these results be explained

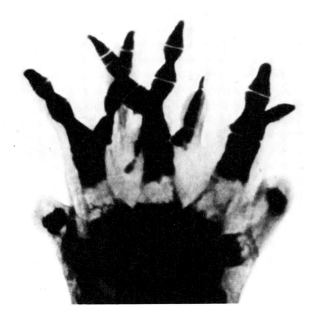

Fig. 7. An example of a newt limb amputated at the webbing level (*c* in Fig. 6). Three toes were removed and up to seven digits (some of which are branched) have regenerated in their place. (Courtesy of Prof. S.V. Bryant.)

in terms of the polar coordinate model, and what is their evolutionary implication?

Stock & Bryant (1981) considered the positional values within the limb as a sequence around the limb circumference (Fig. 8). The limb at this level is not circular but elliptical, with the dorsal and ventral axes flattened. This shape is critical because, when the separations between the digits occur, cells from the dorsal and ventral sides of the foot will contact more readily than cells from the anterior and posterior limb sides (see Holder, 1981). When separation begins in normal development, intercalation of new positional values occurs between contacting dorsal and ventral cells. The model states that these new intercalated positional values will be those normally intervening via the shortest route on the circumferential sequence between contacting cells (see French *et al.*, 1976). Shortest route intercalation between dorsal and ventral cells with specific separation points is shown in Fig. 8. The result of

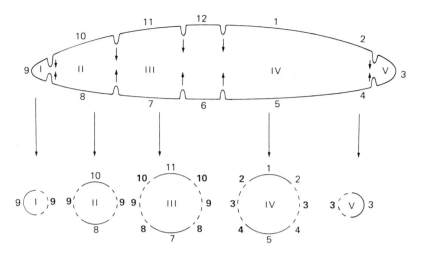

Fig. 8. Diagrammatic representation of the polar coordinate model explanation of the results of Stock & Bryant (1981). Following separation of the digits, shortest route intercalation leads to the creation of a series of symmetrical circles. The number of circles (i.e. digits) will depend upon the points of separation. In this representation the values 12 and 6 are assumed not to interact, although this additional consideration is not a necessary prerequisite for functioning of the model. (From Stock & Bryant, 1981.)

each intercalation path between contacting cells in the figure is a symmetrical circle, which has been shown to have specific properties during distal outgrowth. It has been proposed that, if, during normal outgrowth, cells from different circumferential positions contact each other (see Bryant & Baca, 1978), limb parts bearing symmetrical arrays of positional values comprising two sets of values representing less than half of the complete sequence will lose mid-line structures, taper and cease regeneration (Fig. 9) (Holder *et al.*, 1980; Bryant *et al.*, 1981, 1982). Hence, in normal development, provided that the digital separation points are in the same anatomical positions, a set of five symmetrical outgrowths will form. These outgrowths should taper and cease development with a precise anatomy that is specific for each digit.

The explanation given by Stock & Bryant (1981) for the abnormal digital patterns formed following amputation at the level of digital separation is that the normal positions of digital separation are disrupted by surgery and subsequent healing. When cells from

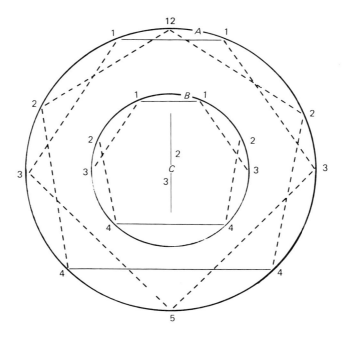

Fig. 9. A property of distally transforming symmetrical circles which bear two sets of positional values which represent less than half of the total set is the loss of mid-line positional values, tapering of the outgrowing structure and the eventual cessation of outgrowth. This process is represented here with distal outgrowth from level *A*. Cell contacts during outgrowth between normally non-adjacent cells (dashed lines) lead to intercalation of new circumferential values via the shortest route. Cells with the same or normally adjacent values will not stimulate division when they contact (solid straight lines). Notice that radial level *B* is missing 12 and 5, and by the stage of formation of level *C* no circumferential values which are normally non-adjacent exist and outgrowth ceases. (See text and Holder *et al.* (1980), for further explanation.)

the dorsal and ventral sides heal after surgery in abnormal places, abnormal symmetrical circumferential sets of positional values are created and novel patterns of toes result.

The evolutionary implications of, or the developmental constraints hinted at by, these simple experiments and their interpretation in terms of a short-range cell–cell interaction model are far reaching. A number of points can be made.

(1) The model predicts that digits will be symmetrical about the antero-posterior axis distal to the point of digital separation. A survey of the hand and foot skeletons of tetrapods and a histological

(a)
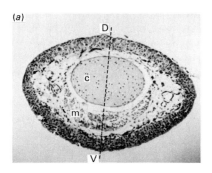

(b)
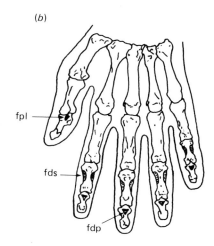

Fig. 10. The digits of tetrapod limbs distal to points of digital separation are symmetrical about the lateral to medial (anterior to posterior) axis. This is illustrated by a transverse section of the finger of an axolotl larva (a) and a drawing of a human hand (b). In (a) the dashed line demarks the dorso-ventral (DV) line of symmetry. In (b) the dotted regions are points of muscle insertions for the flexor digitorum superficialis (fds), flexor digitorum profundus (fdp) and flexor pollicis longus (fpl) muscles. Note the remarkable impression of symmetry of these muscle attachment sites on the phalanges. m, muscle; c, cartilage.

analysis of a few transverse-section series of limbs from the mammals and amphibians bears out this prediction. For the present purpose, this point is illustrated using a section of an axolotl finger and a drawing of a skeleton of a human hand (Fig. 10).

(2) The consequence of symmetrical circles which bear subsets of less than half of the complete circumferential sequence of values is that the structure will taper and eventually cease growing during distal development or regeneration. It is a consequence of the model therefore, that digits will be the most distal part of the limb pattern and will have a tapered appearance.

(3) The lateral digits of the hand or foot bear symmetrical subsets of positional values of the medial digits (see Fig. 8). In consequence, if reduction in digital separation points occurred, medial digits will always remain and be the most complex (Stock & Bryant, 1981). This point is illustrated in Fig. 11, which shows the hand skeletons of a great armadillo and a bat. This arrangement of lateral subsets of medially situated more complete symmetrical

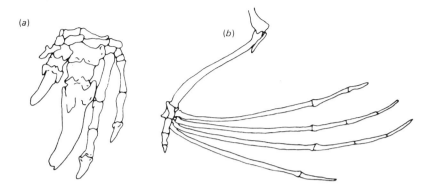

Fig. 11. Drawings of the hand skeleton of the great armadillo (*a*) and the bat (*b*). Note the impressive nature of the most central digits in each case. (After Flower, 1885.)

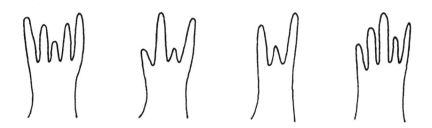

Fig. 12. The polar coordinate model interpretation of digital formation predicts that certain specific digital morphologies will not be formed during development and will therefore not appear in evolution of the tetrapods. Four examples of such forbidden morphologies are shown here.

arrays of positional values places well-defined limitations on possible morphologies and any morphologies which do not fit this pattern should not be found in the tetrapods. Some such forbidden morphologies are shown in Fig. 12.

(4) One of the striking features of the tetrapod limb is the apparent upper limit of five unique digits. Limbs can make more than five digits in abnormal circumstances, such as following a graft of posterior tissue to an anterior site in a developing limb (see for example Tickle *et al.*, 1975) or the polydactylous mutant of the cat (Danforth, 1947), but, when identifiable, the extra digits are always copies of those which are part of the original set of five or

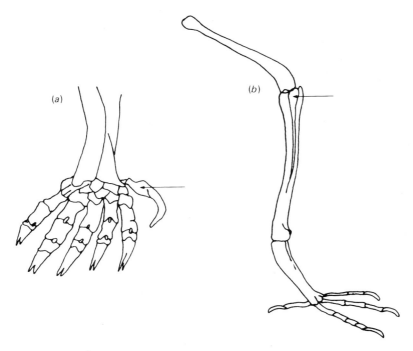

Fig. 13. The changes in specific parts of the limb pattern may occur in very localised limb positions. Two examples of this point are illustrated here. In (a), which is a drawing of the forelimb of a mole, a single carpal bone, the radial sesamoid, has become a functional digital aid while the rest of the carpal bones have remained similar to other common carpal patterns (After Flower, 1885.) In (b), which is the skeleton of a pigeon, the fibula (arrow) has become a short stiletto-shaped bone while the adjacent tibia has remained of standard form.

less. If atavistic digits occur they only do so in cases when the normal limb of the species bears less than five digits (see, for example, Stockard, 1930). The implication of the model in regard to this highly conserved pattern is that the production of greater than four digital separation points produces a series of small symmetrical circles, the development of which will lead to a set of stubby symmetrical outgrowths which are of little use to the animal. As the number of separation points increases the symmetrical circles become ever smaller and the symmetrical digits ever shorter. If more than five large strong digits are needed then the evidence is that other parts of the limb pattern are requisitioned to make novel structures; as is the case in the mole (Fig. 13) or the panda, where one of the wrist bones, the radial sesamoid, has become specialised to have the function of an extra digit. (Flower, 1895; Gould, 1977, 1980.)

Conclusions

It becomes clear from this type of analysis that different models which focus on essentially different cellular mechanisms for establishing patterns in developing tissues exhibit very different anatomical constraints in terms of both development and evolution. Altering a long-range signalling mechanism, such as the ZPA, produces coordinated anatomical variations which affect the whole field. On the other hand, specific alterations in healing modes, for example, cause short-range cell interaction models to produce very much more localised effects.

In order to make this type of conclusion meaningful in evolutionary terms, we are forced to examine the range of limb anatomies produced throughout evolution of the vertebrates and ask which of these two different developmental constraints best fits the picture. The answer, as is discussed in detail below, is that anatomical variations occur, almost exclusively, in localised limb positions. The major exception to this observation is where drastic alterations are apparent, such as in total limb loss, and this is readily explicable in terms of changes in permissive mechanisms, as was discussed earlier in the essay. We must infer that if the long-range signalling system of the ZPA is real then evolution has left it basically untouched and that all anatomical variation that is seen occurs as the result of mutations affecting the cellular interpretation of the positional value produced by the diffusible morphogen. There is no reason *a priori* why this should be so, although if the framework of positional information is meaningful then it must be assumed that some of the alterations to patterns seen in tetrapod limbs were the result of such alterations in the interpretative mechanism. If all the eggs of variation are placed in the interpretative basket, little of any significance can be said about the relationship between evolution and development because nothing is known of the process of cellular interpretation of positional value.

This blind alley is avoided if local cell–cell interaction models are considered. Furthermore, because of the predictive nature of the polar coordinate model, experiments such as the one performed by

Stock & Bryant (1981) can be devised to extend our understanding of the nature of the developmental constraint. The above discussion has focused on the developmental considerations of the nature of the developmental constraint. Possibly as much information can be derived from examining the morphology of limbs which have evolved during the last 200 million years. Apart from the obvious, albeit crucial point concerning the localised nature of the alterations in pattern, several other interesting conclusions can be drawn.

The localised nature of pattern alteration is clearly illustrated by the oddity of the panda's thumb (Fig. 13, and see Gould, 1980). In this case a single wrist bone out of the array of carpals has become specialised into an extra functional digit. Similarly, in the leg of the chicken and the wing of the bat, the loss of an epiphysis of one skeletal element leaves it short and insignificant compared to the bone lying alongside it (see Fig. 11 and 13).

The other striking feature of the vertebrate limb is the remarkable consistency with which some features are maintained. For example, the proximal limb segment, the zeugopod, never has more than one bone in it, a humerus or a femur. Similarly, the stylopod never bears more than two bones, the radius and ulna or the tibia and fibula. The major alterations in number of elements of the skeleton comes in the digits and the wrist or ankle. The number of tarsals in the amphibian and reptile ankle can vary from as few as 2 or 3 to as many as 13 (Schaeffer, 1941; and see Hinchliffe, this volume) and the number of digits in mammals vary from 1 to 5. All of the evolutionary consistencies indicate the nature of the developmental mechanisms whose limitations they reflect. For example, the striking variation in the number and form of digits contrasts markedly with the static presence of only one humerus. This may be because the localised interaction centres at the digital separation points supply a ready source of anatomical variation in terms of a short-range interaction model. This developmental source of variation exists at the site of formation of digits, so it is this region and not that of the humerus which shows so much variation.

It is a pleasure to thank Susan Bryant for allowing me to use data from her work, for giving me some previously unpublished photographs, and for reading an earlier draft of this essay. I also wish to thank Rosie Burton, Julian Lewis and Vernon French for comments on various stages of the manuscript. Special thanks are due to Prof. Angus Bellairs for allowing me access to his excellent reprint collection. Any work of my own mentioned in the text was supported financially by the SRC.

References

Bryant, S. V. & Baca, B. A. (1978). Regenerative ability of double-half and half upper arms in the newt, *Notophthalmus viridescens. J. exp. Zool.* **204**, 307–24.

Bryant, S. V., French, V. & Bryant, P. J. (1981). Distal regeneration and symmetry. *Science* **212**, 993–1002.

Bryant, S. V., Holder, N. & Tank, P. W. (1982). Cell–cell interactions and distal outgrowth in amphibian limbs. *Am. Zool.* **22**, 143–52.

Chevalier, A., Kieny, M. & Mauger, A. (1976). Sur l'origine de la musculature de l'Ailele chez les Oiseaux. *C. r. Acad. Sci. Paris D* **282**, 302–11.

Chevalier, A., Kieny, A., Mauger, A. & Sengel, P. (1977). Developmental fate of the somitic mesoderm in the chick embryo. In *Vertebrate limb and somite morphogenesis*, eds. D. Ede, J. R. Hinchliffe and M. Balls, pp. 421–32. Cambridge University Press: Cambridge.

Christ, B., Jacob, H. J. & Jacob, M. (1974). Uber den Ursprung der Flugelmuskulatur. Experimentelle Unterschungen mit Wachtel- und Huhnerembryonen. *Experientia* **30**, 1446–9.

Christ, B., Jacob, H. J. & Jacob, N. (1977). Experimental analysis of the wing musculature in avian embryos. *Anat. Embryol.* **150**, 171–86.

Danforth, C. H. (1947). Morphology of the feet in polydactyl cats. *Am. J. Anat.* **80**, 143–71.

Fallon, J. F. & Crosby, G. M. (1977). Polarising zone activity in limb buds of amniotes. In *Vertebrate limb and somite morphogenesis*, eds. D. Ede, J. R. Hinchliffe and M. Balls, pp. 55–69. Cambridge University Press: Cambridge.

Fallon, J. F. & Kelley, R. D. (1977). Ultrastructural analysis of the apical ectodermal ridge during vertebrate limb morphogenesis. II. Gap junctions as distinctive ridge structures common to birds and mammals. *J. Embryol. exp. Morphol.* **41**, 223–32.

Felts, W. J. L. (1966). Some functional and structural characteristics of cetacean flippers and flukes. In *1st International symposium on cetacean research – whales, dolphins and porpoises*, ed. K. S. Norris, pp. 255–76. University College Press: Berkeley and Los Angeles.

Flower, W. H. (1885). *An introduction to the osteology of the Mammalia*, 3rd edn. Macmillan: London.

French, V., Bryant, P. J. & Bryant, S. V. (1976). Pattern regulation in epimorphic fields. *Science* **193**, 969–81.

Gans, C. (1975). Tetrapod limblessness: evolution and functional corollaries. *Am. Zool.* **15**, 455–67.

Goel, S. C. & Mathur, J. K. (1977). Morphogenesis in reptilian limbs. In *Vertebrate limb and somite morphogenesis*, eds. D. A. Ede, J. R. Hinchliffe and M. Balls, pp. 386–404. Cambridge University Press: Cambridge.

Gould, S. J. (1977). *Ontogeny and phylogeny*. Belknap: Harvard.

Gould, S. J. (1980). *The panda's thumb: more reflections in natural history*. Norton & Co.: New York.

Hinchliffe, J. R. & Ede, D. (1973). Cell death and the development of limb form and skeletal pattern in normal and wingless chick embryos. *J. Embryol. exp. Morph.* **30**, 753–72.

Hinchliffe, J. R. & Johnson, D. R. (1981). *The development of the vertebrate limb.* Oxford Scientific Publ.: Oxford.

Holder, N. (1981). Pattern formation and growth in the regenerating limbs of urodelean amphibians. *J. Embryol. exp. Morph.* **65** (suppl.), 19–36.

Holder, N., Tank, P. W. & Bryant, S. V. (1980). Regeneration of symmetrical forelimbs in the axolotl, *Ambystoma mexicanum. Devl Biol.* **74**, 302–14.

Iten, L. E. (1982). Pattern specification and pattern regulation in embryonic chick limb buds. *Am. Zool.* **22**, 117–29.

Iten, L. E. & Murphy, D. J. (1980). Pattern regulation in the embryonic chick limb: supernumerary limb formation with anterior (non ZPA) limb bud tissue. *Devl Biol.* **75**, 373–85.

Javois, L. C. & Iten, L. E. (1981). Position of origin of donor posterior chick wing bud tissue transplanted to an anterior host site determines the extra structures formed. *Devl Biol.* **82**, 329–42.

Javois, L. C., Iten, L. E. & Murphy, D. J. (1981). Formation of supernumerary structures by the embryonic chick wing depends on the position and orientation of a graft in a host limb bud. *Devl Biol.* **82**, 343–9.

Jorquera, B. & Pugin, E. (1971). Sur le comportement du mèsoderme et de l'ectoderme du beorgeon de membre dans les échanges entre le poulet et le rat. *Comp. r. Acad. Sci. Paris* **272**, 1522–5.

Jurand, A. (1965). Ultrastructural aspects of early development of the forelimb buds in the chick and the mouse. *Proc. R. Soc. Lond. Ser. B* **162**, 387–405.

Lande, R. (1978). Evolutionary mechanisms of limb loss in tetrapods. *Evolution* **32**, 73–92.

Lewis, J. H. & Wolpert, L. (1976). The principle of non-equivalence in development. *J. theor. Biol.* **62**, 479–90.

MacCabe, J. A. & Parker, B. W. (1976). Polarising activity in the developing limb of the Syrian hamster. *J. exp. Zool.* **195**, 311–17.

Milaire, J. (1957). Contribution à l'étude morphologique et cytochimique des bourgeons de membres chez quelques Reptiles. *Archs. Biol., Liége* **8**, 429–72.

Milaire, J. (1965). Aspects of limb morphogenesis in mammals. In *Organogenesis*, eds. R. Dehaan and H. Ursprung, pp. 283–300. Holt, Rinehart & Winston: New York.

O'Rahilley, R., Gardner, E. & Gray, D. J. (1956). The ectodermal thickening and ridge in the limbs of stage human embryos. *J. Embryol. exp. Morph.* **4**, 254–64.

Pautou, M. P. (1968). Role déterminant du mésoderme dans la différentiation sppécifique de la patte de l'Oiseau. *Archs. Anat. misc. Morph. exp.* **57**, 311–28.

Rahmani, T. M. (1974). Morphogenesis of the rudimentary hindlimbs of the Glass snake (*Ophisaurus apodus*). *J. Embryol. exp. Morph.* **32**, 431–43.

Raynaud, A. (1977). Somites and early morphogenesis of reptile limbs. In *Vertebrate limb and somite morphogenesis*, eds. D. Ede, J. R. Hinchliffe and M. Balls, pp. 373–85. Cambridge University Press: Cambridge.

Renous, S. (1977). Le problème de la rudimentation des membres chez les squamatos contribution à l'étude de la réduction des membres pelviens. *Bull. biol. Fr. Belg.* **61**, 75–89.

Rubin, L. & Saunders, J. W. (1972). Ectodermal–mesodermal interactions in the growth of limb buds in the chick embryo. Constancy and temporal limits of ectodermal induction. *Devl. Biol.* **28**, 94–112.

Saunders, J. W. (1948). The proximo-distal sequence of origin of the parts of the chick wing and the role of the ectoderm. *J. exp. Zool.* **108**, 363–403.

Saunders, J. W. (1972). Developmental control of three dimensional polarity in the avian limb. *Ann. N.Y. Acad. Sci.* **193**, 29–42.

Saunders, J. W. (1977). The experimental analysis of chick limb bud development. In *Vertebrate limb and somite morphogenesis*, eds. D. Ede, J. R. Hinchliffe and M. Balls, pp. 1–24. Cambridge University Press: Cambridge.

Saunders, J. W. & Gasseling, M. T. (1968). Ectodermal–mesenchymal interactions in the origin of limb symmetry. In *Epithelial–mesenchymal interactions*, eds. R. Fleischmayer and R. F. Billingham, pp. 78–97. Williams & Wilkins: Baltimore.

Schaeffer, B. (1941). The morphological and functional evolution of the tarsus in amphibians and reptiles. *Bull. Am. Mus. nat. Hist.* **78**, 395–472.

Smith, J. C. (1980). The time required for positional signalling in the chick wing bud. *J. Embryol. exp. Morph.* **60**, 321–8.

Stébler, R. (1973). Die morphologie der äpikalen epidermis während der frühen Extremitatenent-Wicklung bei Anuran. *Wilhelm Roux Arch. EntwMech Org.* **172**, 131–48.

Stock, G. B. & Bryant, S. V. (1981). Studies of digit regeneration and their implications for theories of development and evolution of vertebrate limbs. *J. exp. Zool.* **216**, 423–33.

Stockard, C. R. (1930). The presence of a factorial basis for character loss in evolution: the atavistic reappearance of digits in mammals. *Am. J. Anat.* **45**, 345–75.

Stocum, D. L. (1975). Outgrowth and pattern formation during limb ontogeny and regeneration. *Differentiation* **3**, 167–82.

Summerbell, D. (1974). A quantitative analysis of the effect of the excision of the AER from the chick limb bud. *J. Embryol. exp. Morph.* **32**, 651–60.

Summerbell, D. (1976). A descriptive study of the rate of elongation and differentiation of the skeleton in the developing chick wing. *J. Embryol. exp. Morph.* **35**, 241–60.

Summerbell, D. (1979). The zone of polarising activity: evidence for a role in normal chick limb morphogenesis. *J. Embryol. exp. Morph.* **50**, 217–233.

Summerbell, D. (1981). The control of growth and the development of pattern across the antero-posterior axis of the chick limb bud. *J. Embryol. exp. Morph.* **63**, 161–80.

Summerbell, D. & Lewis, J. H. (1975). Time, place and positional value in the chick limb bud. *J. Embryol. exp. Morph.* **33**, 621–43.

Summerbell, D., Lewis, J. H. & Wolpert, L. (1973). Positional information in chick limb morphogenesis. *Nature* **224**, 492–6.

Summerbell, D. & Tickle, C. (1977). Pattern formation along the antero-posterior axis of the chick wing bud. In *Vertebrate limb and somite morphogenesis*, eds. D. Ede, J. R. Hinchliffe and M. Balls, pp. 41–57. Cambridge University Press: Cambridge.

Tank, P. W., Carlson, B. M. & Connelly, T. G. (1977). A scanning electron microscope comparison of the development of embryonic and regenerating limbs in the axolotl. *J. exp. Zool.* **201**, 417–30.

Tank, P. W. & Holder, N. (1981). Pattern regulation in the regenerating limbs of urodele amphibians. *Q. Rev. Biol.* **56**, 113–42.

Tarin, D. & Sturdee, A. (1973). Histochemical features of hindlimb development in *Xenopus laevis. J. Anat.* **114**, 101–7.

Thornton, C. S. (1965). Influence of wound skin on blastemal cell aggregation. In *Regeneration in animals and related problems*, eds. V. Kiortis and H. Tampusch, pp. 333–40. North-Holland: Amsterdam.

Tickle, C. (1981). The number of polarising region cells required to specifiy additional digits in the developing chick wing. *Nature* **289**, 295–8.

Tickle, C., Shellswell, G., Crawley, A. & Wolpert, L. (1976). Positional signalling by mouse limb polarising region in the chick wing bud. *Nature* **259**, 396–7.

Tickle, C., Summerbell, D. & Wolpert, L. (1975). Positional signalling and specification of digits in chick limb morphogenesis. *Nature* **254**, 199–202.

Tschumi, P. A. (1957).The growth of the hindlimb bud of *Xenopus laevis* and its dependence upon the epidermis. *J. Anat.* **91**, 149–73.

Wolpert, L. (1969). Positional information and the spatial pattern of cellular differentiation. *J. theor. Biol.* **25**, 1–47.

Wolpert, L. (1971). Positional information and pattern formation. *Curr. Top. devl. Biol.* **6**, 183–224.

Wolpert, L. & Hornbruch, A. (1981). Positional signalling along the antero-posterior axis of the chick wing bud. The effects of multiple polarising region grafts. *J. Embryol. exp. Morph.* **63**, 145–59.

Yntema, C. L. (1968). A series of stages in the development of *Chelydra serpentina. J. Morph.* **125**, 219–52.

Zwilling, E. (1949). The role of epithelial components in the developmental origin of the wingless syndrome of chick embryos. *J. exp. Zool.* **111**, 175–88.

Zwilling, E. (1956). Interaction between limb bud ectoderm in the chick embryo. III. Experiments with a wingless mutant. *J. exp. Zool.* **132**, 241–253.

Zwilling, E. (1959). Interaction between mesoderm and ectoderm in duck–chicken limb bud chimaeras. *J. exp. Zool.* **142**, 521–32.

Zwilling, E. & Hansborough, L. A. (1956). Interaction between limb bud ectoderm and mesoderm in the chick embryo. III. Experiments with polydactylous limbs. *J. exp. Zool.* **132**, 219–39.

Index

Page numbers in italic type refer to figures, plates and tables.

427

Boolean switching rules, 217
brachydactyly, 53
brain, 124–7: embryonic development, 126; epigenetic development, 358–9, *358*
Bruchidius sp. beetle: germ band extension, 145, *146*

canalization, 253; constraint in ciliate patterns, 283, 291; limitations of theory, 200; quantitative genetics, 199–200, *199*, 202
canonical elements of carpus, 100–1, *101*
cell development: molecular jigsaws and proteins, 34
cell sorting, 35
cell-surface patterns of ciliates, *see* patterning, ciliate cell surface
cells: interpretation of positional values, 51–3; response to monotonic gradients, 36, *37*; spatial organization, 34–8; transdetermination, 70
central nervous system: embryonic development, 124–5; induction of, 126–7; *see also* brain
character phylogeny, 20–3
chick: activity of inducers, 371–4, *373*; aortic arches and structural response to function, 334, *335*; epigenetic jaw development, 361–2; induction of chondroid bone, 363–4; Meckel's cartilage chondrogenesis, 361–2, *369*
chick limbs: archetype theories, 107, 109, *109*; development of, 52–3, 332–3, *332–3*, 382–3, 401, (AER in) 402–3, 405, 406, (differential growth) 49–50, (parameters involved) 90, (global signalling model) 407–10, *408*, (and vitamin A) *393*, 394; evolution, 82; growth and genetic control, 116; muscle and axis inversion, 387–8, *387–8*, *390*; prechondrogenic condensations, 104–9, *106–9*; supernumerary, and induction, *385*, 386, *386*, 388–9, *389*, 390
chondroid bone, 363–4
chondroitin-$^{35}SO_4$, autoradiography of, in limb buds, 104–16, (methods) 105
ciliate cell-surface patterns, *see* patterning, ciliate cell surface
ciliates: common ancestry, 279; cortical inheritance, 39, 318; genetic basis of corticotypes, 70; molecular relative to anatomical divergence, 280–3; similarity to multicellular organisms, 280
cladistics: monophyletic taxa, 19
cladograms, 2; and forms of ancestry, 12; of vertebrates, 5, *7*, 24, 25, *25*; and X-trees, 26
clonal analysis: imaginal discs, 231–3; insect

segmentation, 164–5, *165*, 175, 181, 189
clones in ciliates: ciliary pattern evolution, 292–7; perpetuation of heritable units, 70
clones and somaclonal variability, 261
coat colour markings: reaction diffusion model, 206–7, *207*; temperature effect on, 334, *335*
cockroach limbs: grafting and regeneration, 178, *179*, 180; intercalary regeneration, 325–6; patterning, 403
combinatorial model of compartmentation, *204*, 205–8, *207*, 218
common ancestry, 23; ciliates, 279; ontogenetic networks, 139; and relatedness, 18; tetrapod limb, 75, 76, 87, 99, 381; *see also* ancestry
common descent, *3*, 8, *9*; conventions of ancestry, 8–12; hierarchy of taxa, 11, 12
compartmentation in insects, 65; combinatorial model, *204*, 205–8, *207*, 218; a developmental constraint, *204*, 205–6, *207*; genetic factors, 69; pattern formation in imaginal discs, 321–2, 323–4
competence, 36–7, *37*, 41; and gradients in limb bud, 79–81, *80*
conservation of form, 41–3, 399, *400*
constraints: hierarchies of, and morphogenetic fields, 92
constraints, developmental, 44, 45, 400–1; ciliates, 283, 289, 290–1, 292, 306; close packing, 202, 208; compartmentation, *204*, 205–6, *207*; conservation of form, 43, 399, *400*; echinoid morphology, 202; in evolution, 195–221; global signalling scheme, 407; imaginal disc compartments, 321; pattern formation, 53–5; phyllotaxy, 201–2, *201*; polar coordinate model, 416–21; shell patterns, 202–3, *203*, 205
constraints, evolutionary: global signalling scheme, 409–11, *410*, *411*
constraints, selective: ciliates, 282–3, 290–2
contour plots: pattern formation in pentadactyl limb, 76–85, *77*, *82*, *84*, *89*
conventions of ancestry, 8–12, 17–20, 24, 25
Cope's law, 27, 337
correspondence maps, 182, *183*, 184
cortical inheritance, 39, 318
crocodiles and birds in evolution, 103
crystallins in lens evolution, 339–340
cuticular patterns in *Drosophila*, *see* patterning, *Drosophila* cuticle
cyclostomes: lateral lenses, *343*; Meckel's cartilage chondrogenesis, *369*; skeletal tissues, 365
cytoplasmic inheritance and non-nuclear DNA and RNA, 39